ENCYCLOPEDIC DICTIONARY OF

NAMED PROCESSES IN CHEMICAL TECHNOLOGY

SECOND EDITION

ENCYCLOPEDIC DICTIONARY OF

NAMED PROCESSES IN CHEMICAL TECHNOLOGY

SECOND EDITION

Alan E. Comyns

CRC Press
Boca Raton   London   New York   Washington, D.C.

Library of Congress Cataloging-in-Publication Data

Comyns, Alan E.
    Encyclopedic dictionary of named processes in chemical technology
    / Alan E. Comyns.
        p.   cm.
    Includes bibliographical references.
    ISBN 0-8493-1205-1 (alk. paper)
    1. Chemical processes--Dictionaries. I. Title.
TP155.7.C664 1999
660′.281′03—dc21

98-41331
CIP

First Edition originally published by Oxford University Press Inc., New York.

© 1993 by Alan E. Comyns
© 1999 by CRC Press LLC

No claim to original U.S. Government works
International Standard Book Number 0-8493-1205-1
Library of Congress Card Number 98-41331
Printed in the United States of America 1 2 3 4 5 6 7 8 9 0
Printed on acid-free paper

Introduction to the First Edition

The purpose of this dictionary is to provide concise descriptions of those processes in chemical technology which are known by special names which are not self-explanatory.

The chemical industry is notoriously difficult to define. In addition to its obvious role as a producer of 'chemicals' such as sulphuric acid, it may be said to embrace all those industries in which chemical processes are conducted. There is no generally agreed list of such industries, but obvious industries include extractive metallurgy, plastics, paper, ceramics, sewage treatment, and now even electronics. It is this broad spectrum of 'chemical technology' that is addressed in this book. It thus includes the gigantic Bessemer process for making steel, and the microscopic Manesevit process for applying circuits to silicon chips. The only deliberate omission is food chemistry.

The aim has been to include all those processes that are known by special names, of whatever origin. Of course, only a minority of industrial chemical processes are distinguished by the possession of special names, so this book does not attempt to include all of even the more important processes. Overviews of the industry are provided by other books, notably the encyclopædias listed below. Many named processes are included in such works, but only a fraction of the names in the present compilation are to be found in them.

The names are a heterogeneous collection—inventors, companies, institutions, places, acronyms, abbreviations, and obvious corruptions of the chemical nomenclature. Derivations, where known, are indicated in square brackets [ ]. The names of chemicals used in the entries are the traditional names commonly used in industry today.

Criteria for selection of names for inclusion are inevitably subjective, but the intention has been to include all named processes in current commercial use anywhere in the world, and those which have been or are being piloted on a substantial scale. Obsolete processes which have been or might have been important in the past are included too. The coverage is primarily of English names but some foreign names are included.

Process names which combine the name of a company with the name of a chemical, e.g. the Monsanto Acetic Acid process, have mostly been excluded because they are self-explanatory and can be found in the encyclopædias.

Some companies (e.g. Lurgi, Texaco) are best known for one process, even though they may have developed many others; in general, only their most famous one is included here under the company name. Their other processes are included if they have special names. Process names which combine the names of two collaborating companies (e.g. Mobil/Badger) have mostly been included. Company names are usually given in the styles in use at the times of their respective inventions, as given in patent applications.

Where two or more processes have the same name, they are distinguished by numbers in parentheses, e.g. **Parex** (1), **Parex** (2). Because the numbers are not parts of the names they are not emboldened.

Names of chemical reactions have mostly been excluded, being adequately defined in standard chemical texts and in the special chemical dictionaries listed below. There is remarkably little overlap between reaction names and process names: discoverers of chemical reactions seldom develop them into manufacturing processes. However, some generic process names which combine two or more reactions (e.g. oxychlorination, dehydrocyclodimerization) have been included because they are not generally to be found in any dictionaries. These hybrid names are distinguished by being given lower-case initial letters.

The lengths of the entries have been tailored to reflect importance and topicality. In general, important processes in current use have the longest entries; obsolete processes of minor importance have the smallest; and processes of intermediate importance receive correspondingly intermediate space. There are exceptions: some obsolete processes are given more space because of their technical interest or historic importance; some important current processes are given little space because their essential features are still secret.

References have been chosen which document the origins of the processes and which review the latest developments. Sources which are reasonably accessible have been quoted wherever possible. Journal abbreviations are those used by Chemical Abstracts.

Patents quoted are generally those of the countries of the original inventions, or their British or US equivalents. Many processes are described in families of patents; only the first or key ones are quoted here.

Much of the information on current processes comes from commercial literature provided to the author by the companies offering or using them. The author acknowledges their help and solicits additional information as it becomes publicly available, so that future editions of this dictionary may be updated.

Copyright and trade mark laws present special problems for works such as this, protection varying from country to country and from time to time. It would not be feasible to indicate which of the many names in this dictionary are so protected. The inclusion of a protected name, in its registered or commonly used format, has no legal significance and does not indicate that the name has become a generic word.

The appendix provides a key to the chemicals and materials whose manufacture is described by one or more of the named processes. Although the dictionary does not purport to be a comprehensive listing of processes for making particular chemicals or materials, reference to the key will identify those processes which have special names, and reference to these entries will often provide general references which will help to identify the unnamed ones too.

A work such as this, covering a wide spectrum of technologies, and having a historical dimension, can never be complete, even within its own terms of reference. More names, old and new, demand the attention of the author almost daily. More information on the processes is always to be found. In such a task, there comes a time when an author must consider whether the body of information already collected, incomplete though it may be, is sufficient for others to find useful. I hope that I have judged that time correctly.

INTRODUCTION TO THE SECOND EDITION

Many new and improved processes have been invented in the six years that have passed since the completion of the first edition of this work, and many have been commercialized. These processes are now included in this dictionary, and the commercial status of the established processes has been updated wherever possible.

There are many incentives for inventing and improving chemical processes, but today the most powerful incentives are those concerned with improving the environment. Many of the new processes described here are for removing gaseous effluents (SO_x, NO_x, volatile organics), destroying organic residues in water, and minimizing the quantities of waste products in general.

Another group of powerful incentives is concerned with the manufacture of fuels from different raw materials, notably liquid hydrocarbons from natural gas. The removal of lead additives from gasoline has necessitated the introduction of new hydrocarbon formulations and new additives, all made by new processes.

Synthetic organic chemists have simplified many processes, using catalysts to make useful chemicals in one step from basic raw materials such as propane. The gigantic paper and pulp industry has devised many new bleaching processes to replace traditional chlorine bleaching.

These are exciting times for industrial chemists and engineers: their language, exemplified by the new names in this dictionary, reflects this.

NOTES

Corporate designations (e.g. Ltd, PLC, Inc, AG, SA, A/S) have been omitted, except in cases in which confusion with personal names might occur.

Company names listed are generally those that were current when the developments described were made. Because many companies have changed their names (often using letters instead of words) and their owners, many of these names are not current.

Dates of invention given are usually the years in which the relevant patents were filed.

Asteriks denote names that are defined elsewhere in this dictionary.

AUTHOR

Dr. Alan Comyns has had an unusually varied career in academic, government, and industrial research laboratories. He graduated with first class honors in chemistry from the University of London at the age of nineteen. His Ph.D. work, carried out in the Hughes-Ingold school of physical-organic chemistry at University College London, was followed by post-doctoral studies in the United States at Caltech and the University of Wisconsin. He has worked at the Atomic Energy Research Establishment, Harwell (U.K.), British Titan Products (now Tioxide Ltd.), U.K., Westinghouse Electric in Pittsburgh, and National Lead in New Jersey. From 1974 to 1988 he was Product Research Manager, later Chief Scientist, at Laporte Industries in Widnes (U.K.). In the 1980s he was a part-time Visiting Lecturer in Industrial Chemistry at the University of East Anglia. He is now an independent consultant and author, specializing in market studies for inorganic chemicals and materials.

His recent publications include: *Fluoride Glasses* (John Wiley & Sons, Chichester, 1989), still the only monograph in this field; *Titania and the Titanates* (Mitchell Market Reports, London, 1989 and 1992), a 232-page study describing the nonpigmentary uses of titanium dioxide and the inorganic titanates; *Dictionary of Named Processes in Chemical Technology* (Oxford University Press, Oxford, 1993), a key reference work which describes more than 2,000 processes in 337 pages; *Inorganic Peroxides and Peroxy Compounds* (in Kirk-Othmer's Encyclopedia of Chemical Technology, 4th ed.,Vol. 18, 203–229); and *Fillers, Extenders and White Pigments in Western Europe and North America* (Materials Technology Publications, Watford, 1997), a 209-page study, describing over 30 products. He edits, and largely writes, *Focus on Catalysts,* a monthly newsletter published by the Royal Society of Chemistry.

Dr. Comyns was one of the founders of the Applied Solid State Chemistry Group of the Royal Society of Chemistry. He has been Chairman of the Industrial Inorganic Chemicals Group of the Royal Society of Chemistry and Chairman of the Liverpool Section of the Society of Chemical Industry.

His hobbies include collecting antique glass and archaeology.

DEDICATION

---

*Dedicated to the generations of industrial chemists and
engineers whose ingenuity has given us the materials
of civilization.*

CONTENTS

Aachen   *See* DR.

A-B   [Adsorptions-Belebsungsverfahren, German, meaning Adsorption-Activation process]
A two-stage *Activated Sludge process for treating sewage and industrial wastes. The first
stage (A) is highly loaded, the second (B) is low loaded. Such a system can cope with sudden changes in the quantity and quality of effluent feed. Developed in 1983 by B. Bohnke at
the Technical University of Aachen and subsequently engineered by Esmil, UK.

> Horan, N. J., Biological Wastewater Systems, John Wiley & Sons, Chichester, England, 1990, 69.
> Gray, N. F., *Activated Sludge: Theory and Practice,* Oxford University Press, Oxford, 1990, 110.

ABATE   A process for removing hydrogen sulfide from sour gases such as landfill gas. The
gas, containing oxygen in addition to the hydrogen sulfide, is passed through water containing an iron chelate compound, which oxidizes the hydrogen sulfide to elemental sulfur.
Dispersants keep the sulfur in suspension until its concentration reaches 10 percent.
Developed by Dow Chemical, derived from the Dow/Shell *SulFerox process.

> *Chem. Eng. (N.Y.),* 1996, **103**(11), 19.

Abbot-Cox   A method of applying vat dyes to cellulosic textiles in package form. The dispersed dye, with a dispersing agent, is circulated through the package. The dye becomes sub-
stantially transferred to the material by the gradual addition of an electrolyte such as sodium
sulfate. When the dye has been transferred to the fabric, it is reduced in situ. The color is re-
stored by a mild oxidizing agent such as hydrogen peroxide.

> Fischer-Bobsien, C.-H., *Internationales Lexicon Textilveredlung+Grenzgebiete,* Rhenus
> Handelsgesellschaft, Vadus, Liechenstein, 1966, 1123.

ABC   Also called Chiyoda ABC. A process for treating heavy hydrocarbons from tar sands
by *hydrocracking. Piloted by the Chiyoda Chemical Engineering and Construction
Company in the 1980s.

> Bowman, C. W., Phillips, R. S., and Turner, L. R., in *Handbook of Synfuels Technology,* Meyers,
> R. A., Ed., McGraw-Hill, New York, 1984, 5-73.
> Marcos, F. and Rosa-Brussin, D., *Catal. Rev., Sci. Eng.,* **37**(1), 3, 1995.

Ab der-Halden   A continuous process for distilling coal tar. It is operated under reduced
pressure with the heat provided by live, superheated steam. This provides a clean separation
of the products, without cracking. Developed in France in the 1920s by C. Ab der-Halden
who formed the company PROABD to exploit it. PROABD is now a division of BEFS
Technologies, Mulhouse, France, which offers this process and others under the same trade
name. Not to be confused with the Abderhalden reaction in biochemistry.

> British Patents 239,841; 253,935.
> Hoffert, W. H. and Claxton, G., *Motor Benzole: Its Production and Use,* National Benzole
> Association, London, 1957, 38.
> *Mines,* 1957, **12**(53), 223.

Abgas-Turbo-Wascher von Kroll   Not a process, but a piece of equipment for scrubbing
flue-gases with an aqueous suspension of lime. Developed by Walter Kroll GmbH and used
in 14 plants in West Germany in 1986.

ACAR   *See* steelmaking.

4A-CAT [Activity adjustment by ammonia adsorption]   A method for pre-sulfiding and passivating hydrocracking catalysts. Developed by EUROCAT in 1989.

Chauvel, A., Delmon, B., and Hölderich, W. F., *Appl. Catal. A: Gen.,* 1994, **115**, 184.

Accar   A direct reduction ironmaking process, using coal and oil as the reductants. Operated at the OSIL plant at Keonjhar, India, from 1983 to 1987. *See* DR.

Accent [Aqueous carbon compound effluent treatment]   A process for oxidizing organic contaminants in aqueous streams by catalyzed oxidation with sodium hypochlorite. The catalyst is promoted nickel oxide, which retains active oxygen at its surface, as well as adsorbing the organics. Developed by ICI Katalco and first offered in 1998.

Acedox [Acetic oxidation]   A pulp-bleaching process using peracetic acid as the oxidant. Developed by Eka Nobel in 1994 and first commercialized, in combination with *Lignox, in Sweden in 1995.

Acetate   A general name for processes for making cellulose acetate fibers. Cellulose is acetylated, dissolved in acetone, and spun into fibers by injecting through orifices into heated chambers. Cellulose mono-acetate is made by acetylating with a mixture of acetic acid, acetic anhydride, and sulfuric acid as the catalyst. Cellulose tri-acetate is made in a similar fashion, but using perchloric acid as the catalyst, and dry-spinning from a solution in ethanol/ methylene chloride. Cellulose tri-acetate fibers were first made commercially by Courtaulds in London in 1950.

Peters, R. H., *Textile Chemistry,* Elsevier, Amsterdam, 1963, Vol. 1, 187.

Acetex   A vapor-phase process for selectively hydrogenating acetylene in the presence of ethylene. Developed by IFP in France in 1993.

Chem. Eng. News, 1993, **71**(34), 21.
Chauvel, A., Delmon, B., and Hölderich, W. F., *Appl. Catal. A: Gen.,* 1994, **115**, 186.

Acetosolv   A wood pulping and bleaching process which uses hydrogen peroxide and acetic acid. *See* Organosolv.

Eur. Chem. News (Finland Suppl.), 1991, May, 28.

Acetylene Black   A process for making carbon black from acetylene by thermal decomposition at 800 to 1,000°C in refractory-lined, water-cooled retorts.

Kühner, G. and Voll, M., in *Carbon Black Science and Technology,* Donnet, J.-B., Bansai, R. C., and Wang, M.-J., Eds., Marcel Dekker, New York, 1993, 61.
Claasen, E. J., in *Inorganic Chemicals Handbook,* Vol. 2, McKetta, J. J., Ed., Marcel Dekker, New York, 1993, 510.

ACH (1) [Acetone cyanhydrin]   A process for making methyl methacrylate via this intermediate. Acetone reacts with hydrogen cyanide to yield the cyanhydrin. This is then converted to methacrylamide, using concentrated sulfuric acid. Methanolysis of this yields methyl methacrylate. Developed by Röhm GmbH Chemische Fabrik, Germany, and ICI, UK; used in 11 countries in 1990.

Porcelli, R. V. and Juran, B., *Hydrocarbon Process.,* 1986, **65**(3), 39.
Chem. Eng. (N.Y.), 1990, **97**(3), 35.

ACH (2) [Aluminium chlorohydrate]   This is the common name for some types of basic aluminum chloride, but the name has been used also to designate the process by which such a product is made. Several processes are used to make the several commercial aluminum chloride products available, some of which are proprietary. In general it is necessary to introduce an excess of aluminum to a chloride solution, such that the atom ratio of aluminum to chlorine is less than three. The aluminum may be introduced as either the metal or the hydrated oxide.

Acheson (1)   A process for making silicon carbide from sand and coke, in an electric furnace, at 2,200 to 2,400°C:

$$SiO_2 + 3C = SiC + 2CO$$

Invented by E. G. Acheson in Monongahela City, PA, in 1892. He was heating clay and carbon by means of an electric arc, in the hope of making diamond. The hard, crystalline product was called carborundum in the mistaken belief that it was a compound of carbon and corundum (alumina). The process and product were patented in 1893 and made on a small scale in Monongahela City, using the town's electricity supply. In 1895, The Carborundum Company was formed to exploit the process in Niagara, NY, using hydroelectric power from the Falls. This same process is now operated in many countries. The name Carborundum is a registered trademark owned by the Carborundum Company, NY, and used for several of its refractory products, in addition to silicon carbide.

U.S. Patent 492,767.
Szymanowitz, R., *Edward Goodrich Acheson: Inventor, Scientist, Industrialist,* Vantage Press, New York, 1971.
Mühlhaeuser, O., *J. Am. Chem. Soc.,* 1893, **15,** 411.

Acheson (2)   A process for converting carbon articles into graphite, invented by E. G. Acheson in 1895 and commercialized in 1897. This process uses transverse graphitization, unlike the *Castner process, which uses lengthwise graphitization.

U.S. Patents 568,323; 617,979; 645,285.
Szymanowitz, R., *Edward Goodrich Acheson: Inventor, Scientist, Industrialist,* Vantage Press, New York, 1971.

Acid   A process for making sodium perborate by reacting sodium borate ("borax") with sodium peroxide and hydrochloric acid:

$$Na_2B_4O_7 + 4Na_2O_2 + 6HCl + 13H_2O = 2Na_2 [B_2O_4 (OH)_4] \cdot 6H_2O + 6NaCl$$

Operated by the Castner-Kellner Company, Runcorn, England, from 1915 until it was supplanted by the *Duplex (2) process in 1950.

Hardie, D. W. F. and Pratt, J. D., *A History of the Modern British Chemical Industry,* Pergamon Press, Oxford, 1966, 141.

Acid Bessemer   An alternative name for the original *Bessemer steelmaking process in which the furnace is lined with a silica refractory. It is suitable only for ores relatively free from phosphorus.

Acid Open Hearth   The original version of the *Open Hearth process for steelmaking in which the hearth is made of a silica refractory. The process does not remove phosphorus or sulfur, the acid impurities in the iron, so the raw materials must be relatively free from these. Pioneered by C. W. Siemens and F. M. E. and P. Martin at Sireuil, France, in 1864.

British Patent 2,031, 1864.
Barraclough, K. C., *Steelmaking 1850–1900,* The Institute of Metals, London, 1990, 137.

ACIMET   [Acid Methane]   A two-stage, anaerobic digestion process for treating municipal wastewaters. In the first stage, organic matter is decomposed to a mixture of acids, aldehydes, and alcohols. In the second, the carbon in this mixture is anaerobically converted to methane. Invented in 1974 by S. Ghosh and D. L. Klass at the Illinois Institute of Gas Technology (IGT), Chicago. First commercialized in 1991 by IGT and DuPage County, IL, at the Woodridge-Greene Valley Wastewater Treatment Plant.

U.S. Patent 4,022,665.
Ghosh, S., Conrad, J. R., and Klass, D. L., *J. Water Pollut. Control Fed.,* 1975, **47**(1), 30.

ACR [**A**dvanced **C**racking **R**eactor] A *thermal petroleum cracking process, the heat being provided by partial combustion of the feed at 2,000°C. Developed by Chiyoda Chemical Engineering & Construction Company, Kureha Chemical Industry Company, and Union Carbide Corp. in the 1970s. A demonstration plant was operated in Seadrift, TX, from 1979 to 1981.

Ishkawa, T. and Keister, R. G., *Hydrocarbon Process.*, 1978, **57**(12), 109.
Hu, Y. C., in *Chemical Processing Handbook,* Marcel Dekker, New York, 1993, 768.

actiCAT A process for pre-sulfurizing hydrotreating catalysts. Pre-sulfurizing differs from pre-sulfiding in that the products are complex metal oxysulfides, rather than sulfides. A novel organic "matrix" retains the sulfur during the conversion process. Developed by CRI International Inc and offered by that company as a service to the petroleum industry.

Welch, J. G., Poyner, P., and Skelly, R. F., *Oil Gas J.,* 1994, **92**(41), 56.
Blashka, S., Bond, G., and Ward, D., *Oil Gas J.,* 1998, **96**(1), 36.

ACTIFLOW A process for treating raw water. Flocculation of insoluble matter by the addition of a polyelectrolyte takes place within an agitated bed of fine sand. Developed in France by OTV and licensed in the UK through General Water Processes.

Actimag A process for reducing metal ions in aqueous solution by metallic iron. The iron is in the form of particles 1 mm in diameter contained in a fluidized bed and kept in violent agitation by means of an alternating magnetic field. The agitation accelerates the reaction and prevents the adhesion of deposits of reduction products. Demonstrated for reducing the cupric ion to metallic copper, and chromate ion to chromic ion. Developed by Extramet, France, in the 1980s and offered in the United Kingdom by Darcy Products.

European Patent 14,109.
Bowden, P., *Water Waste Treat.,* 1989, **32**(7), 21.
Bowden, P., *Processing,* 1990, 27.

Activated MDEA A version of the *MDEA process for scrubbing acid gases from gas streams, in which the aqueous MDEA solution is regenerated by flashing rather than by stripping. Developed by BASF, Germany in 1971, with the Ralph M. Parsons Co. becoming the sole licensor in most of the Western Hemisphere in 1982. The process is now operated in Europe, Canada, and the United States.

Hydrocarbon Process., 1996, **75**(4), 105.

Activated Sludge A sewage treatment process, developed in the 1920s and soon widely adopted. Based on the aeration of wastewater with flocculating biological growth, followed by separation of the treated wastewater. It removes dissolved and colloidal organic material, suspended solids, some of the mineral nutrients (P- and N-compounds), and some volatile organic compounds. Generally ascribed to H. W. Clark and S. M. de Gage in Massachussetts (1912), followed by E. Arden and M. T. Lockett in Manchester (1914). The first plant was installed in Worcester, England, in 1916.

Arden, E. and Lockett, M. T., *J. Soc. Chem. Ind. (London),* 1914, **33**(10), 523; (23), 1122.
Ganczarczyk, J.J., *Activated Sludge Process: Theory and Practice,* Marcel Dekker, New York, 1983.

ADAM-EVA *See* EVA-ADAM.

Addipol A process for making polypropylene, developed and licensed by Himont, in the United States, and commercialized in 1988. *See also* Spheripol.

Adex   A process for removing heavy metals from phosphoric acid by precipitation of their complexes with 2-ethylhexyl dithiophosphate. Developed by Hoechst, Germany.

Becker, P., *Phosphates and Phosphoric Acid,* 2nd.. ed., Marcel Dekker, New York, 1989, 531.

Adib   A process for extracting isobutene from petroleum fractions by reaction with phenol. The reaction takes place in the gas phase, over an acid catalyst, and yields all the mono-, di-, and tri-butyl phenols. Heating this mixture liberates isobutene; the phenol and the catalyst are recovered for re-use. Piloted in Argentina in the 1980s.

Miranda, M., *Hydrocarbon Process.,* 1987, **66**(8), 51.

Adip   [Possibly an acronym of DIPA, di-isopropanolamine]   A process for removing hydrogen sulfide, mercaptans, carbonyl sulfide, and carbon dioxide from refinery streams by extraction into an aqueous solution of di-isopropanolamine or methyl diethanolamine. Developed and licensed by the Shell Oil Company, Houston, TX. More than 320 units were operating in 1992.

Bally, A. P., *Erdoel Kohle Erdgas Petrochemie,* 1961, **14,** 921.
Hydrocarbon Process., 1975, **54**(4), 79.
Kohl, A. L. and Riesenfeld, F. C., *Gas Purification,* 4th ed., Gulf Publishing Co., Houston, TX, 1985, 41.
Hydrocarbon Process., 1992, **71**(4), 86.

Adkins-Peterson   The oxidation of methanol to formaldehyde, using air and a mixed molybdenum/iron oxide catalyst. Not an engineered process, but the reaction which formed the basis of the *Formox process.

U.S. Patent 1,913,405.
Adkins, H. and Peterson, W. R., *J. Am. Chem. Soc.,* 1931, **53,** 1512.

ADOX   *See* CATOX.

ADU   [Ammonium **di**uranate]   A process for converting uranium hexafluoride into uranium dioxide, for use as a nuclear reactor fuel. The hexafluoride is hydrolyzed in water:

$$UF_6 + 2H_2O = UO_2F_2 + 4HF$$

and the solution treated with ammonia, precipitating ammonium diuranate:

$$2UO_2F_2 + 8HF + 14NH_3 + 3H_2O = (NH_4)_2U_2O_7 + 12NH_4F$$

which is filtered off and reduced with hydrogen. Developed in the United States in the 1950s.

Büchner, W., Schliebs, R., Winter, G., and Büchel, K. H., *Industrial Inorganic Chemistry,* VCH Publishers, Weinheim, Germany, 1989, 581.

ADVACATE   A *flue-gas desulfurization process, similar to *CZD, but using a suspension of fly-ash instead of lime. Developed by the University of Texas, the U.S. Environmental Protection Agency, and Acurex Corporation.

AEROSIL   A process for making sub-micron sized silica, alumina, or titania powders by the flame hydrolysis of the respective chlorides. The chloride vapor is passed through an oxy-hydrogen flame; the reaction is thus a flame hydrolysis, rather than an oxidation, so is to be distinguished from the *Chloride process for making titanium dioxide pigment. Developed by Degussa in 1941 and operated by that company in Reinfelden, Germany.

German Patent 870,242.
Ulrich, G. D., *Chem. Eng. News,* 1984, **62**(32), 22.

AFC   *See* Compagnie AFC.

AGC-21   A process for converting natural gas to liquid fuels in three stages: generation of syngas in a fluidized bed, Fischer-Tropsch synthesis in a slurry bubble column reactor, and hydrocracking. Piloted in 1997 and proposed for installation in Qatar.

Appl. Catal., A: Gen., 1997, **155**(1), N5.

AhlStage   [**Ahl**strom **stage**]   A pulp-bleaching process that economizes on oxidizing agents by first destroying hexenuronic acid derivatives that would otherwise consume them. They are destroyed by hydrolysis with dilute sulfuric acid. Developed by Ahlstrom Machinery Corporation, Finland, in 1996.

Chem. Eng. (N.Y.), 1996, **103**(12), 17.

AHR   [**A**dsorptive **h**eat **r**ecovery]   A vapor-phase process for removing water from other vapors by selective adsorption in a bed of zeolite molecular sieve, and regenerating the adsorbent by passing a noncondensible gas through it at essentially the same temperature and pressure. The heat of adsorption is stored as a temperature rise within the bed, and provides the heat required for desorption. Developed by Union Carbide Corporation for energy-efficient drying of petrochemical streams containing substantial amounts of water, and for drying ethanol for use in motor fuels. Five units have been licensed by UOP.

Garg, D. R. and Ausikaitis, J. P., *Chem. Eng. Prog.,* 1983, **79**(4), 60.
Garg, D. R. and Yon, C. M., *Chem. Eng. Prog.,* 1986, **82**(2), 54.

AH Unibon   A process for hydrogenating aromatic hydrocarbons in petroleum fractions to form aliphatic hydrocarbons. Developed by UOP.

AIAG Neuhausen   An electrolytic process for making aluminum from an all-fluoride melt. Developed by the Société Suisse de l'Aluminium Industrie at Neuhausen, Germany.

Airco   A modification of the *Deacon process for oxidizing hydrogen chloride to chlorine. The copper catalyst is modified with lanthanides and used in a reversing flow reactor without the need for external heat. Developed by the Air Reduction Company from the late 1930s.

U.S. Patents 2,204,172; 2,312,952; 2,271,056; 2,447,834.
Redniss, A., in *Chlorine, Its Technology, Manufacture and Uses,* Sconce, J. S., Ed., Reinhold, New York, 1962, 252.

Airlift Thermofor Catalytic Cracking   Also called Airlift TCC. A continuous catalytic process for converting heavy petroleum fractions to lighter ones. The catalyst granules are moved continuously by a stream of air. Developed by Mobil Oil Corp., United States, and first operated in 1950. See also Thermofor.

Enos, J. L., *Petroleum Progress and Profits,* MIT Press, Cambridge, 1962, Chap. 5.
Unzelman, G. H. and Wolf, C. J., in *Petroleum Processing Handbook,* Bland, W. F. and Davidson, R. L., Eds., McGraw-Hill, New York, 1967, Chap. 3, p 7.

Ajax   An oxygen steelmaking process in which the oxygen is injected into an *open hearth furnace through water-cooled lances. Used at the Appleby-Frodingham steelworks, UK.

Akzo-Fina CFI   A process for improving the quality of diesel fuel by dewaxing, hydrotreating, and hydrocracking. Developed by Akzo Nobel and Fina from 1988.

Absci-Halabi, M., Stanislaus, A., and Qabazard, H., *Hydrocarbon Process.,* 1997, **76**(2), 49.

Albene   [**Al**cohol **bene**zene]   A process for making ethylbenzene from aqueous ethanol and benzene. The aqueous ethanol may contain as little as 30 percent ethanol, such as that obtained by one distillation of liquors from sugar fermentation. The mixed vapors are passed over a catalyst at approximately 350°C. The catalyst ("Encilite-2") is a ZSM-5–type zeolite in which some of the aluminum has been replaced by iron. Developed in India jointly by the

National Chemical Laboratory and Hindustan Polymers; operated commercially by Hindustan Polymers at Vizay, Andhra Pradesh, since 1989.

Indian Patent 157,390.

Alberger   A process for crystallizing sodium chloride from brine. The brine is heated under pressure to 145°C to remove calcium sulfate. Flashing to atmospheric pressure produces fine cubic crystals of sodium chloride, and surface evaporation in circular vessels produces flakes of it. Developed by J. L. and L. R. Alberger in the 1880s. *See also* Recrystallizer.

U.S. Patents 351,082; 400,983; 443,186.

Richards, R. B., in *Sodium Chloride*, D. W. Kaufmann, Ed., Reinhold Publishing, New York, 1960, 270.

Alkar   *See* Alkar.

Alcell   [Alcohol **cell**ulose]   A process for delignifying wood pulp by dissolving it in aqueous ethanol at high temperature and pressure. Developed by Repap Technologies, United States.

Chem. Eng. (N.Y.), 1991, **98**(1), 41.

Alceru   A process for making cellulosic filaments and staple fibres. The cellulose is first dissolved in an aqueous solution of N-methylamine-N-oxide. Developed by Zimmer (Frankfurt) and TITK (Rudolstadt) from 1987. A pilot plant was expected to be built by April 1998.

Chem. Week, 1997, **159**(25), 21.

ALCET   [advanced low-capital ethylene technology]   A process for separating ethylene from the gases made by cracking naphtha. It replaces the conventional cryogenic stages with a proprietary solvent absorption process. Developed by a consortium of Brown & Root, Advanced Extraction Technologies, and Kinetics Technology International but not yet commercialized. A demonstration unit was planned for summer 1996.

Chem. Eng. News, 1994, **72**(29), 6.

Eur. Chem. News, CHEMSCOPE, 1996, **65,** Jun. 8.

Hydrocarbon Process., 1995, **74**(3), 118.

Alco   An early process for thermally polymerizing refinery gases (mainly C_3 and C_4 hydrocarbons) to yield liquid hydrocarbon mixtures, suitable for blending with gasoline. The process was operated without a catalyst, at 480 to 540°C, and 50 atm. Developed by the Pure Oil Company, Chicago, and licensed to Alco Products, United States.

Asinger, F., *Mono-olefins: Chemistry and Technology,* translated by B. J., Hazzard, Pergamon Press, Oxford, 1968, 426.

ALCOA   A process proposed for manufacturing aluminum metal by the electrolysis of molten aluminum chloride, made by chlorinating alumina. It requires 30 percent less power than the *Hall-Héroult process and operates at a lower temperature, but has proved difficult to control. Developed by the Aluminum Company of America, Pittsburgh, in the 1970s and operated in Palestine, TX, from 1976; abandoned in 1985 because of corrosion problems and improvements in the efficiency of conventional electrolysis.

Grjotheim, K., Krohn, C., Malinovsky, M., Matiaskovsky, K., and Thonstad, J., *Aluminium Electrolysis—Fundamentals of the Hall-Hérault process,* CRC Press, Boca Raton, FL, 1982, 17.

Palmear, I. J., in *The Chemistry of Aluminium, Gallium, Indium, and Thallium*, Downs, A. J., Ed., Blackie, London, 1993, 87.

Aldip   *See* metal surface treatment.

Aldol   Also called the Four-step process. A process for converting acetylene to synthetic rubber, used on a large scale in Germany during World War II. A four-step synthesis converted the acetylene to butadiene, and this was then polymerised by the *Buna process. The four steps were:

1. hydration of acetylene to acetaldehyde, catalyzed by sulfuric acid and mercuric sulfate;
2. condensation of acetaldehyde to aldol, using aqueous alkali (the "aldol condensation");
3. hydrogenation of aldol to 1,3-butanediol;
4. dehydrogenation of 1,3-butanediol to 1,3-butadiene, catalyzed by sodium phosphate on coke.

The process was still in use in East Germany in the 1990s.

Fisher, H. L., in *Synthetic Rubber,* Whitby, G. S., Davis, C. C., and Dunbrook, R. F., Eds., John Wiley & Sons, New York, 1954, 121.
Weissermel, K. and Arpe, H.-J., *Industrial Organic Chemistry,* 3rd ed., VCH Publishers, Weinheim, Germany, 1997, 106.

Aldox   [**Ald**olization **OXO**]   A *hydroformylation process for converting olefins having n carbon atoms to aldehydes having (2n + 2) carbon atoms. The olefins are reacted with carbon monoxide and hydrogen, in the presence of an organometallic catalyst. Invented by Esso Research & Engineering Co., United States, in 1954, and operated since 1962 by Humble Oil & Refining Company at Baton Rouge, LA.

British Patents 761,024; 867,799.
Chem. Eng. (N.Y.), 1961, **68**(25), 70
Weissermel, K. and Arpe, H.-J., *Industrial Organic Chemistry,* 3rd ed., VCH Publishers, Weinheim, Germany, 1997, 138.

Alfene   [**Alf**a ol**efene**]   Also spelled Alfen. A process for making higher alpha-olefins. Ethylene is reacted with triethyl aluminum, yielding high molecular weight aluminum alkyls, and these are treated with additional ethylene, which displaces the higher olefins. Developed by the Continental Oil Company.

Chem. Eng. News, 1962, **40**(16), 68, 70.
Acciarri, J. A., Carter, W. B., and Kennedy, F., *Chem. Eng. Prog.,* 1992, **58**(6), 85.
Weissermel, K. and Arpe, H.-J., *Industrial Organic Chemistry,* 3rd ed., VCH Publishers, Weinheim, Germany, 1997, 75.

Alfin   An obsolete process for making synthetic rubber by polymerizing butadiene in pentane solution. The catalyst was an insoluble aggregate of sodium chloride, sodium isopropoxide, and allyl sodium. The name is actually the name of the catalyst, derived from **al**cohol, used to make the sodium *iso*propoxide, and ole**fin,** referring to the propylene used to make the allyl sodium.

Morton, A. A., Magat, E. E., and Letsinger, R. L., *J. Am. Chem. Soc.,* 1947, **69**, 950.
Morton, A. A., *Ind. Eng. Chem.,* 1950, **42**, 1488

Alfol   Also called the Conoco process and the Mühlheim process. The same name is used for the products as well. A process for making linear primary alcohols, from C_2 to C_{28}, from ethylene. The ethylene is reacted with triethyl aluminum, yielding higher alkyl aluminums. These are oxidized with atmospheric oxygen under mild conditions to aluminum alkoxides, which are then hydrolyzed by water to the corresponding alcohols:

$$2AlR_3 + 3O_2 = 2Al(OR)_3$$
$$2Al(OR)_3 + 3H_2O = 6ROH + Al_2O_3$$

Invented by K. Ziegler at the Max Planck Institut für Kohlenforschung, Mühlheim/Ruhr, Germany. Operated in the United States by Conoco since 1962, and in Germany by Condea Chemie since 1964. *See also* Epal.

German Patent 1,014,088.
East German Patent 13,609.
Belgian Patent 595,338.
Ziegler, K., Krupp, F., and Zosel, K., *Angew. Chem.*, 1955, **67**, 425.
Lobo, P. A., Coldiron, D. C., Vernon, L. N., and Ashton, A. T., *Chem. Eng. Prog.*, 1962, **58**(5), 85.
Hydrocarbon Process., 1963, **42**(11), 140.
Weissermel, K. and Arpe, H.-J., *Industrial Organic Chemistry*, 3rd ed., VCH Publishers, Weinheim, Germany, 1997, 75.

AlgaSORB A process for removing toxic heavy metals from aqueous wastes by the use of algae supported on silica gel.

Veglio, F. and Beolchini, F., *Hydrometallurgy*, 1997, **44**, 301.
Veglio, F., and Beolchini, F., and Toro, L., *Ind. Eng. Chem. Res.*, 1998, **37**(3), 1105.

Alkacid A process for removing sulfur compounds from gas streams. All the sulfur compounds are first catalytically hydrogenated to hydrogen sulfide using a cobalt/molybdena catalyst. The hydrogen sulfide is then absorbed in an aqueous solution of an amino acid salt. Heating this solution regenerates the hydrogen sulfide as a concentrate, which is then treated by the *Claus process. Invented by IG Farbenindustrie in 1932; by 1950, 50 plants were operating in Europe, the Middle East, and Japan. *See also* Alkazid.

U.S. Patent 1,990,217.
Lühdemann, R., Noddes, G., and Schwartz, H. G., *Oil Gas J.*, 1959, **57**(32), 100.
Kohl, A. L. and Riesenfeld, F. C., *Gas Purification*, 4th ed., Gulf Publishing, Houston, TX, 1985, 203.

Alkad A process for improving the safety of *alkylation processes using hydrofluoric acid as the catalyst. A proprietary additive curtails the emission of the acid aerosol that forms in the event of a leak. Based on observation of G. Olah in the early 1990s that liquid polyhydrogen fluoride complexes (of amines such as pyridine) depress the vapor pressure of HF above alkylation mixtures. Developed by UOP and Texaco and operated at Texaco's refinery at El Dorado, TX, since 1994. A competing process is *ReVAP, developed by Phillips and Mobil.

U.S. Patent 5,073,674.
Chem. Eng. (N.Y.), 1995, **102**(12), 68.
Sheckler, J. C., Hammershaimb, H. U., Ross, L. J., and Comey, K. R., III, *Oil Gas J.*, 1994, **92**(34), 60.

Alkar [**Alk**ylation of **ar**omatics] Also (incorrectly) spelled Alcar. A catalytic process for making ethylbenzene by reacting ethylene with benzene. The ethylene stream can be of any concentration down to 3 percent. The catalyst is boron trifluoride on alumina. Introduced by UOP in 1958 but no longer licensed by them. Replaced by the *Ethylbenzene process.

Grote, H. W. and Gerald, C. F., *Chem. Eng. Prog.*, 1960, **56**(1), 60.
Hydrocarbon Process., 1963, **42**(11), 141.
Mowry, J. R., in *Handbook of Petroleum Refining Processes*, Meyers, R. A., Ed., McGraw-Hill, New York, 1986, 1–29.

Alkazid A development of the *Alkacid proces. The absorbent is an aqueous solution of the potassium salt of either methylamino propionic acid ("Alkazid M"), or dimethylamino acetic acid ("Alkazid DIK"). Developed by Davy Powergas, Germany. Over 80 plants were operating in 1975.

Bähr, H., *Chem. Fabrik,* 1938, **11**(23/24), 283.
Unzelman, G. H. and Wolf, C. J., in *Petroleum Processing Handbook,* Bland, W. F. and Davidson, R. L., Eds., McGraw-Hill, New York, 1967, 133.
Hydrocarbon Process., 1975, **54**(4), 85.
Speight, J. G., *Gas Processing,* Butterworth Heinemann, Oxford, 1993, 256.

alkylation   Any process whereby an alkyl group is added to another molecule; however, in process chemistry the word is most commonly used to designate a reaction in which an olefin is added to a saturated aliphatic hydrocarbon or an aromatic compound. In the petroleum and petrochemical industries, this term refers to the conversion of a mixture of light olefins and isobutane into a mixture of alkanes suitable for blending into gasoline in order to increase the octane number. An acid catalyst is used. Originally the acid chosen was anhydrous hydrofluoric or sulfuric acid. Proprietary solid acids were introduced in the 1990s which were easier to dispose of. The product is called alkylate. Those alkylation processes having special names which are described in this Dictionary are: Alkar, FBA, Detal, Detergent Alkylate, Mobil-Badger, Stratco, Thoma.

Alkymax   A process for removing benzene from petroleum fractions. They are mixed with light olefin fractions (containing mainly propylene) and passed over a fixed-bed catalyst, which promotes benzene alkylation. The catalyst is solid phosphoric acid (SPA), made by mixing a phosphoric acid with a siliceous solid carrier, and calcining. Invented in 1980 by UOP.

U.S. Patent 4,209,383.
Hydrocarbon Process., 1994, **73**(11), 90.

Allis-Chalmers   *See* DR.

ALMA   [Alusuisse maleic anhydride]   A process for making maleic anhydride by oxidizing *n*-butane, using a fluid bed reactor and a special organic solvent recovery system. The catalyst contains vanadium and phosphorus on iron oxide. Developed jointly by Alusuisse Italia and ABB Lummus Crest. First licensed to Shin-Daikowa Petrochemical Company, Yokkaichi, Japan, in 1988. The world's largest plant plant was built for Lonza in Ravenna, Italy, in 1994.

Budi, F., Neri, A., and Stefani, G., *Hydrocarbon Process.,* 1982, **61**(1), 159.
Arnold, S. C., Suciu, G. D., Verde, L., and Neri, A., *Hydrocarbon Process.,* 1985, **64**(9), 123.
Chem. Eng. *(N.Y.),* 1996, **103**(11), 17.
Weissermel, K. and Arpe, H.-J. *Industrial Organic Chemistry,* 3rd ed., VCH Publishers, Weinheim, Germany, 1997, 370.

Aloton   Also called Büchner. A process proposed for extracting aluminum from clay. Calcined clay is leached with ammonium hydrogen sulfate solution under pressure, and ammonium alum is crystallized from the liquor. Invented by M. Büchner in Hanover-Kleefeld in 1921; piloted in Germany in the 1920s and in Oregon in 1944. It was never commercialized, but provided the basis for the *Nuvalon process which was.

British Patent 195,998.
U.S. Patent 1,493,320.
O'Connor, D. J., *Alumina Extraction from Non-bauxitic Materials,* Aluminium-Verlag, Düsseldorf, 1988, 159.

Alpha   A process for making aromatic hydrocarbons and LPG from C_3–C_7 olefins. The catalyst is a metal-modified ZSM-5 zeolite. Developed by Asahi Chemical Industries and Sanyo Petrochemical and used since 1993 at Sanyo's Mitzushima refinery.

Eur. Chem. News, CHEMSCOPE, 1994, Apr. 7; 1996, Jun. 4.

Alphabutol   Also called IFP-SABIC. A process for dimerizing ethylene to 1-butene. It operates under pressure at 80°C, using a complex Ziegler-Natta catalyst, a titanium alkoxide. Developed by the Institut Français du Pétrole. First operated in Thailand in 1987. Seven plants had been licensed by 1993, of which three were operating.

> Commereuc, D., Chauvin, Y., Gaillard, J., Léonard J., and Andrews, J., *Hydrocarbon Process.,* 1984, **63**(11), 118.
> Hennico, A., Léonard, H. J., Forestiere, A., and Glaize, Y., *Hydrocarbon Process.,* 1990, **69**(3), 73.
> Chauvel, A., Delmon, B., and Hölderich, W. F., *Appl. Catal. A: Gen.,* 1994, **115**(2), 201.
> Weissermel, K. and Arpe, H.-J. *Industrial Organic Chemistry,* 3rd ed., VCH Publishers, Weinheim, West Germany, 1997, 67.

Alplate   *See* metal surface treatment.

Alrak   *See* metal surface treatment.

Alstan   A process for electroplating aluminum by pretreating the surface with a stannate. Developed by M & T Chemicals.

> Di Bari, G. A., *Plat. Surf. Finish.,* 1977, **64**(5), 68.

Alumet   A process for extracting alumina and potassium sulfate from alunite ore (a basic hydrated potassium aluminum sulfate) involving reductive calcination and alkali leaching. Developed and piloted by the Alunite Metallurgical Company, UT, in the mid 1970s but not commercialized. *See also* Kalunite.

> O'Connor, D. J., *Alumina Extraction from Non-bauxitic Materials,* Aluminium-Verlag, Düsseldorf, 1988, 198.

Alumilite   *See* metal surface treatment.

ALUREC   A process for recovering aluminum from residues obtained from the remelting of aluminum scrap. The material is melted in a rotating furnace heated with natural gas and oxygen. Previous processes involved melting with salt. Developed jointly by AGA, Hoogovens Aluminium, and MAN GHH, and offered in 1994.

Alzak   A method for electropolishing aluminum, using fluoroboric acid. Developed by The Aluminum Company of America. *See also* metal surface treatment.

Amalgam   A process for making sodium dithionite by reacting sodium amalgam with sulfur dioxide:

$$2(Hg)Na + 2SO_2 = Na_2S_2O_4 + 2(Hg)$$

> Bostian, L. C., in *Speciality Inorganic Chemicals,* Thompson, R., Ed., Royal Society of Chemistry, London, 1981, 63.

Aman   A process for thermally decomposing metal chloride or sulfate solutions in a spray roaster. Used for recovering hydrochloric acid from iron pickle liquors. Developed by J. J. Aman in Israel in 1954.

> British Patent 793,700.

AMAR   A solvent extraction process for recovering copper. Used in approximately 50 installations worldwide in 1993.

AMASULF   A two-stage process for removing hydrogen sulfide and ammonia from coke oven gas. In the first stage, hydrogen sulfide is removed by scrubbing with aqueous ammonia; the resulting ammonium sulfide solution is heated in another vessel to expel the hydrogen sulfide:

$$H_2S + 2NH_4OH = (NH_4)_2S + 2H_2O$$

$$(NH_4)_2S + H_2O = H_2S + 2NH_4 OH$$

In the second stage, ammonia is removed by scrubbing with water. Developed and licensed by Krupp Koppers, Germany.

AMASULF^{R}PURE   A variation on the *AMASULF process in which only the hydrogen sulfide is recovered, not the ammonia.

aMDEA [activated **M**ethyl **D**iethanolamine]   A process for removing CO_2, H_2S, and trace sulfur compounds from natural gas and *syngas. Developed by BASF and used in more than 90 plants in 1997.

American   Also known as the Wetherill process, and the Direct process. A process for making zinc oxide, in the form of a white pigment, from a zinc oxide ore. The ore is usually franklinite, which is predominately $ZnFe_2O_4$. The ore is mixed with coal and heated in a furnace to approximately 1,000°C, forming zinc vapor in a reducing atmosphere. The vapors pass to a second chamber in which they are oxidized with air, forming zinc oxide and carbon dioxide. *See also* French.

Ames (1)   A process for making uranium by reducing uranium tetrafluoride with calcium or magnesium.

Ames (2)   A wet oxidation process for desulfurizing coal in which the oxidant is oxygen and the sulfur dioxide is absorbed by aqueous sodium carbonate. Developed in the 1970s by the Ames Laboratory of Iowa State University, with funding from the U.S. Department of Energy. *See also* PETC.

   IEA Coal Research, *The Problems of Sulphur,* Butterworths, London, 1989, 20.

AMEX [**Am**ine **ex**traction]   A process for the solvent extraction of uranium from sulfuric acid solutions using an amine extractant:

$$UO_2(SO_4)_2{}^{2-} + 2H^+ + 2B \rightarrow (BH^+)_2 \cdot [UO_2 (SO_4)_2{}^{2-}]$$

The amine (B) is a proprietary mixture of C_8 and C_{10} primary alkylamines dissolved in kerosene. The uranium is stripped from the organic solution with an alkaline stripping solution and precipitated as ammonium diuranate. *See also* Dapex.

   Chem. Eng. News, 1956, **34**(21), 2590.
   Eccles, H. and Naylor, A., *Chem. Ind. (London),* 1987, (6), 174.
   Danesi, P. R., in *Developments in Solvent Extraction,* Alegret, S., Ed., Ellis Horwood, Chichester, England, 1988, 204.

Amine Guard   A process for extracting acid gases from refinery streams by scrubbing with an alkanolamine. Many such processes have been developed, this one was developed by the Union Carbide Corp. and uses monoethanolamine. It has been used to purify hydrogen produced by *steam reforming. In 1990, over 375 units were operating.

   Butwell, K. F., Hawkes, E. N., and Mago, B. F. *Chem. Eng. Prog.,* 1973, **69**(2), 57.
   Butwell, K. F., Kubek, D. J., and Sigmund, P. W., *Chem. Eng. Prog.,* 1979, **75**(2), 75.
   Kubek, D. K. and Butwell, K. F., in *Acid and Sour Gas Treating Processes,* Newman, S. A., Ed., Gulf Publishing, Houston, TX, 1985, 235.
   Hydrocarbon Process., 1992, **71**(4), 86.
   Hydrocarbon Process., 1996, **75**(4), 105.

AMINEX   A process for removing hydrogen sulfide and carbon dioxide from gas and LPG streams, by circulating an aqueous amine solution through bundles of hollow fibers immersed in them. Developed in 1991 by the Merichem Company, Houston, TX.

   Hydrocarbon Process., 1996, **75**(4), 126.

AMISOL   A process for removing sulfur compounds and carbon dioxide from refinery streams by absorption in methanol containing mono- or di-ethanolamine and a proprietary additive. Developed by Lurgi, Germany, in the 1960s and first commercialized in the early 1970s.

Bratzler, K. and Doerges, A., *Hydrocarbon Process.*, 1974, **53**(4), 78.

Kohl, A. L. and Riesenfeld, F. C., *Gas Purification,* 4th ed., Gulf Publishing, Houston, TX, 1985, 871.

Ammonex   An ion-exchange process for continuously purifying the water circuits of electric power generators. Ammonia is used to regenerate the cation exchange resins. Developed by Cochrane Environmental Systems in the 1960s and widely used.

Crits, D. J., in *Ion Exchange Technology,* Naden, D. and Streat, M., Eds., Ellis Horwood, Chichester, England, 1984, 119.

Ammonia-soda   Also called the Solvay process. A process for making sodium carbonate. The basic process was invented and partially developed in the first half of the 19th century by several workers, but the key invention was made by E. Solvay in Belgium in 1861. The first plant was built at Couillet, Belgium, in 1864 and thereafter the process became accepted worldwide, displacing the *Leblanc process. The raw materials are limestone and salt; calcium chloride is a waste product. The overall reaction is:

$$CaCO_3 + 2NaCl = CaCl_2 + Na_2CO_3$$

When carbon dioxide is passed into a nearly saturated solution of sodium chloride containing some ammonia, ammonium bicarbonate is formed. The heart of the process is the exploitation of the equilibrium between this bicarbonate and sodium and ammonium chlorides:

$$NH_4HCO_3 + NaCl \rightleftharpoons NaHCO_3 + NH_4Cl$$

In this system, the least soluble component is sodium bicarbonate, so this crystallizes out. On calcination it yields sodium carbonate and the carbon dioxide is recycled. The ammonia is recovered by adding calcium hydroxide, producing calcium chloride waste and liberating the ammonia for re-use:

$$2NH_4Cl + Ca(OH)_2 = 2NH_3 + CaCl_2 + 2H_2O$$

British Patent 3,131 (1863).

Wood, R. D. E., in *Industrial Inorganic Chemicals: Production and Use,* Thompson, R., Ed., Royal Society of Chemistry, Cambridge, 1995, 128.

ammoxidation   The catalytic oxidation of a mixture of an aliphatic hydrocarbon and ammonia to give an alkyl cyanide:

$$2RCH_3 + 2NH_3 + 3O_2 = 2RCN + 6H_2O$$

This was a development of the *Andrussov process by which methane yields hydrogen cyanide. In one important version, propylene and ammonia yield acrylonitrile:

$$2CH_2=CH-CH_3 + 2NH_3 + 3O_2 \rightarrow 2CH_2=CH-CN + 6H_2O$$

Invented and developed independently in the late 1950s by D.G. Stewart in the Distillers Company, and R. Grasselli in Standard Oil of Ohio. The former used a tin/antimony oxide catalyst; the latter bismuth phosphomolybdate on silica. Today, a proprietary catalyst containing depleted uranium is used. *See also* Erdölchemie, OSW, Sohio.

Another variation is the catalytic oxidation of toluene with ammonia to produce benzonitrile. Such a process has been developed and is offered for license by Nippon Shokubai Kagaku Kogyo Company. Their plant is in Himeji, Japan.

U.S. Patent 2,904,580.

Wiseman, P., *Chem. Br.,* 1987, **23**, 1198.

Hydrocarbon Process., 1987, **66**(11), 66.

Sokolovskii, V. D., Davydov, A. A., and Ovsitser, O. Yu., *Catal. Rev. Sci. Eng.,* 1995, **37**(3), 425.

ammoximation   The conversion of an aldehyde or ketone to its oxime by treatment with ammonia and hydrogen peroxide:

$$R\text{--}CHO + NH_3 + H_2O_2 = R\text{--}CH{=}NOH + 2H_2O$$

Cyclohexanone is thus converted to cyclohexanone oxime, an intermediate in the manufacture of Nylon-6. The catalyst is titanium silicalite-2. Commercialized by Enichem who built a 12,000 ton/year plant in Porta Marghera in 1994.

Reddy, J. S., Sivasanker, S., and Ratnasamy. P., *J. Mol. Catal.* 1991, **69,** 383.

Chem. Br., 1995, **31**(2), 94.

Amoco   Amoco Chemicals Company, a subsidiary of Amoco Corporation, formerly Standard Oil Company (IN), is best known in the chemicals industry for its modification of the *Mid-Century process for making pure terephthalic acid. *p*-Xylene in acetic acid solution is oxidized with air at high temperature and pressure. Small amounts of manganese, cobalt, and bromide are used as catalysts. The modification allows the use of terephthalic acid, rather than dimethyl terephthalate, for making fiber. The process can also be used for oxidizing other methylbenzenes and methylnaphthalenes to aromatic carboxylic acids. *See also* Maruzen.

Spitz, P. H., *Petrochemicals, the Rise of an Industry,* John Wiley & Sons, New York, 1988, 327.

Weissermel, K. and Arpe, H.-J. *Industrial Organic Chemistry,* 3rd ed., VCH Publishers, Weinheim, Germany, 1997, 396.

AMV   A modified process for making ammonia, invented by ICI and announced in 1982. It uses a new catalyst and operates at a pressure close to that at which the synthesis gas has been generated, thereby saving energy. Construction licenses have been granted to Chiyoda Corporation, Kvaerne, and Mannesman. In 1990 it was operated in the CIL plant in Ontario, Canada and then in Henan Province, China.

European Patent 49,967.

Livingston, J. G. and Pinto, A., *Chem. Eng. Prog.,* 1983, **79**(5), 62.

Chem. Eng. (Rugby, Eng.), 1990, 21.

Hydrocarbon Process., 1991, **70**(3), 134.

ANAMET   [**Ana**erobic **met**hane]   An anaerobic biological process for treating industrial effluents containing relatively high concentrations of organic matter. The microorganisms are removed in a lamella separator in which they slide down inclined plates. Developed by Purac, Sweden, which had installed more than 50 plants by 1992, mostly in the food industry.

Anatread   A hydrometallurgical process for extracting copper from a sulfide ore with ferric chloride solution.

Ancit   *See* carbonization.

Andco-Torrax   A process for making a fuel gas by the partial oxidation of organic wastes in a vertical shaft furnace. The residue is removed as a liquid slag from the base of the furnace.

Sixt, H., *Chem. Ing. Tech.,* 1981, **53**(11), 844.

Kirk-Othmer's Encyclopedia of Chemical Technology, 3rd ed., Vol 13, John Wiley & Sons, New York, 1981, 195.

Andrussov   A process for making hydrogen cyanide by reacting ammonia, methane and air at approximately 1,000°C over a platinum/rhodium catalyst:

$$2NH_3 + 3O_2 + 2CH_4 = 2HCN + 6H_2O$$

The product gases are freed from ammonia by scrubbing with sulfuric acid and the hydrogen cyanide is then absorbed in water or diethanolamine. Invented in 1930 by L. Andrussov at IG Farbenindustrie, Germany.

U.S. Patent 1,934,838.

Andrussov, L., *Ber. Dtsch. Chem. Ges.,* 1927, **60**, 2005.

Andrussov, L., *Angew. Chem.,* 1935, **48**, 593.

Andrussov, L., *Chem. Ing. Tech.,* 1955, **27**, 469.

Dowell, A. M., III, Tucker, D. H., Merritt, R. F., and Teich, C. I., in *Encyclopedia of Chemical Processing and Design,* McKetta, J. J. and Cunningham, W. A., Eds., Marcel Dekker, New York, 1988, **27**, 7.

Anglo-Jersey   A paraffin isomerization process, catalysed by aluminum trichloride supported on bauxite. Developed by the Anglo Iranian Oil Company and Standard Oil Development Company.

Perry, S. F., *Trans. Am. Inst. Chem. Eng.,* 1946, **42**, 639 (*Chem. Abstr.,* **40,** 6792).

Asinger, F., *Paraffins, Chemistry and Technology,* translated by B. J. Hazzard, Pergamon Press, Oxford, 1968, 708.

Angus Smith   *See* metal surface treatment.

Aniline   *See* Laux.

Anortal   [**Anort**hosite **al**uminium]   A process for extracting alumina from anorthosite ore (a calcium aluminosilicate) by leaching with hydrochloric acid, precipitating aluminum trichloride hexahydrate, and calcining this. Developed and piloted by I/S Anortal in Norway in the late 1970s but not commercialized.

Gjelsvik, N., *Light Met. Met. Ind.,* 1980, 133.

O'Connor, D. J., *Alumina Extraction from Non-bauxitic Materials,* Aluminium-Verlag, Düsseldorf, 1988, 127.

Anox   An integrated water treatment process for removing organic contaminants. The energy is obtained by burning the biogas generated in the process, which contains approximately 70 percent methane. Developed by W. D. Evers; a demonstration plant was built in France in 1979.

Evers, W. D., *Chimia,* 1979, **33**(6), 217.

ANTHANE/ANODEK   [**An**aerobic **Methane**/**An**aerobic **O. de** Konickx]   A process for generating methane by the anaerobic fermentation of industrial organic wastes. Invented by the Institute of Gas Technology, Chicago; engineered by the Studiebureau O. de Konickx, Belgium, and commercialized since 1977.

Anthracine   *See* carbonization.

AO   [**A**utoxidation, or Air **O**xidation, or Anthraquinone oxidation]   A process for making hydrogen peroxide from hydrogen and oxygen (air) by cyclic oxidation/reduction of an alkyl anthraquinone solution (the working solution). Invented by H.-J. Riedl and G. Pfleiderer in Germany in the mid-1930s; piloted by IG Farbenindustrie in Ludwigshaven during World War II, and commercialized in the UK and United States during the 1950s. Now virtually the sole manufacturing process.

$$\text{Anthraquinone} + H_2 \xrightarrow[\text{catalyst}]{\text{heterogeneous}} \text{Anthraquinol}$$

$$\text{Anthraquinol} + O_2 \rightarrow \text{Anthraquinone} + H_2O_2$$

The anthraquinone derivative is usually 2-ethyl- or 2-pentyl-anthraquinone. The solvent is usually a mixture of two solvents, one for the quinone and one for the quinol. The hydrogenation catalyst is usually nickel or palladium on a support. The hydrogen peroxide is produced at a concentration of 20 to 40 percent and is concentrated by distillation.

U.S. Patents 2,215,856; 2,215,883.
German Patent 671,318.
Bertsch-Frank, B., Dorfer, A., Goor, G., and Süss, H. U., in *Industrial Inorganic Chemicals: Production and Use,* Thompson, R., Ed., Royal Society of Chemistry, Cambridge, 1995, 176.
Goor, G., in *Catalytic Oxidations with Hydrogen Peroxide as Oxidant,* Strukul, G., Ed., Kluwer Academic Publishers, Dordrecht, England, 1993, 13.

A/O   A modification of the *Activated Sludge process, designed to maximize the removal of phosphate ion. Developed by Air Products & Chemicals.

Bowker, R. P. G. and Stensel, H. D., *Phosphorus Removal from Wastewater,* Noyes Data, Park Ridge, NJ, 1990, 21.

AOD   [Argon oxygen decarburization]   A steelmaking process in which a mixture of oxygen and argon is injected into molten iron to reduce the carbon content. Developed by the Union Carbide Corporation in the mid 1970s. By 1989, 90 percent of the stainless steel made in the United States was made with this process.

Isalski, W. H., *Separation of Gases,* Clarendon Press, Oxford, 1989, 9.

APAC   A coal gasification combined cycle process that produces fuel gases, acetylene, and electricity. Limestone is added, which produces calcium carbide, in turn used to generate acetylene by reaction with water. Operated at the Acme power plant, Sheridan, WY.

APOL   [Alkaline pressure oxidation leaching]   A process for extracting gold from refractory ores, developed by Davy McKee (Stockton, UK).

Appleby-Frodingham   A process for removing hydrogen sulfide and organic sulfur compounds from coke-oven gas by absorption on iron oxide particles in a fluidized bed at 350°C. The absorbent is regenerated with air at a higher temperature, and the resulting sulfur dioxide is used to make sulfuric acid. Invented by L. Reeve and developed by the South Western Gas Board at Exeter, UK, in the 1950s and operated at the Appleby-Frodingham steelworks.

British Patent 719,056.
Reeve, L., *J. Inst. Fuel,* 1958, **31,** 319.
Claxton, G., *Benzoles, Production and Uses,* National Benzole & Allied Products Association, London, 1961, 210.
Kohl, A. L. and Riesenfeld, F. C., *Gas Purification,* 4th ed., Gulf Publishing, Houston, TX, 1985, 479.

Aquaclaus   A modification of the *Claus process in which hydrogen sulfide is removed from water by reaction with sulfur dioxide. Developed by Stauffer Chemical Company and operated by the Heflin Oil Company, in Queen City, TX.

Hayford, J. S., *Hydrocarbon Process.,* 1973, **52**(10), 95.
Sulphur, 1974, (111), 48.
Chem. Eng., (N.Y.), 1984, **91**(13), 150.

Aquaconversion   A process for converting heavy crude petroleum oils into lighter products which are more easily converted into more valuable products in oil refineries. Intended for use at the well head rather than the oil refinery. Developed by Foster Wheeler USA Corporation, Intevep, and UOP from 1998.

Hydrocarbon Process., 1997, **76**(12), 36.

AQUAFINING   A process for extracting water and other nonsulfur-containing contaminants from petroleum fractions by the use of a proprietary bundle of hollow fibers called a

FIBER-FILM contactor. Developed by Merichem Company, Houston, TX, and used in 11 installations in 1991.

> *Hydrocarbon Process.,* 1992, **71**(4), 120.
> *Hydrocarbon Process.,* 1996, **75**(4), 126.

Aquarrafin   A wastewater treatment process using activated carbon in fixed beds. Developed by Lurgi.

Arbiter   Previously known as the Sherritt-Gordon ammonia process. A process for leaching copper from sulfide concentrates, using ammoniacal ammonium sulfate solution at 85°C and relying on air oxidation. Copper is produced from the leachate by solvent extraction and electrowinning. Sulfur is recovered as ammonium sulfate. Operated on a large scale by the Anaconda Copper Company in Montana from 1974 to 1979. See Sherritt-Gordan.

> Kuhn, M. C., Arbiter, N., and Kling, H. *Can. Inst. Min. Met. Bull.,* 1974, **67**, 62.
> Arbiter, N., *New Advances in Hydrometallurgy,* Institute of Gas Technology, Chicago, 1974.

Arc   *See* Berkland-Eyde.

Arco   A process for making isobutene by dehydrating *t*-butanol. The reaction takes place in the gas phase at 260 to 273°C, 14 bar, in the presence of an alumina-based catalyst.

> Weissermel, K. and Arpe, H.-J., *Industrial Organic Chemistry,* 3rd ed., VCH Publishers, Weinheim, Germany, 1997, 70.

ARDS   A process for upgrading petroleum residues by catalytic hydrogenation.

> *Hydrocarbon Process.,* 1997, **76**(2), 50.

Arex   A process for removing aromatic hydrocarbons from petroleum streams by extraction with 1-methyl piperidone (N-methyl caprolactam) at 60°C. Developed by Leuna Werke, Germany.

> *Chem. Tech., (Leipzig),* 1977, **29**, 573.

Arge   [Arbeitsgemeinschaft]   A version of the *Fischer-Tropsch process, using a fixed catalyst bed. It converts *synthesis gas to a mixture of gasoline, diesel fuel, and waxes. The catalyst is made by adding sodium carbonate solution to a solution of mixed iron and copper nitrates, binding the resulting precipitate with potassium silicate, and reducing it with hydrogen. Used in the *SASOL plant in South Africa since 1955 and being considered for use in New Zealand in 1992. Developed by Ruhr Chemie-Lurgi.

> Mako, P. F. and Samuel, W. A., in *Handbook of Synfuels Technology,* Meyers, R. A., Ed., McGraw-Hill, New York, 1984, 11.

Aris   A process for the hydrocatalytic isomerization of C_8 fractions. Developed by Leuna-Werk and Petrolchemische Kombinate Schwedt in 1976. The catalyst is platinum deposited in a mixture of alumina and natural mordenite.

> Weissermel, K. and Arpe, H.-J. *Industrial Organic Chemistry,* 3rd ed., VCH Publishers, Weinheim, Germany 1997, 331.

Armco   A direct reduction ironmaking process which used natural gas as the reductant. Operated in Houston, TX, from 1972, dismantled in 1982. *See* DR.

Armour (1)   A continuous soapmaking process developed by the Armour Company in 1964.

> Potts, R. H. and McBride, G. W. *Chem. Eng. (N.Y.),* 1950, **57**(2), 124.

Armour (2)   A process for separating fatty acids by fractional crystallization from acetone.

> Potts, R. H. and McBride, G. W., *Chem. Eng. (N.Y.),* 1950, **57**(2), 124.

ARODIS A process for converting light aromatic hydrocarbons to diesel fuel. It involves hydrodealkylation and hydrogenation. Developed by the University of New South Wales and BHP Research.

Jiang, C. J., Trimm, D. L., Cookson, D., Percival, D., and White, N., in *Science and Technology in Catalysis,* Izumi, Y., Aral, H., and Iwamoto, M., Eds., Elsevier, Amsterdam, 1994, 149.

Arofining A process for removing aromatic hydrocarbons from petroleum fractions by catalytic hydrogenation to naphthenes. Developed by Labofina, France, and licensed by Howe-Baker Engineers.

Hydrocarbon Process., 1970, **49**(9), 205.

Aroforming A process for making aromatic hydrocarbons from aliphatic hydrocarbons. Based on the Aromizing process. Developed by Salutec, Australia, and IFP, France. A demonstration unit with capacity of 500 bbl/day was being designed in 1994.

Mank, L., Shaddick, R., and Minkkinen, A., *Hydrocarbon Technol. Internat.,* 1992, 69.
Eur. Chem. News, CHEMSCOPE, 1994, 7.
Eur. Chem. News, 1994, **62**(1648), 18.

Aromax (1) A catalytic process for converting light paraffins to benzene and toluene, using a zeolite catalyst. Developed by Chevron Research & Technology Company. Installations were planned for Mississippi, Thailand, and Saudi Arabia.

Aromax (2) Also known as Toray Aromax. A chromatographic process for separating *p*-xylene from its isomers. Similar to the *Parex (1) process, it operates in the liquid phase at 200°C, 15 bar. Developed in 1971 by Toray Industries, Japan.

U.S. Patent 3,761,533.
Otani, S., *Chem. Eng. (N.Y.),* 1973, **80**(21), 106.
Weissermel, K. and Arpe, H.-J. *Industrial Organic Chemistry,* 3rd ed., VCH Publishers, Weinheim, Germany, 1997, 322.

Aromex A process for removing aromatic hydrocarbons from petroleum reformate by extraction with diglycolamine (also called [2-(2-aminoethoxy) ethanol], and DGA). Developed by Howe-Baker Engineers. *See also* Econamine.

Jones, W. T. and Payne, V., *Hydrocarbon Process.,* 1973, **52**(3), 91.
Bailes, P. J., in *Handbook of Solvent Extraction,* Lo, C. C., Baird, M. H. I., Hanson, C., Eds., John Wiley & Sons, Chichester, England, 1983, Chap. 18.2.4.

Aromizing A petroleum reforming process for converting aliphatic to aromatic hydrocarbons. Developed by the Institute Français du Pétrole.

Bonnifay, P., Cha, B., Barbier, J.-C., Vidal, A., Jugin, B., and Huin, R., *Oil Gas J.,* 1976, **74**(3), 48.

Arosat [Aromatics saturation] A *hydroprocessing process developed by C-E Lummus.

Arosolvan A solvent extraction process for removing aromatic hydrocarbons from petroleum mixtures, using N-methyl pyrrolidone (NMP) containing 12 to 14 percent water at 20 to 40°C. Developed by Lurgi, and first used commercially in Japan in 1961.

Eisenlohr, K.-H., *Erdoel Kohle,* 1963, **16**, 530.
Eisenlohr, K.-H. and Grosshaus, W., *Erdoel Kohle,* 1965, **18**, 614.
Oil Gas J., 1966, **64**(29), 83.
Müller, E., *Chem. Ind. (London),* 1973, 518.

Arosorb A process for extracting aromatic hydrocarbons from refinery streams using a solid adsorbent, either silca gel or activated alumina. Developed by the Sun Oil Company in 1951. California Research Corporation developed a similar process.

Harper, J. I., Olsen, J. L., and Shuman, F.R., Jr., *Chem. Eng. Prog.,* 1951, **48**(6), 276.

Davis, W. H., Harper, J. I., and Weatherly, E. R., *Pet. Refin.,* 1952, **31**(5), 109.

Unzelman, G. H. and Wolf, C. J., in *Petroleum Processing Handbook,* Bland, W. F. and Davidson, R. L., Eds., McGraw-Hill, New York, 1967, 106.

ARS [Advanced Recovery System] An integrated set of engineering modifications for upgrading catalytic crackers for making ethylene, developed jointly by Mobil Chemical Co. and Stone and Webster Engineering Corp. The first plant was planned for an AMOCO plant in S. Korea for completion in 1994.

Eur. Chem. News, 1990, **54**(1434), 22.

Arseno A process for extracting gold from arsenic-containing ores, developed by Arseno Processing. Similar to the *Cashman process.

Yannopoulos, J. C., *The Extractive Metallurgy of Gold,* Van Nostrand Reinhold, New York, 1991, 103.

ART [Asphalt Residuum Treating] A process for converting heavy petroleum fractions into more easily processed liquid fractions. Developed by Engelhard Corp. and offered by the MW Kellogg Co. Three units were operating in 1996.

Hydrocarbon Process., 1996, **75**(11), 121.

Arthur D. Little *See* DR.

ASAM [Alkali-Sulfite Anthraquinone Methanol] A process for delignifying wood pulp. Wood chips are digested in sodium hydroxide or sodium carbonate solution, and sodium sulfite is added to remove the lignin. Methanolic anthraquinone is used as a catalyst. Invented by the Department of Chemical Wood Technology at the University of Hamburg. Further developed by the University with Kraftanlage Heidelburg, and demonstrated at a plant of Feldmühle in Düsseldorf, Germany, in 1990.

Chem. Eng. (N.Y.), 1991, **98**(1), 37.

Patt, R., Kordsachia, O., and Schubert, H.-L., in *Environmentally Friendly Technologies for the Pulp and Paper Industries,* Young, R.A. and Akhar, M., Eds., John Wiley & Sons, New York, 1998, 101.

ASARCO [American Smelting and Refining Company] This large metallurgical company has given its name to a *flue-gas desulfurization process in which the sulfur dioxide is absorbed in dimethylaniline and subsequently desorbed at a higher temperature. Operated in California, Tennessee, and Norway.

Fleming, E.P. and Fitt, T.C., *Ind. Eng. Chem.,* 1950, **42**(11), 2253.

Kohl, A. L. and Riesenfeld, F. C., *Gas Purification,* 4th ed., Gulf Publishing, Houston, TX, 1985, 382.

ASCOT [Asphalt coking technology] A process combining de-asphalting and decoking, offered by Foster-Wheeler, United States.

U.S. Patent 4,686,027.

ASEA-SKF *See* steelmaking.

Ashcroft-Elmore A process for extracting tin from its ores. The ore is mixed with coke and calcium chloride and heated in a rotary kiln to 800°C. Stannous chloride, formed by the reaction:

$$SnO_2 + C + CaCl_2 = CaO + CO + SnCl_2$$

volatilizes and is condensed in water. The aqueous condensate is neutralized and electrolyzed. Invented by E.A. Ashcroft and S. Elmore and operated in Thailand from 1941 to 1949.

British Patents 302,851; 602,245; 602,246; 602,247.

Wright, P. A., *Extractive Metallurgy of Tin,* 2nd. ed., Elsevier, Amsterdam, 1982, 175.

ASR Sulfoxide   [Alberta Sulfur Research]   A process for removing residual sulfur dioxide and hydrogen sulfide from the tail gases from the *Claus process by wet scrubbing with a solution containing an organic sulfoxide. Elemental sulfur is produced. It had not been piloted in 1983.

ASVAHL   A process combining *HDM, *HDN, and *HDS. Developed by ELF, IFP, and Total in the 1980s. Piloted in France in 1983.

Chauvel, A., Delmon, B., and Hölderich, W. F., *Appl. Catal. A: Gen.,* 1994, **115,** 186.

Atgas   [Applied Technology Coporation gasification]   A coal gasification process in which powdered coal and limestone, mixed with steam and oxygen, are injected into a bath of molten iron at 1,400°C. The product gas is a mixture of hydrogen and carbon monoxide, and the sulfur is converted to a calcium sulfide slag. Piloted by the Applied Technology Corporation in the 1970s but not fully developed.

Hebden, D. and Stroud, H. J. F., in *Chemistry of Coal Utilization,* 2nd Suppl. Vol., Elliott, M. A. Ed., John Wiley & Sons, New York, 1981, 1739.

ATOL   [Atochem polymerization]   A gas-phase process for making polyethylene. Developed by Atochem and first commercialized in 1991. It uses a *Ziegler-Natta catalyst containing titanium and magnesium halides. First commercialized at Gonfreville, France, in 1991.

Chauvel, A., Delmon, B., and Hölderich, W. F., *Appl. Catal. A: Gen., 1994,* **115,** 180.

ATR (1)   [Autothermal reforming]   A process for making CO-enriched *syngas. It combines partial oxidation with adiabatic *steam-reforming. Developed in the late 1950s for ammonia and methanol synthesis. Further developed in the 1990s by Haldor Topsoe.

Christensen, T. S. and Primdahl, I. I., *Hydrocarbon Process.,* 1994, **73**(3), 39.

ATR (2)   [Autothermal reforming]   A process for making nitrogen-diluted *syngas, suitable for use in the *Fischer Tropsch process. Developed by Syntroleum in 1989.

Oil Gas J., 1997, **95**(25), 18.

Atrament   *See* metal surface treatment.

ATS   [Ammonium thiosulfate]   A process for removing residual sulfur dioxide from *Claus tail gas by absorption in aqueous ammonia to produce ammonium sulfite and bisulfite. Addition of hydrogen sulfide from the Claus unit produces saleable ammonium thiosulfate. Developed by the Pritchard Corporation and first operated by the Colorado Interstate Gas Company at Table Rock, WY.

Zey, A., White, S., and Johnson, D., *Chem. Eng. Prog.,* 1980, **76**(10), 76.

Attisholz   A process for recovering valuable waste products from the *sulfite process for making wood pulp. This includes methanol, cymenes, and furfural. Developed in Switzerland.

AUC   [Ammonium uranyl carbonate]   A process for converting uranium hexafluoride into uranium dioxide for use as a nuclear reactor fuel. The hexafluoride vapor, together with carbon dioxide and ammonia, are passed into aqueous ammonium carbonate at 70°C, precipitating ammonium uranyl carbonate:

$$UF_6 + 5H_2O + 10NH_3 = (NH_4)_4[UO_2(CO_3)_3] + 6NH_4F$$

The precipitate is filtered off, washed, and calcined in hydrogen in a fluidized bed. Developed by Nukem at Hanau, Germany.

Büchner, W., Schliebs, R., Winter, G., and Büchel, K. H., *Industrial Inorganic Chemistry,* VCH Publishers, Weinheim, Germany, 1989, 581.

Auger A process for chlorinating benzene to chlorobenzene, catalyzed by metallic iron. Invented by V.E. Auger in 1916 and operated in France and Italy in the early 20th century.

French Patent 482,372.

Ellis, C., *The Chemistry of Petroleum Derivatives,* The Chemical Catalog Co., New York, 1934, 765.

AuPLUS A hydrometallurgical process for extracting gold. Addition of calcium peroxide to the cyanide leaching liquor increases the rate and amount of gold extracted.

AUROBAN A catalytic process for *hydrotreating and converting the asphaltenes in residual oils and heavy crude oils into lighter products. Developed by UOP.

Bowman, C. W., Phillips, R. S., and Turner, L. R., in *Handbook of Synfuels Technology,* Meyers, R. A., Ed., McGraw-Hill, New York, 1984, 73.
Hydrocarbon Process., 1997, **76**(2), 45.

AUSCOKE *See* carbonization.

Autofining A fixed-bed catalytic process for removing sulfur compounds from petroleum distillates. This process uses a conventional cobalt/molybdenum hydrodesulfurization catalyst but does not require additional hydrogen. Developed by The Anglo-Iranian Oil Company in 1948.

British Patent 670,619.
U.S. Patent 2,574,449.
McKinley, J. B., in *Catalysis,* Emmett, P. H., Ed., Reinhold, New York, 1957, 405.
Hydrocarbon Process., 1964, **43**(9), 186.
Unzelman, G. H. and Wolf, C. J., in *Petroleum Processing Handbook,* Bland, W. F. and Davidson, R. L., Eds., McGraw-Hill, New York, 1967, 42.

AUTO-PUREX G A process for removing carbon dioxide from air by *PSA, using alumina as the sorbent. Developed by Marutani Chemical Plant & Engineering Company, Japan.

Suzuki, M., in *Adsorption and Ion Exchange: Fundamentals and Applications,* LeVan, M. D., Ed., American Institute of Chemical Engineers, New York, 1988, 121.

Autopurification A wet-scrubbing process for removing hydrodrogen sulfide from coke-oven gas. The scrubbing liquor was an ammoniacal suspension of ferric ammonium ferrocyanide. The process was developed by ICI, Billingham, UK, in the 1930s and 40s, but was abandoned in 1947.

Smith, F. F. and Pryde, D. R., *Chem. Ind. (London),* 1934, **12**, 657.
Craggs, H. C. and Arnold, M. H. M., *Chem. Ind. (London),* 1947, **66**, 571,590.

Avaro [Aviation aromatics] A process for increasing the aromatics content of gasoline by *thermal reforming in the presence of low molecular weight hydrocarbons. Used at the Shell refinery in Curacao during World War II.

Avco An electric arc process for making acetylene from coal and hydrogen. The arc in hydrogen is rotated by a magnetic field in order to spread it out and thus make better contact with the coal passing through. Developed by V. J. Krukonis at the Avco Corporation in the early 1970s with support from the U.S. Office of Coal Research. Piloted at the rate of 55 kg/hr but not yet commercialized.

Gannon, R. E., Krukonis, V. J., and Schoenberg, T., *Ind. Eng. Chem. Prod. Res. Dev.,* 1970, **9**, 343.

AVM [Atelier de Vitrification de Marcoule] A continuous process for immobilizing radioactive waste by incorporation in a borosilicate glass. Developed at Marcoule, France, in 1972, based on the earlier *PIVER process. In 1988, two larger vitrification plants were being designed for installation at La Hague, France. A modified form of AVM is used at the THORP nuclear fuel reprocessing plant at Springfields, England.

> Lutze, W., *Radioactive Waste Forms for the Future,* Lutze, W. and Ewing, R. C., Eds., North-Holland, Amsterdam, 1988, 10, 133.

Axorb   A process for removing carbon dioxide and hydrogen sulfide from gases by scrubbing with an aqueous solution of potassium carbonate containing proprietary additives.

Ayers   An early process for making carbon black from oil. The air for combustion is injected tangentially into the furnace, producing swirl, and the atomized oil is injected into this. Invented by J. W. Ayers and developed by Phillips Petroleum Company.

> U.S. Patents 2,292,355; 2,420,999.
> Shearon, Jr., W. H., Reinke, R. A., and Ruble, T. A., in *Modern Chemical Processes,* Vol. 3, Reinhold Publishing, Washington, 1954, 45.

B

Babcock and Wilcox   The Babcock and Wilcox company developed a number of processes but is perhaps best known for its coal gasification process. This uses a single-stage, two-zone gasifier. In the lower zone coal and recycled char are contacted with oxygen (or air) and steam at 1,650 to 1,855°C and molten slag is removed from its base. The upper zone is cooled to 900°C. A commercial scale oxygen-blown plant was operated in West Virginia in the 1950s, and a pilot-scale air-blown plant was operated in Ohio in the 1960s. *See also* BiGas.

> Hebden, D. and Stroud, H. F. G., in *Chemistry of Coal Utilization,* 2nd. Suppl. Vol., Elliott, M. A., Ed., John Wiley & Sons, New York, 1981, 1724.

Babcock W-D   *See* Woodall-Duckham.

BACFOX   [Bacterial film oxidation]   A process for regenerating hydrometallurgical leach liquors by bacterial oxidation of their iron from Fe^{2+} to Fe^{3+}. Developed by Mathew Hall Ortech Company and used in South Africa.

> *Eng. Min. J.,* 1978, **179**(12), 90.
> Jackson, E., *Hydrometallurgical Extraction and Reclamation,* Ellis Horwood, Chichester, England, 1986, 68.

Bachmann   A process for making the explosive RDX. Hexamethylene tetramine is nitrated in acetic acid solution, using a mixture of ammonium nitrate and acetic anhydride. Invented by W. E. Bachmann at the University of Michigan during World War II. *See also* KA, Woolwich.

> Bachmann, W. E. and Sheehan, J. C., *J. Am. Chem. Soc.,* 1949, **71**, 1842.
> *Kirk-Othmer's Encyclopedia of Chemical Technology,* 4th ed., Vol. 9, John Wiley & Sons, New York, 1991–1998, 583.

Backus   [Backhaus]   A process for purifying carbon dioxide obtained by fermentation, using activated carbon. The carbon beds are reactivated with steam. Invented in 1924 by A. A. Backhaus at the U.S. Industrial Alcohol Company. *See also* Reich.

> U.S. Patents 1,493,183; 1,510,373.

Bacus   A scrubbing process for removing olefins and carbon monoxide from gas streams. The adsorbent is a solution of a copper compound that is not deactivated by traces of water. Developed by D. Haase of Herr Haase, Nixon, TX.

Chem. Eng. (N. Y.), 1995, **102**(3), 19.

Baekeland   A process for making organic polymers by reacting phenols with formaldehyde. Based on an observation by A. von Bayer in 1872 and developed into an industrial process by L. H. Baekeland from 1905 to 1909. It was used to make Bakelite, one of the first commercial plastics. The first industrial manufacture began in Germany in 1910.

von Bayer, A., *Ber. Dtsch. Chem. Ges.,* 1872, **5**, 280.
Baekeland, L. H., *Ind. Eng. Chem.,* 1909, **1**, 149.
Kirk-Othmer's Encyclopedia of Chemical Technology, 4th ed., Vol. 18, John Wiley & Sons, New York, 1991–1998, 603.

BAF   [Biological aerated filter]   A generic type of sewage treatment process in which the biological medium is supported on a porous matrix. See BIOBEAD, BIOCARBONE, BIO-FOR, BIOPUR, BIOSTYR, COLOX, CTX, FAST, SAFe, STEREAU.

Stephenson, T., Mann, A., and Upton, J., *Chem. Ind. (London),* 1993, (14), 533.

Balbach   A variation of the *Moebius process for electrolytically removing gold from silver, in which the anodes are placed horizontally in wooden trays lined with canvas to retain the slimes. The silver is deposited as crystals on graphite cathodes at the base of the cell. Invented by E. Balbach. *See also* parting, Parkes, Thum, Wohlwill.

Balke   A process for making niobium by reducing niobium pentoxide with carbon in a carbon crucible, *in vacuo,* at 1,800°C.

Banox   *See* metal surface treatment.

Bardenpho   A modification of the *Activated Sludge process, designed for the removal of high levels of nitrogen and phosphorus.

Horan, N. J., *Biological Wastewater Treatment Systems,* John Wiley & Sons, Chichester, England, 1990, 234.

Bardet   Also called Samica. A process for expanding mica in order to make it into paper. It is partially dehydrated by heating and the hot product is quenched in alkaline water. After drying, it is immersed in dilute sulfuric acid, which generates gas between the layers, forcing them apart. In this expanded condition it can easily be made into a paper.

Barffing   *See* metal surface treatment.

Bari-Sol   A petroleum *dewaxing process using solvent extraction by a mixture of dichloroethane and benzene.

Unzelman, G. H., and Wolf, C. J., in *Petroleum Processing Handbook,* Bland, W. F. and Davidson, R. L., Eds., McGraw-Hill, New York, 1967, 96.

Barium   A process for making hydrogen peroxide by reacting barium peroxide with sulfuric or phosphoric acid:

$$BaO_2 + H_2SO_4 = BaSO_4 + H_2O_2$$

The barium peroxide was made using the *Brin process. The barium was recovered by reducing the sulfate with carbon, and then converting this barium sulfide to the oxide via the carbonate:

$$BaSO_4 + 2C = BaS + 2CO_2$$

Phosphoric acid was sometimes used instead of sulfuric acid. The process was first operated in Berlin in 1873 by the Schering Company. In the United Kingdom it was first operated in 1888 by B. Laporte & Company. It was progressively replaced by the electrolytic process developed between 1908 and 1932. Also in the United Kingdom, Laporte Chemicals abandoned the barium process in 1950.

Wood, W. S., *Hydrogen Peroxide,* Royal Institute of Chemistry, London, 1954, 2.

Schumb, W. C. Satterfield, C. N., and Wentworth, R. L., *Hydrogen Peroxide,* Reinhold Publishing, New York, 1955, 14, 115.

Pascal, P., Ed., *Nouveau Traité de Chimie Minérale,* Vol. 13, Masson et Cie., Paris, 1960, 528.

Barton   A process for making black lead monoxide by atomizing molten lead in air.

BASF/CAN   *See* CAN.

BASF/Scholven   A pretreatment process for benzole, the product formed by hydrogenating hard coal. The benzole is hydrogenated at 300 to 400°C using a molybdenum or cobalt/molybdenum catalyst. The product is a mixture of aromatic hydrocarbons, suitable for separation by a variety of physical processes. The process was invented by BASF in 1925 and adopted by Scholven-Chemie in 1950. Eleven plants in France and Germany subsequently adopted the process.

Jäckh, W., *Erdoel Kohle,* 1958, **11**, 625.

Reitz, O., *Erdoel Kohle,* 1959, **12**, 339.

Muder, R., *Chemistry of Coal Utilization,* Suppl. Vol., Lowry, H. H., Ed., John Wiley & Sons, New York, 1963, 647.

Bashkirov   A process for making aliphatic alcohols by oxidizing paraffins. The reaction is conducted in the presence of boric acid, which scavenges the hydroperoxide intermediates. Borate esters of secondary alcohols are formed as intermediates and then hydrolyzed. Developed in the USSR in the 1950s and now operated there and in Japan.

Bashkirov, A. N., *et al.,* in *The Oxidation of Hydrocarbons in the Liquid Phase,* Emanuel, N. M., Ed., Pergamon Press, Oxford, 1965, 183.

Basic Bessemer   *See* Thomas.

Basic Open Hearth   A version of the *Open Hearth process for steelmaking in which the hearth is made from calcined dolomite (calcium and magnesium oxides). The sulfur and phosphorus impurities in the raw materials are converted to basic slag, which is separated from the molten steel. First operated in 1882 at Alexandrovsky, near St. Petersburg, Russia, and at Le Creusot, France. It was the major steelmaking process in the world in the first half of the 20th century. *See* Thomas.

Barraclough, K. C., *Steelmaking 1850–1900,* The Institute of Metals, London, 1990, 247.

Basset   *See* DR.

Batenus   A series of processes, including solvent extraction and ion exchange, for recovering metals from scrap batteries. Developed by Pira, Germany, in 1993.

Chem. Eng. (N.Y.), 1993, **100**(11), 21.

Battersea   A pioneering *flue-gas desulfurization process, operated at Battersea power station, London, from 1931 until the station was closed. The flue-gases were washed with water from the River Thames whose natural alkalinity was augmented by chalk slurry. One of the problems of this process was cooling of the stack gases, which caused the plume to descend on the neighborhood.

J. Air Pollut. Control Assoc., 1977, **27**, 948.

Kohl, A. L. and Riesenfeld, F. C., *Gas Purification,* 4th ed., Gulf Publishing Co., Houston, TX, 1985, 302.

Rees, R. L., *J. Inst. Fuel,* 1953, **25**, 350.

Bayer A process for making pure alumina hydrate from bauxite, used principally as a raw material for the manufacture of aluminum metal. The ore is digested with hot sodium hydroxide solution, yielding a solution of sodium aluminate. Insoluble impurities are separated off in the form of red mud and the solution is then nucleated with alumina hydrate from a previous batch, causing alumina trihydrate to precipitate:

$$Al_2O_3{\cdot}3H_2O + 2NaOH = 2NaAlO_2 + 4H_2O$$
$$2NaAlO_2 + 4H_2O = Al_2O_3{\cdot}3H_2O + 2NaOH$$

The hydrate is dehydrated by calcination. Invented by K. J. Bayer in Russia in 1887 and now universally used, with minor variations depending on the nature of the ore. The German company Bayer AG was not involved in this invention.

German Patents 43,977, 1887; 65,604, 1892.

Misra, C., *Industrial Alumina Chemicals,* American Chemical Society, Washington, D.C., 1986, 31.

Hudson, L. K., *Production of Aluminium and Alumina,* Burkin, A. R., Ed., John Wiley & Sons, Chichester, England, 1987, 13.

McMichael, B., *Ind. Miner. (London),* 1989, (267), 19.

Gupta, C. K. and Mukherjee, T. K., *Hydrometallurgy in Extraction Processes,* Vol. 1, CRC Press, Boca Raton, FL, 1990, 129.

Bayer-Bertrams A process for concentrating and purifying waste sulfuric acid by distillation.

Büchner, W., Schliebs, R., Winter, G., and Büchel, K. H., *Industrial Inorganic Chemistry,* VCH Publishers, Weinheim, Germany, 1989, 117.

Bayer-Hoechst A gas-phase process for making vinyl acetate from ethylene and acetic acid, using a supported palladium catalyst. Developed jointly by Bayer and Hoechst. In 1991, nearly 2 million tonnes per year of vinyl acetate were made by this process.

Weissermel, K. and Arpe, H.-J. *Industrial Organic Chemistry,* 3rd ed., VCH Publishers, Weinheim, Germany, 1997, 230.

Bayer ketazine A process for making hydrazine by the reaction of sodium hypochlorite with ammonia in the presence of acetone. Acetone azine is an intermediate. Never commercialized. *See also* Raschig (1).

Büchner, W., Schliebs, R., Winter, G., and Büchel, K. H., *Industrial Inorganic Chemistry,* VCH Publishers, Weinheim, Germany, 1989, 48.

BCD [Base-Catalyzed Decomposition] A process for decomposing hazardous organochlorine compounds by treating them in oil at 300 to 350°C with sucrose and a high-boiling solvent. Developed by the U.S. Environmental Protection Agency in 1991.

Kawahara, F. K. and Michalakos, P. M., *Ind. Eng. Chem. Res.,* 1997, **36**(5), 1580.

Beacon A process for recovering carbon from coal gasification. Developed in 1991 by TRW.

Chem. Eng. (N.Y.), 1984, **91**(13), 157.

Beavon [Beavon Sulfur Removal] Also called BSR. A process for removing residual sulfur compounds from the effluent gases from the *Claus process. Catalytic hydrogenation over a cobalt/molybdena catalyst converts carbonyl sulfide, carbon disulfide, and other organic sulfur compounds to hydrogen sulfide, which is then removed by the *Stretford process. A variation (**BSR/MDEA**), intended for small plants, uses preliminary scrubbing with methyl diethanolamine. Developed by the Ralph M. Parsons Company and Union Oil Company of California in 1971. In 1992, more than 15 plants were operating in the United States and Japan. *See also* SCOT.

Beavon, D. K., *Chem. Eng. (N.Y.),* 1971, **78**(28), 71.

Kohl, A. L. and Riesenfeld, F. C., *Gas Purification,* 4th ed., Gulf Publishing, Houston, TX, 1985, 739.

Hydrocarbon Process., 1996, **75**(4), 106.

Béchamp   A process for reducing organic nitro-compounds to amines, using iron, ferrous salts, and acetic acid. Invented by A. J. Béchamp in 1854 and still used for making certain aromatic amines.

Béchamp, A. J., *Ann. Chim. Phys.,* Ser. 3, 1854, **42**, 186.

Béchamp, A. J., *Annalen,* 1854, **92**, 401.

Becher   A process for making synthetic rutile (a titanium concentrate), from ilmenite ore. The ore, typically containing 58 percent titanium dioxide, is first roasted with coal and 0.2 to 0.5 percent of elemental sulfur at 1,200°C in a specially designed rotary kiln. This reduces the iron in the ilmenite to the metallic state. After cooling, partially burnt coal and ash is removed from the product by screening and magnetic separation. The reduced ilmenite is then agitated in water containing ammonium chloride as a catalyst, and air is blown through. This converts the metallic iron to a flocculent precipitate of iron oxides, which are then removed by wet classification. The synthetic rutile product contains approximately 93 percent of titanium dioxide.

The process was invented and developed in Australia in the 1960s, initially by R. G. Becher at the Mineral Processing Laboratories of the West Australian Government Chemical Centre, and later by Western Titanium. In 1998 it was operating in three companies at four sites in Western Australia. Most of the beneficiate is used for the manufacture of titanium pigments; some is used in titanium metal production and in welding rod coatings. Annual production in 1997 was approximately 600,000 tonnes.

Australian Patent 247,110.

Bechtel CZD   *See* CZD.

Behr   An obsolete process for separating the drying and nondrying constituents of bodied oils by selective precipitation. The oils are treated with low molecular weight alcohols or ketones, with a medium solvent power, which will not dissolve compounds of molecular weight greater than 900. The polymerized oil settles out, while the nonpolymeric nondrying constituents remain dissolved.

Beilby   A process for making potassium cyanide by passing ammonia gas over a molten mixture of potassium carbonate, potassium cyanide, and carbon:

$$K_2CO_3 + C + 2NH_3 = 2KCN + 3H_2O$$

The fused product is poured into molds. The process was invented by G. T. Beilby in Scotland and first used in 1891; in 1900 it was replaced by the *Castner process.

British Patent 4,820, 1891.

Kirk-Othmer's Encyclopedia of Chemical Technology, 4th ed., Vol. 7, John Wiley & Sons, New York, 1991–1998, 765.

Beja   A process for extracting gallium from sodium aluminate solution, as used in the *Bayer process, by means of successive carbonations. Developed by Pechiney in 1946.

Palmear, I. J., in *The Chemistry of Aluminium, Gallium, Indium, and Thallium,* Downs, A. J., Ed., Blackie, London, 1993, 88.

Wilder, J., Loreth, M. J., Katrack, F. E., and Agarwal, J. C., in *Inorganic Chemicals Handbook,* Vol. 2., McKetta, J. J. Ed., Marcel Dekker, New York, 1993, 942.

Belgian   A process for making zinc from zinc oxide by reducing it with carbon in a retort. Invented in 1810.

Bemberg   An alternative name for the *Cuprammonium process for making artificial silk, named after J. P. Bemberg who commercialized the process in the early 1900s in several countries.

Moncrieff, R. W., *Man-made Fibres,* 6th ed., Newnes-Butterworths, London, 1975, 224.

Bender   A continuous process for oxidizing mercaptans in petroleum fractions to disulfides, using a lead sulfide catalyst in a fixed bed. Developed and licensed by Petrolite Corp.; 98 units were operating in 1990.

Waterman, L. C. and Wiley, R. A., *Pet. Refin.,* 1955, **34**(9), 182.
Unzelman, G. H. and Wolf, C. J., in *Petroleum Processing Handbook,* Bland, W. F. and Davidson, R. L., Eds., McGraw-Hill, New York, 1967, 127.
Hydrocarbon Process., 1996, **75**(4), 108.

Benfield   [**Ben**son and **Field**]   A process for removing carbon dioxide, hydrogen sulfide, and other acid gases from industrial gas streams by scrubbing with hot aqueous potassium carbonate containing activators:

$$K_2CO_3 + H_2S \rightleftharpoons KHS + KHCO_3$$

Invented by H. E. Benson in 1952 and then developed with J. H. Field at the U.S. Bureau of Mines. First licensed by the Benfield Corporation of Pittsburgh, subsequently acquired by the Union Carbide Corporation, and now licensed by UOP. The current UOP version includes new solution activators and incorporates zeolites or membrane processes for complete separation of acid gases and minimal loss of product gases. More than 650 plants were operating in 1996. Variations include the Benfield HiPure process and the Benfield LoHeat process. *See also* Carsol, CATACARB, Giammarco-Vetrocoke, HiPure.

U.S. Patent 2,886,405.
British Patent 725,000.
Benson, H. E., Field, J. H., and Jimeson, R. M., *Chem. Eng. Prog.,* 1954, **50**(7), 356.
Benson, H. E., Field, J. H., and Haynes, W. P., *Chem. Eng. Prog.,* 1956, **52**(10), 433.
Kohl, A. L. and Riesenfeld, F. C., *Gas Purification*, 4th ed., Gulf Publishing, Houston, TX, 1985, 211.
Bartoo, R. K., in *Acid and Sour Gas Treating Processes*, Newman, S. A., Ed., Gulf Publishing, Houston, TX, 1985, 342.
Hydrocarbon Process., 1996, **75**(4), 108.

Bengough-Stuart   *See* metal surface treatment.

Benilite   [**Ben**eficiation of **i**lmenite]   Also called the Wah Chang process. A process for increasing the titanium content of ilmenite by extracting some of the iron with hot hydrochloric acid. The ore is pretreated by reduction in a rotary kiln. The extraction takes place in a rotating spherical iron pressure vessel. The hydrochloric acid is recovered for re-use by the *Woodhall-Duckham process. The process originated with the Wah Chang Corp. in the 1960s; the patent by J. H. Chen being filed in 1969. It was further developed by the Benilite Corporation of America, Corpus Christi, TX, now owned by Hitox Corporation of America. The product is used as a feedstock for the *Chloride process and as a pigment. The process is now operated by Hitox Corporation in Corpus Christi, by the Kerr-McGee Corporation in Mobile, AL and in three locations in India.

British Patent 1,262,401.
German Patent 2,004,878.

U.S. Patent 3,825,419.

Robinson, S. M., *Polym. Paint Colour J.*, 1986, **176,** 754.

BenSat [**Ben**zene **sat**uration]   A process for removing benzene from C_5 to C_6 petroleum fractions by selective hydrogenation to cyclohexane. Developed by UOP, based on its *HB Unibon process, and first offered for license in 1991.

> *Hydrocarbon Process.*, 1996, **75**(11), 94.
> Sullivan, D. K., in *Handbook of Petroleum Refining Process*, Meyers, R. A., Ed., McGraw-Hill, New York, 1997, 9.3.

Bensmann   A process for recovering lubricating oils by treatment with a strong acid, followed by fuller's earth. Invented in 1926 by N. Bensmann in Germany.

> German Patent 472,184.
> Bensmann, N., *Oel Kohle*, 1933, **1,** 159.

Benson   A process for converting methane to ethylene, developed by Hydrocarbon Research, CA.

> *Eur. Chem. News*, 1982, **39**(1049), 31.

Benzoraffin   A hydrofining process for treating naphtha fractions derived from coal. It is a fixed-bed, gas-phase process using a cobalt/molybdenum oxide catalyst. Developed jointly by BASF, Veba-Chemie, and Lurgi, Ground 1960.

> *Ullmann's Encyclopedia of Industrial Chemistry*, Vol. A12, 5th ed., VCH Publishers, Weinheim, Germany, 1989, 284.

Benzorbon   A process for separating and recovering benzene from coke-oven gas and town gas by adsorption on activated carbon. Developed in 1930 by Lurgi.

Bergbau-Forschung   Also called BF. A process for separating nitrogen from air by selective adsorption over activated carbon in a *PSA system. Developed by Bergbau-Forschung (now Bergwerksverband), Germany. Licensed by Nitrox, UK, which uses this process in its laboratory gas supply equipment.

> U.S. Patent 4,572,723.
> European Patent 132,756.
> British Patent 2,152,834.
> Knoblauch, K., *Chem. Eng. (N.Y.)*, 1978, **85**(25), 87.

Bergbau-Forschung/Uhde   A *flue-gas desulfurization process that uses a movable bed of hot coke. Operated in a power station in Arzberg, Germany, since 1987.

> Wieckowska, J., *Catal. Today*, 1995, **24**(4), 453.

Bergius (1)   A coal liquifaction process (also called hydroliquifaction), invented in Germany in 1913 by F. Bergius and subsequently developed by IG Farbenindustrie. The inventor, together with C. Bosch, was awarded the Nobel Prize for chemistry for this invention in 1931. A pilot plant was operated at Rheinau near Mannheim between 1921 and 1927. The first commercial plant was built at Leuna in 1927. Twelve plants of this type provided much of the aviation fuel used by Germany in World War II. After the war, the process was further developed by the U.S. Bureau of Mines. The process is essentially one of hydrogenation at high pressures and temperatures, catalyzed by an iron oxide catalyst. In Germany, the catalyst was the red mud waste from the *Bayer aluminum process. *See also* Bergius Pier.

> German Patents 301,231; 304,348.
> British Patent 18,232 (1914).
> Bergius, F., *J. Gasbeleucht*, 1912, **54,** 748.
> Storch, H. H., in *Chemistry of Coal Utilization*, 2nd. Suppl. Vol., Lowry, H. H., Ed., John Wiley & Sons, New York, 1945, 1750.

James, L. K., Ed. *Nobel Laureates in Chemistry, 1901–1992*, American Chemical Society and Chemical Heritage Foundation, Washington, D.C., 1993, 192.

Bergius (2)   Also known as Bergius-Willstäter-Zechmeister. A process for extracting sugar from wood by hydrolyzing the cellulose with concentrated hydrochloric acid. Lignin remains undissolved. Developed in the 1920s.

Willstäter, R. and Zechmeister, L., *Ber. Dtsch. Chem. Ges.*, 1913, **40**, 2401.
Bergius, F., *Ind. Eng. Chem.*, 1937, **29**, 247.

Bergius-Pier   An improved version of the *Bergius (1) process in which the activity of the catalyst was increased by treatment with hydrofluoric acid. Invented by H. Pier and others in the 1930s and used in Germany during World War II.

U.S. Patents 2,154,527; 2,194,186.
Pier, M., *Angew. Chem.*, 1938, **51**, 603.
Pier, M., *Z. Elektrochem. Angew. Phys. Chem.*, 1953, **57**, 456.

Bergius-Rheinau   A *saccharification process in which wood is hydrolyzed with concentrated hydrochloric acid to produce sugars for subsequent fermentation to ethanol.

Kent, J. A., Ed., *Riegel's Handbook of Industrial Chemistry*, 9th ed., Van Nostrand Reinhold, New York, 1992, 255.

Bernardini   A process for separating fatty acids by fractional crystallization.

Coppa-Zuccari, G., *Oleagineux*, 1971, **26**, 405 (*Chem. Abstr.*, **75**, 128522).

Bertrand   A microbiological process for oxidizing aldoses to ketoses. Used in the manufacture of ascorbic acid.

Pigmann, W. W. and Goepp, R. M., *Chemistry of the Carbohydrates*, Academic Press, New York, 1948, 90.

Bertrand Thiel   A variation of the Basic *Open Hearth steelmaking process, suitable for ores rich in silicon and phosphorus. Two hearths are used; by varying the quantities of lime added to each it is possible to concentrate most of the silicon and phosphorus in the slag from the first. Developed by E. Bertrand and O. Thiel at Kladno, Bohemia, in 1894; subsequently adopted at Hoesch, Germany (hence the alternative name for the process).

Bertrand, E., *J. Iron Steel Inst. (London)*, Pt. 1, 1897, 115.
Barraclough, K. C., *Steelmaking 1850–1900*, The Institute of Metals, London, 1990, 281.

Berzelius   A tin smelting process operated by Berzelius Metalhütten at Duisberg-Wanheim, Germany. A mixture of oxide ore, coal, and sodium carbonate is fed continuously into a rotating tubular kiln having a constriction at the discharge end. The molten metal product collects in the sump at the end and the slag, which forms an upper layer, and flows off.

Mantell, C. L., *Tin: Its Mining, Production, Technology, and Application*, Reinhold, New York, 1949; published in facsimile by the Hafner Publishing, New York, 1970, 141.

Bessemer   Also called the Pneumatic process. A steelmaking process invented by H. Bessemer in Sheffield, England, in 1855 while experimenting with the manufacture of gun barrels. He noticed that a draught of air decarbonized iron bars that were lying on the rim of his furnace. Commercialized by associates of Bessemer in Sheffield in 1858 and soon widely adopted. Molten pig iron is contained in a Bessemer converter, which is a pear-shaped vessel with a double bottom, lined with silica, and capable of being tilted on a horizontal axis. Compressed air is blown through the base of the converter, oxidizing most of the dissolved carbon. A similar process was developed in the United States by W. Kelly at around the same time. Kelly made his invention in 1851 but delayed applying for his patent until 1857;

although his priority was recognized by the U.S. Patent Office, his name has not become associated with the process.

Bessemer, H., British Patents 2,768 (1855); 630 (1856).

Kelley, W., U.S. Patent 17,628 (1857).

Allen, J. A., *Studies in Innovation in the Steel and Chemical Industries*, A. M. Kelley, New York, 1968.

Barraclough, K. C., *Steelmaking 1850–1900*, The Institute of Metals, London, 1990, 39,127.

Dennis, W. H., *A Hundred Years of Metallurgy*, Gerald Duckworth, London, 1963, 98.

Bethanising   *See* metal surface treatment.

Bethell   Also known as the Full-cell process. A method for impregnating timber with a creosote preservative. The wood is first degassed under partial vacuum and then impregnated under a pressure of up to 10 atm. *See also* Rueping.

Betterton (1)   A process for removing bismuth from lead, A calcium-magnesium alloy is added to the molten lead. The bismuth concentrates in an upper layer and is skimmed off.

Betterton (2)   A process for removing zinc from lead by reaction with gaseous chlorine at 400°C. The resulting zinc chloride floats on the molten metal and is skimmed off. Invented in 1928 by J. O. Betterton at the American Smelting & Refining Company.

Dennis, W. H., *A Hundred Years of Metallurgy*, Gerald Duckworth, London, 1963, 192.

Betterton-Kroll   A metallurgical process for removing bismuth from lead. Metallic calcium and magnesium are added to the molten lead, causing precipitation of the sparingly soluble $CaMgBi_2$.

Betts   An electrolytic process for refining lead and recovering silver and gold from it. The electrolyte is a solution of lead fluosilicate and hydrofluosilicic acid. The other metals collect as a slime on the anode and are retained there. Developed by A.G. Betts in 1901, first operated at Trail, British Columbia, in 1903, and now widely used in locations having cheap electric power.

Powell, A. R., *The Refining of Non-ferrous Metals,* The Institute of Mining and Metallurgy, London, 1950.

Dennis, W. H., *A Hundred Years of Metallurgy,* Gerald Duckworth, London, 1963, 196.

Bextol   A catalytic *hydrodealkylation process using an oxide catalyst.

Roebuck, A. K. and Evering, B. L., Ind. Eng. Chem., 1958, **50,** 1135.

Ballard, H. D., Jr., in *Advances in Petroleum Chemistry and Refining,* Vol. 10, McKetta, J. J., Ed., Interscience Publishing, New York, 1965, 219.

BF   *See* Bergbau-Forschung, carbonization.

BFL   *See* carbonization.

BFR   [Bleach filtrate recycle]   A process for almost completely recycling the waste liquors from pulp-bleaching using chlorine dioxide. The key to the process is the separation of the sodium sulfate from potassium chloride by crystallization. Developed by Champion International Corporation, Sterling Pulp Chemicals, and Wheelabrator Technologies. It was proposed for installation in North Carolina in 1997.

Chem. Eng. *(N.Y.),* 1996, **103**(9), 27.

BF/Uhde   [Bergbau-Forschung]   A *flue-gas treatment which removes sulfur dioxide and nitrogen oxides from power station effluent gases. Sulfur dioxide is first adsorbed on activated coke (made by the German company Bergbau-Forschung). Nitrogen oxides are then converted to elemental nitrogen by the *SCR process. The system was first demonstrated at the Arzberg power station, Germany, in 1988. Engineered by Uhde.

Bianchi An early entrained-flow coal gasification process.

Biazzi A continuous, two-phase process for nitrating polyols to form nitrate esters, used as explosives. The nitrating agent is a mixture of anhydrous nitric and sulfuric acids. Used for making nitroglycerine, triethylene glycol nitrate, butanetriol trinitrate, and trimethylolethane trinitrate. Invented by M. Biazzi in Switzerland in 1935. More than 110 plants were sold, worldwide; many are still operating. Today, Biazzi SA is known also for its hydrogenation process and many other aromatic processes. *See also* Meissner, Gyttorp.

BicarboneR A biological sewage treatment process, using bacteria contained in a fixed bed. Developed by Compagnie Général des Eaux, France, and used in the United States and Japan.

Bifilm A sewage treatment process utilizing two aerated biological filters filled with granular plastic media. Developed by Biwater, UK, and piloted in 1994.

Bi-Gas [**Bi**tuminous **Gas**] A coal gasification process using a two-stage, entrained-flow slagging gasifier. Developed by Bituminous Coal Research. A 120-tons-per-day pilot plant was built in 1976 at Homer City, PA, under sponsorship from the U.S. Energy Research and Development Administration and the American Gas Association.

> Hegarty, W. P. and Moody, B. E., *Chem. Eng. Prog.,* 1973, **69**(3), 37.
> *Hydrocarbon Process.,* 1975, **54**(4), 119.
> Hebden, D. and Stroud, H. J. F., in *Chemistry of Coal Utilization,* 2nd. Suppl. Vol., Elliott, M. A., Ed., John Wiley & Sons, New York, 1981, 1722.

Binax A process for removing carbon dioxide from methane from landfill by washing with water; absorption takes place under pressure and desorption is done by an air blast at atmospheric pressure. Piloted at a wastewater treatment plant at Modesto, CA, in 1978.

> Henrich, R. A., *Energy Biomass Wastes,* 1983, 879,916 (Chem. Abstr., **99,** 125564).
> Kohl, A. L. and Riesenfeld, F. C., *Gas Purification,* 4th ed., Gulf Publishing, Houston, TX, 1985, 263.

Bio3 A process for oxidizing organic wastes in industrial wastewater by a combination of ozone, hydrogen peroxide, and UV radiation. Intended to be integrated with the *Vitox process by using the excess of oxygen from the Bio3 process in the Vitox process.

> *Eur. Chem. News,* 1994, **61**(1607), 27.

BIOBEAD A *BAF process offered by Brightwater.

> Stephenson, T., Mann, A., and Upton, J., *Chem. Ind. (London),* 1993, 14, 533.

Biobed An anaerobic digestion system for treating industrial organic wastes, developed from the *Biothane system. It uses an expanded granular sludge bed and a three-phase separation system; these features enable it to handle ten times the hydraulic load of a similar Biothane system. Developed by Gist-Brocades NV (Delft) and licensed in the UK through Babcock Water Engineering Ltd. Twenty systems were operating, worldwide, in 1996.

> *Water Waste Treat.,* 1996, **39**(5), 27.

Biobor HSR [**Bio**logical **Bor**sig **H**ubstrahl**r**eaktor] A high-intensity biological process for treating concentrated effluents from farms and food processing. Compressed air forces the waste rapidly up a tower containing moving, perforated discs. Developed by Borsig, Berlin, in the early 1990s. *See also* Deep Shaft.

BIOCARBONE A *BAF process offered by OTV.

> Stephenson, T., Mann, A., and Upton, J., *Chem. Ind., (London),* 1993, (14), 533.

BIOCLAIM   A process for removing toxic metals from aqueous systems by use of bacteria supported on poly(ethylenimine) glutaraldehyde.

> Brierly, C. L., *Geomicrobiol.,* 1990, **8,** 201.
> Vegio, F., Beolchini, F., and Toro, L., *Ind. Eng. Chem. Res.,* 1998, **37**(3), 1107.

Bio-Claus   A biological desulfurization process for removing sulfur dioxide from waste gases. There are four stages: scrubbing with a solvent to produce a sulfite solution, biological reduction of the sulfite to sulfide, biological oxidation of the sulfide to elemental sulfur, and separation of the sulfur. Developed by Stork Engineering and first announced in 1997.

> *Chem. Eng. (Rugby, Engl.),* 1997, (638), 24.

BIOFIX   A process for removing toxic heavy metals from aqueous wastes by the use of peatmoss and a polysulfone.

> Veglio, F. and Beolchini, F., *Hydrometallugy,* 1997, **44**, 301.
> Vegio, F., Beolchini, F., and Toro, L., *Ind. Eng. Chem. Res.,* 1998, **37**(3), 1105.

BIOFOR   A *BAF process offered by Degremont.

> Stephenson, T., Mann, A., and Upton, J., *Chem. Ind., (London)* 1993, (14), 533.

Biogas   A process for generating fuel gas from sewage. Developed by EMS-Inventa and installed in Switzerland and Italy between 1979 and 1983. *See* BIOGAS.

BIOGAS   A process for generating methane from wastes and biomass. Developed by the Institute of Gas Technology, Chicago, since 1970, and demonstrated at Walt Disney World, Orlando, FL, in 1990. *See also* Biogas.

BIOHOCH   An aerobic wastewater treatment process. Optimimum reactor design and a proprietary air injection system achieve 50 percent of the energy consumption of similar systems. Fifty systems had been installed worldwide by 1994. Licensed by Brown-Minneapolis Tank and Hoechst-Uhde Corporation.

> *Hydrocarbon Process.,* 1993, **72**(8), 92.

BIOKOP   A process for treating liquid effluents containing wastes from organic chemical manufacture. It combines aerobic fermentation, in special reactors known as BIOHOCH reactors, with treatment by powdered activated carbon. Developed originally for treating the effluent from the Griesheim works of Hoechst, it was engineered by Uhde and is now offered by that company. *See also* PACT.

BIOPUR   A *BAF process offered by Sulzer and John Brown.

> Stephenson, T., Mann, A., and Upton, J., *Chem. Ind. (London),* 1993, (14), 533.

Biostil   A continous fermentation process for making ethanol. Developed by Chematur Engineering AB, Sweden.

Bio-SR   A process for removing hydrogen sulfide from gas streams. Developed by NKK Industries, Japan. It uses a solution of unchelated iron, regenerated microbiologically.

> *Oil Gas J.,* 1994, **92**(21), 58.
> Quinlan, M. P., Echterhof, L. W., Leppin, D., and Meyer, H. S., *Oil Gas J.,* 1997, **95**(29), 54.

BIOSTYR   [Biological polystyrene]   A biological system for treating efluents containing dissolved organic matter. The microbiological organisms are trapped within rigid, lighter-than-water porous polystyrene granules. The effluent flows upward through a bed of these granules and air is injected at the base of the bed. Developed in France by OTV and licensed in the UK through General Water Processes.

> Stephenson, T., Mann, A., and Upton, J., *Chem. Ind. (London),* 1993, (14), 533.

Biothane An anaerobic digestion system for treating industrial organic wastes. The reactor contains an upflow sludge blanket and is operated at approximately 35°C, with the heat provided by burning some of the product gas which contains 70 percent methane. It is usually necessary to add nutrients such as urea and iron. Developed in the early 1970s in The Netherlands by Centrale Suiker Maatschappij; in 1984, Gist-Brocades (Delft) acquired the rights and subsequently licensed the process in the United Kingdom to Esmil. In 1990, more than 70 units had been built, worldwide, for a variety of industries. *See* Biobed.

Water Waste Treat., 1991, **34**(5), 24.

BIOX [biological oxidation] A general term for effluent treatment processes employing biological oxidation, such as the *Activated Sludge process.

Capps, R. W., Matelli, G. N., and Bradford, M. L., *Hydrocarbon Process.*, 1993, **72**(12), 81.

Birkeland-Eyde Also known as the **Arc** process. A process for making nitric acid by passing air through an electric arc, forming nitric oxide, oxidizing this with air, and absorbing the resulting oxides of nitrogen in water:

$$N_2 + O_2 = 2NO$$

$$2NO + O_2 = 2NO_2$$

$$2NO_2 + H_2O = HNO_2 + HNO_3$$

$$3HNO_2 = HNO_3 + 2NO + H_2O$$

The arc is spread into a disc by an electromagnet. Operated in Norway, using hydroelectric power, from 1905 to 1930, when it was made obsolete by the ammonia oxidation process. In the United States it was first used in 1917.

Sherwood Taylor, F., *A History of Industrial Chemistry,* Heinemann, London, 1957, 428.
Haber, L. F., *The Chemical Industry 1990–1930,* Clarendon Press, Oxford, 1971, 86.
Davies, P., Donald, R. T. Ed., and Harbord, N. H., in *Catalyst Handbook,* 2nd. ed., Twigg, M. V., Wolfe Publishing, London, 1989, 470.

BiRON A biological process for removing iron from public water supplies. Developed in the UK by Biwater Europe Ltd and piloted in 1994 at a water treatment plant in Ipswich.

Alani, S., *Water Waste Treat.*, 1994, **37**(6), 50.

Bischof An obsolete process for making white lead (basic lead carbonate), invented by G. Bischof around 1900. Lead monoxide was reduced by the carbon monoxide in water-gas to form black lead suboxide, oxidized in damp air to lead hydroxide, slurried in dilute acetic acid, and carbonated with carbon dioxide. Piloted in Willsden, London, and commercialized by L. Mond in the Brimsdown White Lead Company. A sample of Bischof's white lead was used by Holman Hunt in his painting "Light of the World," now in St. Paul's Cathedral, London.

Cohen, J. M., *The Life of Ludwig Mond,* Methuen, London, 1956, 211.

Bischoff A *flue-gas desulfurization process. A slurried mixture of lime and limestone is sprayed into the gas in a spray tower. The calcium sulfite in the product is oxidized by air to calcium sulfate. Used in Europe in the 1980's. Lurgi Bishoff is a part of the Lurgi group. The process is offered by Lentjes, Germany, a subsidiary of Lurgi.

IEA Coal Research, *The Problems of Sulphur,* Butterworths, London, 1989, 100.

Black ash One of the two processes comprising the *Leblanc process for making sodium carbonate; the other is the Salt-cake process. The heart of the process was a rotating kiln made of cast iron, known as a revolver. Invented by G. Elliot and W. Russel in St. Helens,

England, in 1853. The name has been used also for a process for extracting barium from barium sulfate.

British Patent 887 (1853).

Black oil conversion   *See* RCD Unibon.

Blanc   A process for extracting alumina from leucite ore (a potassium aluminosilicate) in which the ore is leached with hydrochloric acid, aluminum trichloride is crystallized, and this is calcined to alumina. Invented by G.A. Blanc in 1921 and used in Italy from 1925 to 1943.

U.S. Patent 1,656,769.
O'Connor, D. J., *Alumina Extraction from Non-bauxitic Materials,* Aluminium-Verlag, Düsseldorf, 1988, 85.

Blaugas   An early thermal cracking process for making liquid petroleum gas from petroleum. Developed by the German company Blau in Augsburg from 1905. Not to be confused with blue gas (see Water gas).

BLISS   [Butylene isomerization system]   A process for isomerizing *n*-butenes to isobutene. Piloted by Texas Olefins Company in Houston, TX, 1990 to 1992.

Chem. Mark. Rep., 1992, **241**(18), 7.

Bloomery   The earliest process for making iron from iron ore, operated from around 1500 BC until the blast furnace was invented around 1500 AD. The ore is heated with charcoal in a furnace blown by bellows; the product, known as "bloom," is a composite of iron particles and slag. When this is hammered, the slag is expelled to the surface and a lump of relatively pure iron remains. *See also* Catalan.

Barraclough, K. C., *Steelmaking Before Bessemer, Vol. 1, Blister Steel,* The Metals Society, London, 1984, 15.

Blueprint   A reprographic process, based on the photochemical reduction of ferric salts. Paper is impregnated with an aqueous solution of potassium ferricyanide, ammonium ferric citrate, and a gum binder. Exposure to light generates Prussian blue, and then Turnbull's blue. Developed in the 19th century and widely used for copying line drawings until superseded by versions of the *Diazo process and subsequently by xerographic processes. *See* reprography.

Blumenfeld   An early version of the *Sulfate process for making titanium dioxide pigment, in which the nucleation of the precipitation of titania hydrate is accomplished by dilution under controlled conditions. Invented by J. Blumenfeld, a Russian working in London in the 1920s.

U.S. Patents 1,504,669; 1,504,671; 1,504,672.
Barksdale, J., *Titanium: Its Occurrence, Chemistry, and Technology,* 2nd. ed., Ronald Press, New York, 1966, 278.

BMA   [Blausaure Methan Ammoniak]   *See* Degussa.

BMS   A process for removing mercury from the effluent from the *Castner-Kellner process. Chlorine is used to oxidize metallic mercury to the mercuric ion, and this is then adsorbed on activated carbon impregnated with proprietary sulfur compounds. Developed by Billingsfors Bruks, Sweden.

Rosenzweig, M.D., *Chem. Eng. (N.Y.),* 1975, **82**(2), 60.

BOC Isomax   [Black Oil Conversion]   *See* Isomax.

BOC Unibon   A process for upgrading petroleum residues by catalytic hydrogenation. *See also* RCD Unibon.

Absci-Halabi, M., Stanislaus, A., and Qabazard, H., *Hydrocarbon Process.* 1997, **76**(2), 50.

Bofors   A process for nitrating organic compounds by mixing the nitrating acid and the substrate rapidly in a centrifugal pump.

Boivan-Loiseau   A process for purifying cane sugar. Calcium hydroxide is added to the syrup, and carbon dioxide passed through it. The precipitated calcium carbonate removes some of the coloring impurities.

Watson, J. A., *A Hundred Years of Sugar Refining,* Tate & Lyle Refineries, Liverpool, 1973, 80.

Boliden (1)   An obsolete process for reducing sulfur dioxide to elemental sulfur, using hot coke. Operated in Sweden by Boliden from 1933 to 1943. *See also* Trail, RESOX.

Katz, M. and Cole, R. J., *Ind. Eng. Chem.,* 1950, **42**, 2258.

Boliden (2)   A lead extraction process in which a sulfide ore, mixed with coke, is smelted in an electric furnace, air jets forming vortices between the electrodes. Discontinued in 1988 in favor of the *Kaldo process, using a rotating furnace.

Boliden/Norzink   A method for removing mercury vapor from zinc smelter off-gases by scrubbing with a solution of mercuric chloride:

$$Hg + HgCl_2 = Hg_2Cl_2$$

The precipitated mercurous chloride separates as a sludge. In the original process, some of this mercurous chloride was chlorinated to mercuric chloride for re-use. In a later version of the process, all the mercuric chloride is electrolytically converted to elemental mercury and chlorine. As of 1994 the electrolytic version had been installed in three plants. Developed by Boliden, Sweden, and Norzink, Norway, and now offered for license by Boliden Contech.

Bolkem   [Boliden]   A process for removing mercury from the sulfuric acid from metallurgical smelters. Addition of aqueous sodium thiosulfate causes mercuric sulfide to be precipitated. Operated at Helsingborg, Sweden, since 1974.

O. Sundström, *Sulphur,* 1975, (116), 37.

Bonderizing   *See* metal surface treatment.

Borchers-Schmidt   *See* metal surface treatment.

Borstar   A catalytic process for polymerizing ethylene. Use of two reactors, a loop reactor and a gas-phase reactor, allows better control of molecular weight distribution. The loop reactor operates under super-critical conditions to avoid bubble formation. Either *Ziegler-Natta or metallocene catalysts can be used. The first commercial unit was installed in Porvoo, Finland, in 1995.

Eur. Chem. News, 1995, **64**(1688), 37; 1996, **65**(1709), 45.
Chem. Eng. (N.Y.), 1995, **102**(11), 17.

BOSAC   [Bofors Sulfuric Acid Concentrator]   A process for recovering sulfuric acid from the production of nitro-compounds. Spent acid is concentrated by distillation, using a heat exchanger with externally heated silica tubes. Developed by Bofors Nobel Chemikur, Sweden.

Douren, L., *Making the Most of Sulfuric Acid,* More, A. I., Ed., British Sulphur, London, 1982, 317.

Bosch   A regenerative, two-stage process for reducing carbon dioxide to oxygen:

$$CO_2 + 2H_2 = 2H_2O + C$$

$$2H_2O = 2H_2 + O_2$$

The first step is conducted over an iron catalyst at 700°C; the second by electrolysis. Proposed for use during prolonged space travel.

Sacco, A., Jr. and Reid, R. C., *Carbon,* 1979, **17,** 459.

Bosch-Meiser   A process for making urea from ammonia and carbon dioxide under high temperature and pressure. Invented by C. Bosch and W. Meiser in 1920.

U.S. Patent 1,429,483.

Boss   A development of the *Washoe process for extracting silver from sulfide ores, invented in 1861. The chloride solution from the Washoe process was passed continuously through a series of amalgamation pans and settlers.

Bouchet-Imphy   *See* DR.

Bower-Barff   *See* metal surface treatment.

BP-Hercules   *See* Hercules-BP.

BPR   [By-Product Recycle]   A process for recycling the chlorine-containing by products from the manufacture of vinyl chloride, 1,2-dichloroethane, and other chlorinated hydrocarbons. Combustion with oxygen converts 90 percent of the chlorine to anhydrous hydrogen chloride, and 10 percent to aqueous hydrochloric acid. Developed by BASF and licensed by European Vinyl Corp.

Eur. Chem. News, 1990, 26.

Bradshaw   A soapmaking process in which glyceryl esters (fats) are first converted to their corresponding methyl esters by transesterification, and then hydrolyzed. Invented by G. B. Bradshaw.

U.S. Patent 2,360,844.

Brassert   A modified blast furnace process for making iron. Less limestone is added than in the usual process, so the product contains more sulfur. The molten iron product is mixed with sodium carbonate, with which the ferrous sulfide reacts yielding sodium sulfide and iron oxide. These float on the molten iron and are skimmed off.

Braun   A variation on the classic ammonia synthesis process in which the synthesis gas is purified cryogenically. Widely used since the mid 1960s.

Isalski, W. H., *Separation of Gases,* Clarendon Press, Oxford, 1989, 155.

Brennstoff-Technik   *See* carbonization.

Bretsznajder   A process for extracting aluminum from clays and a variety of aluminous wastes by continuous digestion with concentrated sulfuric acid in an autoclave. Developed in Poland in the 1980s but not yet commercialized.

O'Connor, D. J., *Alumina Extraction from Non-bauxitic Materials,* Aluminum-Verlag, Dusseldorf, 1988, 160.

Bridgman   Also called Bridgman-Stockbarger. A process for growing large single crystals. The material is contained in a cylindrical ampoule that is slowly lowered through a temperature gradient.

Stockbarger, D. C., *J. Opt. Soc. Am.,* 1927, **14,** 448.
Vere, A. W., *Crystal Growth; Principles and Progress,* Plenum Press, New York, 1987, 67.

Brin   An obsolete process for making oxygen. Barium monoxide was roasted in air to produce barium peroxide, which was roasted at a higher temperature to produce oxygen:

$$2BaO + O_2 = 2BaO_2$$

$$2BaO_2 = O_2 + 2BaO$$

The process was developed in France by the brothers L. and A. Brin, based on a chemical reaction discovered by J. B. J. D. Boussingault in 1851. Boussingault used temperature cycling between 537 and 926°C, which caused a progressive loss in the efficiency of the barium oxide. The Brin brothers used pressure cycling at a lower temperature, which overcame this problem. The process was first operated commercially by Brin Frères et Cie. in France in 1879, but was abandoned in favor of air liquification at the beginning of the 20th Century. In England, Brins Oxygen Company was incorporated to operate the process in 1886, changing its name to the British Oxygen Company in 1906.

> Smith, W., *J. Soc. Chem. Ind., London,* 1885, **4**, 568.
> *Chem Br.,* 1997, **33**(5), 58.

Brittania   A process for removing silver from lead, operated by Brittania Refined Metals in England, using ore from the Mount Isa mine in Australia. After initial concentration by the *Parkes process, and removal of the zinc by vacuum distillation, the mixture, which contains silver (70 percent), lead, and some copper is treated in a bottom blown oxygen cupel in which lead and copper are removed by the injection of oxygen through a shielded lance.

> Barrett, K. R. and Knight, R. P., *Silver–Exploration, Mining and Treatment.* The Institute of Mining and Metallurgy, London, 1988.

Brodie   A naphthalene crystallization process.

BSR   *See* Beavon.

BSR/MDEA   *See* Beavon.

Bubiag   An early two-stage coal gasification process.

Bucher   A process for making sodium cyanide by the reaction between sodium carbonate and coke:

$$Na_2CO_3 + 4C + N_2 = 2NaCN + 3CO$$

Invented by J. E. Bucher in 1924.

> U.S. Patent 1,091,425.
> Sittig, M., *Sodium, Its Manufacture, Properties and Uses,* Reinhold Publishing, New York, 1956, 223.

Buchner   *See* Aloton.

Budenheim   *See* CFB.

Bueb   A process for removing cyanogen from coal gas by scrubbing with a concentrated aqueous solution of ferrous sulfate. A complex sequence of reactions occurs, involving the ammonia and hydrogen sulfide which are always present in coal gas; the final product is a blue mud containing ammonium ferrocyanide. This mud is boiled with lime, thereby expelling the ammonia, and several further steps result in the formation of saleable sodium or potassium ferocyanide. Invented by J. Bueb in 1898 and operated in the United States in the early 1900s.

> German Patent 112, 459, 1898.
> Muller, M. E., *J. Gas Lighting,* 1910, **112**, 851.
> Hill, W. H., in *Chemistry of Coal Utilization,* Vol. 2, Lowry, H. H., Ed., John Wiley & Sons, New York, 1945, 1104.

Bufflex [from **Buff**elsfontein (South Africa), **ex**traction]   A process for extracting uranium from its ores, using a solution of an amine (Alamine 336). It was developed in South Africa, and was later replaced by Purlex. *See also* Eluex.

> Eccles, H. and Naylor, A., *Chem. Ind. (London)*, 1987, (6), 174.

Bullard-Dunn   *See* metal surface treatment.

Buna [**Bu**tadien **n**atrium]   The name has been used for the product, the process, and the company VEB Chemische Werke Buna. A process for making a range of synthetic rubbers from butadiene, developed by IG Farbenindustrie in Leverkusen, Germany, in the late 1920s. Sodium was used initially as the polymerization catalyst, hence the name. **Buna S** was a copolymer of butadiene with styrene; **Buna N** a copolymer with acrylonitrile. The product was first introduced to the public at the Berlin Motor Show in 1936. Today, the trade name Buna CB is used for a polybutadiene rubber made by Bunawerke Hüls using a *Ziegler-Natta type process.

> German Patent 570, 980.
> Morris, P. J. T., *The American Synthetic Rubber Research Program,* University of Pennsylvania Press, Philadelphia, 1989, 7.

Burgess   A coal liquifaction process piloted by the U.S. Office of Synthetic Fuels at Louisiana, MO, in 1949.

> Kastens, M., Hirst, L. L., and Chaffe, C. C., *Ind. Eng. Chem.*, 1949, **41**, 870.

Burkheiser   Also known as the sulfite-bisulfite process. A complicated process for removing hydrogen sulfide and ammonia from coal gas by absorption in an aqueous solution containing ammonia, iron oxide, and elemental sulfur. The end products are sulfur and ammonia. Invented by K. Burkheiser in 1907 and developed in Germany in the early 1900s.

> German Patents 212, 209; 215, 907; 217, 315; 223, 713.
> Kohl, A. L. and Riesenfeld, F. C., *Gas Purification,* 4th ed., Gulf Publishing, Houston, TX, 1985, 488.

Burton   The first commercial process for thermally cracking heavy petroleum fractions to obtain gasoline. Invented in 1912 by W. M. Burton at Standard Oil (Indiana) and operated commercially from 1913 through the 1920s. *See also* Dubbs.

> U.S. Patent 1, 049, 667.
> Enos, J. L., *Petroleum Progress and Profits,* MIT Press, Cambridge, MA, 1962, Chap. 1.
> Achilladelis, B., *Chem. Ind. (London)*, 1975, 19 Apr., 337.

BUTACRACKING   A process for converting *iso*-butanes to *iso*-butene, which can then be converted to gasoline-blending components such as methyl *t*-butyl ether. Developed by Kinetics Technology International.

> Monfils, J. L., Barendregt, S., Kapur, S. K., and Woerde, H. M., *Hydrocarbon Process.*, 1992, **71**(2), 47.

Butamer [**Buta**ne iso**mer**ization]   A process for converting *n*-butane into *iso*-butane; conducted in the presence of hydrogen over a dual-functional catalyst containing a noble metal. Developed by UOP and licensed worldwide since 1959. In 1992, more than 55 units had been licensed.

> *Hydrocarbon Process.*, 1990, **69**(4), 73.
> Cusher, N. A., in *Handbook of Petroleum Refining Processes,* Meyers, R. A., Ed., McGraw-Hill, New York, 1997, 5-39.

BUTENEX   A process for separating several C_4 components from C_4 hydrocarbon streams by extractive distillation using Butenex, a proprietary extraction agent. Piloted by Krupp Koppers in 1987. Several plants were being engineered in 1994.

> *Chem. Week,* 1987, **141**(9), 24.

Butesom   *See* C$_4$ Butesom.

Butex   A process for separating the radioactive components of spent nuclear fuel by solvent extraction from nitric acid solution, using diethylene glycol dibutyl ether (also called Butex, or dibutyl carbitol) as the solvent. Developed by the Ministry of Supply (later the UK Atomic Energy Authority) in the late 1940s. Operated at Windscale from 1952 until 1964 when it was superseded by the *Purex process.

> Martin, F. S. and Miles, G. L., *Chemical Processing of Nuclear Fuels,* Butterworth Scientific Publications, London, 1958, 102.
> Howells, G. R., Hughes, T. G., Mackey, D. R., and Saddington, K., in *Proc. 2nd. U.N. Internat. Conf. Peaceful Uses of Atomic Energy,* United Nations, Geneva, 1958, **17,** 3.
> British Nuclear Fuels PLC, *Nuclear Fuel Processing Technology,* 1985.

Butomerate   A catalytic process for isomerizing *n*-butane to isobutane. Developed by the Pure Oil Company.

> *Hydrocarbon Process.,* 1964, **43**(9), 173.
> Unzelman, G. H. and Wolf, C. J., in *Petroleum Processing Handbook,* Bland, W. F. and Davidson, R. L., Eds., McGraw-Hill, New York, 1967, 3–51.

BWHP   *See* Woodall-Duckham

BYAS   [**By**pass **a**mmonia **s**ynthesis]   An economical process for expanding existing ammonia synthesis plants by introducing the additional natural gas at an intermediate stage in the process. The additional nitrogen in the air, which has also to be introduced, is removed by *PSA. Developed and offered by Humphreys and Glasgow, UK.

C

CAA   [**C**uprous **a**mmonium **a**cetate]   A general process for separating alkenes, di-alkenes, and alkynes from each other by extraction of their cuprous complexes from aqueous cuprous ammonium acetate into an organic solvent. Exxon used it for separating C$_4$ fractions containing low concentrations of butadiene. The liquid–liquid extraction processes for butadiene have all been replaced by extractive distillation processes.

> U.S. Patents 2,369,559; 2,429,134 (Jasco): 2,788,378 (Polymer Corp.); 2,847,487; 2,985,697; 3,192,282 (Esso).
> Morrell, C. E., Palz, W. J., Packie, J. W., Asbury, W. C., and Brown, C.L., *Trans. Am. Inst. Chem. Eng.,* 1946, **42,** 473.
> Weissermel, K. and Arpe, H.-J. *Industrial Organic Chemistry,* 3rd ed., VCH Publishers, Weinheim, Germany, 1997, 108.

CAB   *See* steelmaking.

CADRE   A process for removing and oxidizing volatile organic compounds from gas streams. The compounds are adsorbed on a fixed bed of carbon and then desorbed by a stream of hot air or inert gas. Developed by Vard International, a division of Calgon Carbon Corporation.

> *Hydrocarbon Process.,* 1993, **72**(8), 77.

CAFB [Chemically active fluidized-bed] A coal-gasification process intended for producing gas for power generation. Coal particles are injected into a shallow bed of lime particles that trap the sulfur dioxide. The bed particles are regenerated in a second fluidized bed, releasing the sulfur dioxide. Developed in the 1970s by the Esso Petroleum Company, UK, but not commercialized.

> Hebden, D. and Stroud, H. J. F., *Chemistry of Coal Gasification,* 2nd. Suppl. Vol., Elliott, M. A., Ed., John Wiley & Sons, New York, 1981, 1668.

Calcilox A process for converting calcium sulfate/sulfite wastes from *flue-gas desulfurization into a disposable, earthy material, by use of a proprietary inorganic additive made from blast furnace slag. Developed by Dravo Corporation of Pittsburgh, PA.

> Labovitz, C. and Hoffman, D. C., *Toxic and Hazardous Waste Disposal,* Pojasek, R. B., Ed., Ann Arbor Science Publishers, Ann Arbor, MI, 1979, Vol. 1, Chap. 5.
> IEA Coal Research, *The Problems of Sulphur,* Butterworths, London, 1989, 127.

Calcor A process for making carbon monoxide from natural gas or liquid petroleum gas. It combines *steam reforming with carbon dioxide recovery or recycle. Designed and licensed by Caoric GmbH. Five commercial plants have been installed as of 1992.

> Teuner, S. *Hydrocarbon Process.,* 1985, **64**(5), 106.
> *Hydrocarbon Process.,* 1992, **71**(4), 90.

Calmet A process for extracting gold from its ores by pressure cyanidation. Developed by the Calmet Corporation in 1983.

> Yannopoulos, J. C., *The Extractive Metallurgy of Gold,* Van Nostrand Reinhold, New York, 1991, 164.

Calorising Also spelled Calorizing. A proprietary process for protecting the surface of iron or steel by applying a layer of aluminum. Several methods of application may be used: dipping, spraying, or chemical reaction with aluminum chloride. *See also* metal surface treatment.

Calsinter A process for extracting aluminum from fly ash and from *flue-gas desulfurization sludge. The ash is sintered with calcium carbonate and calcium sulfate at 1,000 to 1,200°C and then leached with sulfuric acid. Developed at Oak Ridge National Laboratory, United States in 1976, but not known to have been piloted.

> Felker, K., Seeley, F., Egan, Z., and Kelmers, D., *CHEMTECH,* 1982, **12**(2), 123.
> O'Connor, D. J., *Alumina Extraction from Non-bauxitic Materials,* Aluminium-Verlag, Düsseldorf, 1988, 102,262.

CAN [calcium ammonium nitrate] A process for making calcium ammonium nitrate fertilizer. Developed by BASF (hence the alternative process name: BASF/CAN) and engineered by Uhde.

CANDID A process for making adiponitrile by reductive dimerization of acrylonitrile. Invented by ICI in 1976 and piloted in the United Kingdom from 1986, but not commercialized.

> U.K. Patents 1,546,807; 1,547,431.
> *Eur. Chem. News,* 1990, **54**(1420), 38.

CANMET Hydrocracking A process for demetallizing and converting heavy oils or refinery residues, in the presence of hydrogen and a proprietary additive, into naphtha, middle distillates, and gas oil. Originally developed in the 1970s by the Canada Center for Mineral and Energy Technology (CANMET), a division of the Department of Energy Mines and Resources, Canada, the process is now licensed by Partec Lavalin and Petro-Canada Products. A large demonstration plant, designed and built by Partec Lavalin in Petro-

Canada's Montreal oil refinery, has operated successfully since 1986. A variation for treating extracts from tar sands was developed by Petro-Canada Exploration and the Department of Energy Mines and Resources and piloted in Canada in the 1980s. Another variation, for making diesel fuel from vegetable oils, was piloted in Vancouver in 1992.

> Canadian Patents 1,094,492; 1,202,588; 1,151,579.
> U.S. Patents 4,963,247; 4,969,988.
> Menzies, M. A., Silva, A. E., and Denis, J. M., *Chem. Eng. (N.Y.),* 1981, **88**(4), 46.
> Silva, A. E., Rohrig, H. K., and Dufresne, A. R., *Oil Gas J.,* 1984, **82**(13), 81.
> *Chem. Eng. (N.Y.),* 1992, **99**(5), 21.

Cansolv A *flue-gas desulfurization process based on the selective absorption of sulfur dioxide in certain "amine-based" organic solvents. Developed by Union Carbide Corporation; the first plant was planned for startup at Newburgh, IN, in 1994. *See* HS.

> *Chem. Mark. Rep.,* 1991, 11 Nov., 5; 9 Dec., 29.

CAPTOR A modification of the *Activated Sludge sewage treatment system, in which the micro-organisms are retained in a reticulated polyether foam. Invented in 1978 at UMIST, Manchester, and developed by Simon-Hartley, UK.

> British Patent 2,006,181.
> Cooper, P. F., *Topics in Wastewater Treatment,* Sidgewick, J. M., Ed., Blackwell Scientific, Oxford, 1985, 48.

CAR [Combined autothermal reforming] A *reforming process for making *syngas from light hydrocarbons, in which the heat is provided by partial oxidation in a section of the reactor. Developed by Uhde and commercialized at an oil refinery at Strazske, Slovakia, in 1991.

> *Chem. Eng. (N.Y.),* 1992, **99**(5), 33.
> Babik, A. and Kurt, J., *Oil Gas J.,* 1994, **92**(12), 66.

Carbacell [Carbamate cellulose] A process for making rayon filament and staple fibre. Cellulose is reacted with urea in an inert organic solvent at a high temperature to yield cellulose carbamate. This process avoids the environmental problems caused by carbon disulfide in the viscose process. Developed by Zimmer in the 1990s and piloted in Germany and Poland. Commercialization is expected by 1999.

> *Chem. Week,* 1997, 159(**25**), 21.

carbochlorination A general name for processes that convert metal oxides to chlorides by heating them with carbon in a chlorine atmosphere. *See* Chloride.

Carbo-Flo An integrated process for treating small volumes of effluent containing agrochemicals or other waste organic materials. Flocculation by proprietary chemicals is used, followed by sand filtration, and activated carbon treatment. Developed by ICI in the mid-1980s.

> Harris, D. A., Johnson, K. S., and Ogilvy, J. M. E., *Env. Protect. Bull.,* No. 017, Institution of Chemical Engineers, Rugby, 1992, 23.

carbonation Any process using carbon dioxide as a reactant. Most commonly used to designate the production of calcium carbonate by passing kiln gases through an aqueous suspension of calcium hydroxide.

carbonization A general term for the heat treatment of coal to produce gases, industrial cokes, and domestic smokeless fuels. Many such processes have been developed and many of them have special names. Most of these are outside the scope of this work, but the principle ones are: **Ancit, Anthracine, Anthracoke, AUSCOKE** [Australian **coke**], **BF** [Bergbau Forschung], **BFL** [Bergbau Forschung Lurgi], **Brennstoff-Technik, Carbocite, Carbolux, Carmaux, CCC** [Consolidation Coal Company], **Charfuel, Clean Coke, Coppee** [Coppee-Houillières du Basin du Nord et du Pas de Calais], **CRIJ** [Coal Research Institute of Japan], **Delayed coking, DKS** [Didier Keihan Sumitomo], **Dr. C. Otto, EBV** [Eschweiler Bergwerks-Verein], **FMC** [Food Machinery and Chemical Corp.], **Formcoke, GI, HBN** [Houillières du Bassin du Nord], **HBNPC** [Houillières du Bassin du Nord et du Pas de Calais], **IGI, INICHAR** [Institut National de l'Industrie Charbonnière], **INIEX** (A development of the previous item), **ISCOR** [Iron and Steel Industrial Corporation], **Krupp-Lurgi, NCB** [National Coal Board], **NIPR** [National Institute for Pollution and Resources], **OZIOLE** (same as HBN), **Sapozhnikov, Schenk-Wenzel, Stamicarbon, Stevens, Synthracite, Taciuk, Weber, Wisner.**

Denig, F., in *Chemistry of Coal Utilization,* Vol. 1, Lowry, H. H., Ed., John Wiley & Sons, New York, 1945, 774.

Wilson, P. J., Jr. and Clendenin, J. D., in *Chemistry of Coal Utilization,* Suppl. Vol., Lowry, H. H., Ed., John Wiley & Sons, New York, 1945, 395.

Schinzel, W., in *Chemistry of Coal Utilization,* 2nd. Suppl. Vol., Elliott, M. A., Ed., John Wiley & Sons, New York, 1981, Chap. 11.

Carbonyl *See* Mond nickel.

Carbosolvan One of the several processes for absorbing carbon dioxide from gases, using hot potassium carbonate solution. *See also* Benfield, Carsol, CATACARB, Giammarco-Vetrocoke, Hi-Pure.

Linsmayer, S., *Chem. Tech. (Leipzig),* 1972, **24**(2), 74.

Carbotherm *See* DR.

carburetted water gas *See* water gas.

Carburol An early thermal process for cracking petroleum.

Asinger, F., *Mono-olefins: Chemistry and Technology,* translated by B. J. Hazzard, Pergamon Press, Oxford, 1968, 322.

Carinthian An obsolete lead smelting process, first operated at Bleiberg, Carinthia, Austria.

Carix An ion-exchange process for purifying water, in which regeneration is accomplished with carbon dioxide. Developed in the 1980s by Kernforschungszentrum, Karlsruhe.

Ullmann's Encyclopedia of Industrial Chemistry, 5th ed., Vol. A14, VCH Publishers, Weinheim, West Germany, 1989, 442.

Carl Still (1) A *hydrofining process.

Claxton, G., *Benzoles, Production and Uses,* National Benzole & Allied Products Association, London, 1961, 452.

Carl Still (2) A process for removing hydrogen sulfide from coke oven gas by scrubbing with aqueous ammonia, itself derived from coke oven gas. Developed in the 1970s by Firma Carl Still, Germany. Operated at the ARMCO steel mill at Middleton, OH. *See also* Diamox, Still.

British Patent 1,348,937.

Kohl, A. L. and Riesenfeld, F. C., *Gas Purification,* 4th ed., Gulf Publishing, Houston, TX, 1985, 171.

Carom [**Car**bide **arom**atics extraction] A two-stage process for removing aromatic hydrocarbons from petroleum refining streams. In the first stage, the aromatics are removed by liquid–liquid extraction with a proprietary solvent (a mixture of polyalkylene glycols and a glycol ether) at ambient temperature. In the second stage, the aromatics are stripped from the solvent by steam distillation. Developed by Union Carbide Corporation; first commercialized in 1986, and now licensed by UOP.

U.S. Patents 4,498,980; 5,022,981.

Caron A process for extracting nickel and cobalt from lateritic ores by reductive roasting, followed by leaching with ammoniacal ammonium carbonate solution in the presence of oxygen. Developed by M. H. Caron at The Hague in the 1920s and used in Cuba (where the location of the mine is named Nicaro, after the metal and the inventor) and in Australia.

U.S. Patent 1,487,145.
Gupta, C. K. and Mukherjee, T. K., *Hydrometallurgy in Extraction Processes,* Vol. 1, CRC Press, Boca Raton, FL, 1990, 144.

Carpenter-Evans A catalytic process for removing organic sulfur compounds from synthesis gas by hydrogenation to hydrogen sulfide, which is absorbed by iron oxide. The hydrogenation catalyst is nickel sub-sulfide, Ni_3S_2. Invented by E. V. Evans and C. C. Carpenter in England around 1913 and operated in three commercial plants.

Carpenter, C. C., *J. Gas Light,* 1913, **122,** 1010; 1913, **123,** 30.
Evans, E. V., *J. Soc. Chem. Ind. (London),* 1915, **34,** 9.
Kohl, A. L. and Riesenfeld, F. C., *Gas Purification,* 4th ed., Gulf Publishing, Houston, TX, 1985, 724.

Carrousel An unconventional aerobic treatment system for sewage and industrial effluents, providing efficient oxygenation, mixing, and quiescent flow in an elliptical aeration channel fitted with baffles. Developed in The Netherlands by DHV Raagevend Ingenieursbureau B.V., and licensed in the United Kingdom by Esmil.

Carsol A process for removing carbon dioxide from gas streams by scrubbing with aqueous potassium carbonate. *See also* Benfield, CATACARB, Giammarco-Vetrocoke, Hi-pure.

Carter Also known as **H.T.S. Carter.** A proces for making basic lead carbonate pigment (white lead). Lead monoxide, in a slowly revolving drum, is moistened and sprayed with acetic acid. Carbon dioxide is then introduced. Carbonation is subsequently completed in a separate vessel. *See also* Dutch, Thompson-Stewart.

Dunn, E. J., Jr., in *Treatise on Coatings,* 3(1), Meyers, R. R. and Long, J. S., Eds., Marcel Dekker, New York, 1975, 333.

Carus A process for making potassium permanganate by reacting manganese dioxide with molten potassium hydroxide, in air. Invented by M. B. Carus in 1958 and operated by the Carus Chemical Company at La Salle, IL.

U.S. Patents 2,848,537; 2,940,821; 2,940,822; 2,940,823; 3,172,830.

CAS *See* steelmaking.

Casale The first synthetic ammonia process, designed by L. Casale, who founded Ammonia Casale of Lugano, Switzerland, in 1921. *See* Claude (1).

Vancini, C. A. *Synthesis of Ammonia,* translated by L. Pirt, Macmillan Press, Basingstoke, England, 1971, 245.
Nitrogen, 1996, (223), 25.

Cashman A high-pressure process for extracting gold from arsenic-bearing ores, concentrates, and flue dusts.

Yannopoulos, J. C., *The Extractive Metallurgy of Gold,* Van Nostrand Reinhold, New York, 1991, 103.

Castner (1)   A process for making graphite articles, invented by H. Y. Castner in 1893. It uses lengthwise graphitization, unlike the *Acheson process, which uses transverse graphitization.

U.S. Patent 572,472.

Castner (2)   A process for making sodium cyanide. Sodamide is first made by passing ammonia gas over molten sodium. The molten sodamide is then poured over red-hot charcoal, which converts it first to sodium cyanamide and then to sodium cyanide:

$$2Na + 2NH_3 = 2NaNH_2 + H_2$$

$$2NaNH_2 + C = Na_2CN_2 + 2H_2$$

$$Na_2CN_2 + C = 2NaCN$$

Invented by H. Y. Castner in 1894. Operated first at Frankfurt-am Main, in 1899, and thereafter in several other counties, until abandoned in the 1960s in favor of the *neutralization process.

British Patents 12,218; 12,219 (1894).

Castner (3)   A process for making sodium by reducing sodium hydroxide with iron carbide:

$$6NaOH + FeC_2 = 2Na + 2Na_2CO_3 + 3H_2 + Fe$$

Invented by H. Y. Castner and operated by the Aluminium Company at Oldbury, England in 1888, in order to supply sodium for the manufacture of aluminum. It was abandoned soon afterward when the *Hall-Hérault process for aluminum was developed.

Hardie, D. W. F., *A History of the Chemical Industry in Widnes,* Imperial Chemical Industries, Widnes, England, 1950, 184.

Sittig, M., *Sodium, Its Manufacture, Properties and Uses,* Reinhold Publishing, New York, 1956.

Castner (4)   A process for making sodium by electrolyzing fused sodium hydroxide. Used in the United Kingdom from the early 1900s until 1952.

Hardie, D. W. F. and Pratt, J. D., *A History of the Modern British Chemical Industry,* Pergamon Press, Oxford, 1966, 90.

Castner-Kellner   Also called the Chor-Alkali process. A process for making chlorine and sodium hydroxide by the electrolysis of aqueous sodium chloride in a cell having a mercury cathode. Invented independently in 1892 by H. Y. Castner, an American chemist working in Birmingham, England, and K. Kestner in Austria. First operated in the United States at Saltville, VA, in 1896; and in England by the Castner Kellner Alkali Company in Runcorn in 1897. Of major importance worldwide in the first half of the 20th century. Concerns over mercury pollution caused by effluents from this process caused the abandonment of many plants in the 1970s, to be replaced by various diaphragm-based electrolytic processes. *See* Diaphragm cell.   2nd. Suppl. Vol.,

Fleck, A., *Chem. Ind. (London),* 1947, **66,** 515.

Sittig, M., *Sodium, Its Manufacture, Properties and Uses,* Reinhold Publishing, New York, 1956, 21.

Hardie, D. W. F., *Electrolytic Manufacture of Chemicals from Salt,* Oxford University Press, London, 1959, 19.

MacMullin, R. B., in *Chlorine, Its Manufacture, Properties and Uses,* Sconce, J. S., Ed., Reinhold Publishing, New York, 1962, 127.

Hocking, M. B., *Modern Chemical Technology and Emission Control,* Springer-Verlag, Berlin, 1984, 141.

Cataban   A process for removing small amounts of hydrogen sulfide from industrial gas streams by oxidation, in aqueous solution, to elemental sulfur. The oxidant is the ferric ion, in a proprietary chelated form. The solution is regenerated by atmospheric oxygen. The overall reactions are:

$$2Fe^{3+} + H_2S = 2Fe^{2+} + 2H^+ + S$$

$$2Fe^{2+} + \frac{1}{2}O_2 + 2H^+ = 2Fe^{3+} + H_2O$$

Developed by Rhodia, New York.

Davis, J. C., *Chem. Eng. (N.Y.)*, 1972, **79**(11), 66.
Kohl, A. L. and Riesenfeld, F. C., *Gas Purification*, 4th ed., Gulf Publishing, Houston, TX, 1985, 518.

CATACARB [**Cata**lyzed removal of **carb**on dioxide] A process for removing carbon dioxide and hydrogen sulfide from gas streams by absorption in hot potassium carbonate solution containing a proprietary catalyst. Developed and licensed by Eickmeyer and Associates, KS, based on work at the U.S. Bureau of Mines in the 1950s. More than a hundred plants were operating in 1997. *See* also Benfield, Carsol, Hi-pure, Giammarco-Vetrocoke.

U.S. Patents 3,851,041; 3,932,582.
Eickmeyer, A. G., *Chem. Eng. Progr.*, 1962, **58**(4), 89.
Kohl, A. L. and Riesenfeld, F. C., *Gas Purification*, 4th ed., Gulf Publishing, Houston, TX, 1985, 236.
Gangriwala, H. A. and Chao, I.-M., in *Acid and Sour Gas Treating Processes*, Newman, S. A., Ed., Gulf Publishing, Houston, TX, 1985, 370.
Hydrocarbon Process., 1992, **71**(4), 91.

Catadiene [**Cata**lytic buta**diene**] Also spelled Catadien. A version of the *Houdry process for converting mixtures of butane isomers into butadiene by dehydrogenation over an alumina/chromia catalyst. Another version converts propane to propylene. Rapid coking of the catalyst necessitates use of several reactors in parallel, so that reactivation can be carried out continuously. Developed by Houdry and first operated at El Segundo, CA, in 1944. By 1993, 20 plants had been built worldwide. Now licensed by ABB Lummus Crest.

Unzelman, G. H. and Wolf, C. J., in *Petroleum Processing Handbook*, Bland, W. F. and Davidson, R. L., Eds., McGraw-Hill, New York, 1967, 149.
Craig, R. G. and Dufallo, J. M., *Chem. Eng. Prog.*, 1979, **75**(2), 62.
Craig, R. G. and Spence, D. C., in *Handbook of Petroleum Refining Processes*, Meyers, R. A., Ed., McGraw-Hill, New York, 1986, 4-3.
Hydrocarbon Process., 1991, **70**(3), 142.
Weissermel, K. and Arpe, H.-J. *Industrial Organic Chemistry*, 3rd ed., VCH Publishers, Weinheim, Germany, 1997, 110.

Catalan An early iron-making process in which selected ores were reduced with charcoal and the slag was expelled from the product by hammering while hot. *See also* Bloomery.

Catalloy A gas-phase process for making olefin co-polymers, using *Ziegler-Natta catalysts. It uses a series of three gas-phase reactors to which monomer is progressively added. The properties of the product can be varied according to the monomer grades used. Developed by Himont and first commercialized in 1990. Now operated by a joint venture of Montell Polyolefins and Japan Polyolefins. *See also* Hivalloy.

Chem. Week, 1992, 13 May, 53.
Plastiques Modernes et Elastomères, 1997, 4(9), 31.

Catalyst A process for treating industrial off-gases by catalytic oxidation over a titania/vanadia catalyst. Offered by BASF.

Oil Gas J., 1990, 1 Oct. (Suppl.), 51.

Catalytic Condensation   Also known colloquially as CATCON. A process for oligomerizing olefins, or alkylating aromatic hydrocarbons with olefins. The catalyst is a solid containing free or combined phosphoric acid. Developed by UOP.

> *Chem. Eng. (Rugby, England)*, 1991, (489), 12.

catalytic cracking   A process used in petroleum refining to convert high-boiling hydrocarbon fractions to lower-boiling fractions suitable for use as gasoline. The first catalysts were natural clays; these were later replaced by synthetic sodium alumino-silicates, and then by zeolites, notably zeolite Y. Those processes with special names which are described in this dictionary are: Airlift, ARS, Catarole, Demet, Dynacracking, FCC, Houdresid, Houdriflow, Houdriforming, Houdry, MDDW, MSCC, Orthoflow, R2R, SBA-HT, Suspensoid, TCC, Thermofor, THERMOCAT, Ultra-Orthoflow, Veba-Combi-Cracking.

> Unzelman, G. H. and Wolf, C. J., in *Petroleum Processing Handbook,* Bland, W. F. and Davidson, R. L., Eds., McGraw-Hill, New York, 1967, Chap. 3.1.
> *The Petroleum Handbook,* 6th ed., Elsevier, Amsterdam, 1983, 284.

Catalytic Dewaxing   Also called CDW. A *hydrocracking process for removing waxes (linear aliphatic hydrocarbons) from petroleum streams by converting them to lower molecular weight hydrocarbons. The catalyst is a synthetic mordenite. Developed by BP; two units were operating in 1988.

> Hardgrove, J. D., *Encyclopedia of Chemical Processing and Design,* McKetta, J. J. and Cunningham, W. A., Eds., Marcel Dekker, New York, 1982, **15**, 346.
> *Hydrocarbon Process.,* 1988, **67**(9), 69.

catalytic distillation   A generic term for processes in which the packing of a distillation column is also a catalyst for the reaction. Developed by CDTECH. In 1996, 54 units were in operation, making several ethers for use as fuel additives. *See also* CD-Cumene, CDE-THEROL, CD-HDS, CDHydro, RWD.

> U.S. Patent 4,443,559.
> *Hydrocarbon Process.,* 1996, **75**(11), 113.

catalytic hydrogenation   In the context of the coal and petroleum industries this term can mean either the conversion of aromatic compounds to alicyclic compounds, or of olefins to saturated aliphatic hydrocarbons. Such processes with special names which are described in this dictionary are BASF/Scholven, COIL, Hydra, Hydrobon, Hydropol, Lignol, Normann, SHP, Unifining, Unionfining.

catalytic reforming   A process for converting linear aliphatic hydrocarbons into a mixture of branched-chain aliphatic hydrocarbons, and aromatic hydrocarbons, in order to increase the octane rating of the product. First operated in 1940 at the refinery of the Pan American Oil Refining Corporation in Texas City, using a molybdena/alumina catalyst. Further developed by V. Haensel at Universal Oil Products in the 1940s and still widely used. The catalyst is usually a platinum metal supported on alumina. Hydrogen is a by-product. Those processes with special names which are described in this dictionary are: Aromizing, Catforming, Cycloversion, Hydroforming, Iso-Plus Houdriforming, Magnaforming, Orthoforming, Platforming, Powerforming, Rheniforming, SBK, Silamit P3, Sinclair-Baker, SMDS, Sovaforming, SSC, STAR, UGI, Ultraforming.

> *Chem. Br.,* 1981, Nov., 536.
> Little, D. M., *Catalytic Reforming.* PennWell Publishing, Tulsa, OK, 1985.
> Menon, P. G. and Paal, Z., *Ind. Eng. Chem. Res,* 1997, **36**(8), 3282.

Catarole   Also spelled Catarol. A process for making aromatic hydrocarbons and olefins by cracking petroleum fractions over copper turnings. Invented by C. Weizmann in England in

1940 and developed by Petrochemicals, which used it from 1947 in its refinery at Carrington, UK, to make ethylene, propylene, and a range of aromatic hydrocarbons.

British Patents 552,115; 552,216; 575,383.
Steiner, H., *J. Inst. Pet.,* 1947, **33**, 410.
Weizmann, C., Bergmann, E., Huggett, W. E., Steiner, H., Sulzbacher, M., Parker, D., Michaelis, K. O., Whincup, S., and Zimkin, E., *Ind. Eng. Chem.,* 1951, **43**, 2312.
Claxton, G., *Benzoles, Production and Uses,* National Benzole & Allied Products, London, 1961, 90.
King, R., *Chem. Ind. (London),* 1989, 309.

Catasulf A catalytic process for converting hydrogen sulfide in gas streams to elemental sulfur. The gas, to which a stoichiometric quantity of air or oxygen has been added, is passed over the hot catalyst. Invented in 1983 by BASF. One plant had been built as of 1990.

U.S. Patent 4,507,274.
Hydrocarbon Process., 1992, **71**(4), 91.

CATAZONE [**cata**lyzed o**zone**] A process for removing traces of organic compounds from groundwater by catalyzed oxidation with ozone. The catalyst is titanium dioxide, and hydrogen peroxide may be added as well.

Masten, S. J. and Davies, S. H. R., in *Environmental Oxidants,* Nriagu, J. O. and Simmons, M. S., Eds., John Wiley & Sons, New York, 1994, 534.

CATCON [**Cat**alytic **Con**densation] *See* Catalytic Condensation.

Catforming [**Cat**alytic re**forming**] A *catalytic reforming process using a platinum catalyst on a silica/alumina support. Developed by the Atlantic Refining Company and first operated in 1952.

Unzelman, G. H. and Wolf, C. J., in *Petroleum Processing Handbook,* Bland, W. F. and Davidson, R. L., Eds., McGraw-Hill, New York, 1967, 3-27.
Little, D. M., *Catalytic Reforming,* PennWell Publishing, Tulsa, OK, 1985, xv.

Cativa Not a process but a catalyst for making acetic acid by the carbonylation of methanol. It contains iridium acetate with promoters. Developed by BP Chemicals at Hull, UK and announced in 1996. Used first in Texas City, TX, and planned for use in Malaysia and in Hull.

Chem. Eng. (N.Y.), 1996, **103**(8), 23.
Process Eng., 1996, **77**(7), 21.

CATnap A process for passivating *hydrotreating and *hydrocracking catalysts in petroleum refining. A proprietary mixture of high molecular weight aromatic compounds is added to the catalyst reactors; this forms an inert film on the surfaces of the catalyst and the hardware, preventing oxidation when the catalyst is discharged. Developed in the 1980s by Kashima Engineering Company and Softard Industries.

Appl. Catal., 1993, **102**(2), N19.

CATOFIN [**CAT**alytic Ole**FIN**] A version of the *Houdry process for converting mixtures of $C_3 - C_5$ saturated hydrocarbons into olefins by catalytic dehydrogenation. The catalyst is chromia on alumina in a fixed bed. Developed by Air Products & Chemicals; owned by United Catalysts, which makes the catalyst, and licensed through ABB Lummus Crest. Nineteen plants were operating worldwide in 1991. In 1994, seven units were used for converting isobutane to isobutylene for making methyl *t*-butyl ether for use as a gasoline additive.

Craig, R. G. and White, E. A., *Hydrocarbon Process.*, 1980, **59**(12), 111.

Wett, T., *Oil Gas J.*, 1985, **83**(35), 46.

Craig, R. G. and Spence, D. C., in *Handbook of Petroleum Refining Processes*, Meyers, R. A., Ed., McGraw-Hill, New York, 1986, 4-3.

Hydrocarbon Process., 1991, **70**(3), 185.

Hu, Y. C., in *Chemical Processing Handbook*, Marcel Dekker, New York, 1993, 806.

Chem. Week, 1994, **154**(10), 38.

CATOX [**Cat**alytic **ox**idation] A process for removing organic solvents and other odorous compounds from gaseous effluents, using catalytic oxidation over a proprietary catalyst. The original catalyst was described as a non-precious metal, mixed-oxide catalyst, known as CK-302; the next generation of catalyst contained both a base metal and a noble metal. Licensed and supplied by Haldor Topsoe and used in the food processing, lacquering, printing, and chemical industries. Related processes are *REGENOX, which uses regenerative heat exchange, and *ADOX.

The name is also used by Nippon Shokubai Company as a trade name for distributed digital control systems for chemical plant.

CAT-OX [**Cat**alytic **ox**idation] An adaptation of the *Contact process for making sulfuric acid, using the dilute sulfur dioxide in flue-gases. A conventional vanadium pentoxide catalyst is used. Developed by Monsanto Enviro-Chemical Systems, and operated in Pennsylvania and Illinois in the early 1970s.

Stites, J. G., Jr., Horlacher, W. R., Jr., Bachover, J. L., Jr., and Bartman, J. S., *Chem. Eng. Prog.*, 1969, **65**(10), 74.

Miller, W. E., *Chem. Eng. Prog.*, 1974, **70**(6), 49.

Kohl, A. L. and Riesenfeld, F. C., *Gas Purification*, 4th ed., Gulf Publishing, Houston, TX, 1985, 410.

IEA Coal Research, *The Problems of Sulphur*, Butterworths, London, 1989, 124.

Speight, J. G., *Gas Processing*, Butterworth Heinemann, Oxford, 1993, 278.

Catoxid [**Cat**alytic **oxid**ation] A process for destroying organic chlorine compounds, especially from the production of vinyl chloride, by catalytic oxidation in a fluid bed. Developed by the B. F. Goodrich Company in Akron, OH.

German Patents 2,531,981; 2,532,027; 2,532,043; 2,532,052; 2,532,075.

Benson, J. S., *Hydrocarbon Process.*, 1979, **58**(10), 107.

Catpoly A polymerization process for making linear olefins for use in making ethers for use as gasoline additives. Developed by UOP, later supplanted by IFP's *Polynaphta Essence process.

Cattstil A catalytic process for making high-octane gasoline from off-gases from a catalytic cracker. It can also be used for making a mixture of benzene, toluene, and xylenes from the off-gases from a *steam reformer. A demonstration plant was operated at a refinery in the mid-western United States in 1988 to 1989. Developed and offered for license by Chemical Research and Licensing Company, Houston, TX.

Chem. Eng. (N.Y.), 1989, **96**(9), 41.

causticization Also called the Lime-soda process. A general name for the generation of sodium hydroxide by reacting sodium carbonate with calcium hydroxide:

$$Na_2CO_3 + Ca(OH)_2 = 2NaOH + CaCO_3$$

The process is operated at 80 to 90°C with a slight excess of the calcium hydroxide. This was the only method used for making sodium hydroxide after the invention of the *Leblanc process, and before the introduction of the *Castner-Kellner process around 1890. The process is still used when the demands for chlorine and sodium hydroxide from the Castner-

Kellner process are unbalanced, and for the regeneration of waste pulping liquors in the manufacture of pulp and paper by the *Kraft process. *See also* Löwig.

> Hocking, M. B., *Modern Chemical Technology and Emission Control,* Springer-Verlag, Berlin, 1985, 129.
> *Chem. Ind. (London),* 1991, 455.

Cazo [Spanish, a copper vessel]   An ancient process for extracting silver from sulfide ores. The ore was boiled in a copper pot with salt and water; addition of mercury gave silver amalgam. The copper served as the reducing agent. Described in 1640 by A. A. Barba who claimed that it had been operated since 1590. Around 1800 it developed into the Fondon process, in which the raw materials were ground together.

> Mellor, J. W., *Comprehensive Treatise on Inorganic and Theoretical Chemistry,* Longmans, Green & Co., London, 1923, **3,** 303.
> Dennis, W. H., *A Hundred Years of Metallurgy,* Gerald Duckworth, London, 1963, 285.

CBA [Cold bed adsorption]   A variation of the *Claus process in which the sulfur product is desorbed from the catalyst by a side stream of hot gas from the main process. Developed by AMOCO Canada Petroleum Company and operated in Alberta.

> Goddin, G. S., Hunt, E. B., and Palm, J. W., *Hydrocarbon Process.,* 1974, **53**(10), 122.
> Nobles, J. E., Palm, J. W., and Knudtson, D. K., *Hydrocarbon Process.,* 1977, **56**(7), 143.
> Kohl, A. L. and Riesenfeld, F. C., *Gas Purification,* 4th ed., Gulf Publishing, Houston, TX, 1985, 449.

C$_4$ Butesom [**But**ene **isom**erization]   A process for isomerizing linear butenes to isobutene, catalyzed by a zeolite. The isobutene is intended for use as an intermediate in the production of ethers for use as fuel additives. Developed by UOP in 1992. *See also* C$_5$ Pentesom.

CCC   *See* carbonization.

CCG [**C**atalytic **C**oal **G**asification]   A generic name. All such processes require very cheap catalysts. Exxon Engineering Corporation developed such a process in the 1980s, which used a catalyst based on potassium carbonate. Tohoku University, Japan developed another process using iron salts deposited on coal.

> Gallagher, J. E., Jr., and Euker, C. A., Jr., *Energy Res.,* 1980, **4,** 137.
> Hirsch, R. L., Gallagher, J. E., Jr., Lessard, R. R., and Wesselhoft, R. D., *Science,* 1982, **251,** 121.
> Ohtsuka, Y. and Asami, K., *Catalysis Today,* 1997, **39,** 111.

CCL [**C**atalytic **c**oal **l**iquids]   A catalytic process developed by Gulf Oil Corporation. The main objective is the production of clean-burning liquid fuels for power plants.

> Crynes, B. L., in *Chemistry of Coal Utilization,* Elliott, M. A., Ed., John Wiley & Sons, New York, 1981, 2005.

CCOP [**C**hlorine-**c**atalyzed **o**xidative **p**yrolysis]   A process for converting methane into a mixture of ethylene and acetylene. Invented by the Illinois Institute of Technology, Chicago, and under development by Dow Chemical Company in 1991.

> *Chem. Eng. (N.Y.),* 1990, **97**(2), 17.

CCR Platforming [**C**ontinuous **C**atalyst **R**egeneration]   A development of the *Platforming process in which the catalyst is moved continuously through the stacked reactors into a catalyst regeneration section. Developed by UOP in 1970.

CD-Cumene   A process for making cumene for subsequent conversion to phenol and acetone. The cumene is made by catalytic alkylation of benzene with propylene in a *catalytic distillation reactor. Developed in 1995 by CDTech.

   Chem. Eng. (N.Y.), 1995, **102**(8), 22.

CDP   A process for destroying dioxins and polychlorinated biphenyls by treatment with a polyethylene glycol and sodium peroxide in a fixed catalyst bed. Developed by Sea Marconi Technologies, Turin, Italy. *See also* KPEG.

   Italian Patents 1,999,283; 2,221,583; 2,444,383.

CDETHEROL   A process for making ethers (MTBE, TAME, ETBE) from alcohols. Combines the *Etherol process with *catalytic distillation.

   Hydrocarbon Process., 1994, **73**(11), 104.

CDHydro   [Catalytic distillation **hydro**genation]   A process for hydrogenating diolefins in butylene feedstocks. It combines hydrogenation with fractional distillation. Developed by CDTECH, a partnership between Chemical Research & Licensing Company and ABB Lummus Crest. The first plant was built at Shell's Norco, LA, site in 1994. Ten units were operating in 1997.

   Hydrocarbon Process., 1996, **75**(11), 129.

CDW   *See* Catalytic Dewaxing.

CEC   [Chisso Engineering Company]   A process for removing oxides of nitrogen from flue-gases by scrubbing with an aqueous solution containing ferrous ion and ethylenediamine tetra-acetic acid (EDTA). An iron nitrosyl compound is formed. Developed by Chisso Engineering Company, Japan, and piloted in France and Japan.

   Kohl, A. L. and Riesenfeld, F. C., *Gas Purification*, 4th ed., Gulf Publishing, Houston, TX, 1985, 371.

Celanese LPO   [Liquid **p**hase **o**xidation]   A process for making acetic acid by oxidizing n-butane in the liquid phase, catalyzed by cobalt acetate. Developed by Hoechst Celanese and operated in the United States and The Netherlands. *See also* DF.

   Weissermel, K. and Arpe, H.-J. *Industrial Organic Chemistry* 3rd ed., VCH Publishers, Weinheim, Germany, 1997, 172.

Celdecor   A process for making paper from straw or bagasse. The fiber is digested in aqueous sodium hydroxide and bleached with chlorine. The essential feature is that the alkali and chlorine are used in the proportions in which they are made by the *Chlor-Alkali process.

   Grant, J., *Cellulose Pulp and Allied Products*, Leonard Hill, London, 1958, 356.

Celobric   An anaerobic wastewater treatment process, suitable for treating high concentrations of organic substances. Developed by Hoechst Celanese Chemicals Group and the Badger Company, and used in seven installations in 1993.

   Hydrocarbon Process., 1993, **72**(8), 92.

cementation   An alchemical term for any reaction that takes place in the solid state. Two examples follow. The name is used also in hydrometallurgy for the electrochemical process by which one metal replaces another in aqueous solution, for example, metallic iron causing precipitation of metallic platinum from a platinum chloride solution (also called "footing").

Cementation (1)   The earliest known process for making steel from iron. It originated in the Iron Age but the first written description was published in Prague in 1574 and the earliest known ironworks using the process was that in Nuremburg, Germany, in 1601. The process was operated in Europe between its induction and around 1950 when it was finally replaced by the *Bessemer process. In the United States it was operated from the mid-18th century, initially in Connecticut. Wrought iron bars were stacked in alternate layers with lump charcoal in a conical brick kiln and fired for 5 to 12 days until the correct carbon content had been achieved. The product was known as "blister steel," because of the appearance of its surface.

> Barraclough, K. C., *Steelmaking Before Bessemer, Vol. 1, Blister Steel,* The Metals Society, London, 1984, 48.

Cementation (2)   An obsolete process for making brass by alloying copper with zinc, introduced as the vapor.

Centaur   A process for reducing sulfur dioxide emissions from sulfuric acid plants. An activated carbon with both absorptive and catalytic properties is used. The technology uses fixed beds of Centaur carbon to oxidize sulfur dioxide to sulfuric acid in the pores of the carbon. The sulfuric acid is recovered as dilute sulfuric acid, which is used a make-up water in the sulfuric acid production process. Developed by Calgon Carbon Corporation in the 1990s. Calgon Carbon and Monsanto Enviro-Chem operated a Centaur pilot plant at an existing sulfuric acid facility in 1996.

> *Eur. Chem. News,* 1996, **66**(1723), 19.
> *Sulphur,* 1996, Jul-Aug (245), 13.

Central-Prayon   A wet process for making phosphoric acid, similar to the *Prayon process but using extra sulfuric acid in the crystallization of the gypsum in order to minimize losses of phosphoric acid. Developed jointly by the Central Glass Company (Japan) and Société de Prayon (Belgium).

> Gard, D. R., in *Encyclopedia of Chemical Processing and Design,* McKetta, J. J. and Cunningham, W. A., Eds., Marcel Dekker, New York, 1990, **35**, 456.

Centripure   *See* De Laval Centripure.

CEP   [Catalytic extraction processing]   A process for destroying hazardous wastes by reaction with a molten metal at high temperature. Invented in 1989 by C. Nagel at the Massachusetts Institute of Technology and developed in the early 1990s by Molten Metals Technology, Waltham, MA. The company filed for bankruptcy in 1997.

> Pierce, A. and Chanenchuck, C., *Poll. Prevent.,* 1992, **2**(4), 69.
> *Chem. Eng. News,* 1993, **71**(39), 9.
> *Chem. Eng. (Rugby, England),* 1996, (605), 15.
> *Chem. Week,* 1997, **159**(47), 5.

CER   [Chlorination with Energy Recovery]   A process for making 1,2-dichloroethane by reacting ethylene with chlorine in the presence of a catalyst based on the tetrachloroferrate complex. Developed by Hoechst, Germany, in 1989. *See also* HTC.

CERNOX   [Ceramic NO_x]   A process for destroying NO_x by reaction with ammonia, catalyzed by a zeolite. Developed by Steuler Industrie Werke in the 1950s for treating vapors from nitric acid baths used for pickling stainless steel. The zeolite was developed for this process by Mobil Corporation and is still proprietary. The process was introduced to the market in 1982, and by 1988 100 units had been installed in Europe alone.

> Grove, M. and Sturm, W., *Ceram. Eng. Sci. Proc.,* 1989, **10**(3-4), 325.

Cerny   A process for crystallizing calcium nitrate tetrahydrate, used as a fertilizer, by cooling the saturated solution with drops of cold petroleum.

> Bamforth, A. W., *Industrial Crystallization.* Leonard Hill, London, 1965.
> Mullin, J. W., *Crystallization,* 2nd. ed., Butterworths, London, 1972, 311.

Cerphos [Centre d'Études et de Recherche de Phosphates Minéraux]   A process for making gypsum, suitable for use as plaster, from the waste from the *wet process for making phosphoric acid.

Cetus   A four-stage fermentation process for making propylene oxide from glucose. The product is obtained as a dilute aqueous solution. Developed by Cetus Corporation in the 1970s but not commercialized.

> U.S. Patents 4,246,347; 4,284,723.
> Weissermel, K. and Arpe, H.-J. *Industrial Organic Chemistry,* 3rd ed., VCH Publishers, Weinheim, Germany, 1997, 274.

CFB (1)   [Chemische Fabrik Budenheim]   A process for removing cadmium from phosphoric acid by extracting with a solution of a long-chain amine in a hydrocarbon. Developed and operated by the German company of that name.

> German Patents 3,218,599; 3,327,394.
> Becker, P., *Phosphates and Phosphoric Acid,* Marcel Dekker, New York, 1989, 528.

CFB (2)   [Circulating fluid bed]   A metallurgical roasting process, developed originally for the aluminum industry and now used for many other nonferrous ores. Offered by Lurgi.

CFI   A catalytic process for simultaneously dewaxing and hydrogenating gas oil. Developed by Nippon Ketjen, Fina, and Akzo Nobel.

> *Jap. Chem. Week,* 1995, **36**(1823), 8.

CGA   [Coal gold agglomeration]   A process for separating gold particles from aqueous slurries of finely ground ores by contacting them with slurries of coal particles in oil. The gold particles adhere to the coal particles. When the coal particles have become saturated with gold they are separated and the coal is burnt off. The process was developed as an alternative to the environmentally unacceptable mercury and cyanidation processes. Invented by BP Minerals and developed in the early 1990s by Davy McKee (Stockton).

CGCC   [Coal Gasification, Combined Cycle]   A general name for processes that both gasify coal and produce electricity. One such process has been operated by Louisiana Gasification Technology, at Plaquemine, LA, since 1987. *See also* APAC.

> *Chem. Mark. Rep.,* 1990, 19 Nov., 7.

Chamber   An obsolete but previously very important process for making sulfuric acid. Invented by J. Roebuck in Birmingham, England in 1746, although the patent was not filed in Scotland until 1771. Progressively improved during the 19th century, and finally abandoned everywhere in favor of the *Contact process by 1980. Essentially it was the gas-phase oxidation of sulfur dioxide to sulfur trioxide, catalyzed by oxides of nitrogen, conducted in a lead-lined chamber, followed by dissolution of the sulfur trioxide in water.

> Clow, A. and Clow, N. L., *The Chemical Revolution,* The Batchworth Press, London, 1952, 140.
> Imperial Chemical Industries, *Sulphuric Acid: Manufacture and Use,* Kynoch Press, Birmingham, England, 1955, 16.
> Campbell, W. A., in *Recent Developments in the History of Chemistry,* Russell, C. A., Ed., Royal Society of Chemistry, London, 1985, 243.
> Roebuck, P., *Chem. Br.,* 1996, **32**(7), 38.

Champion   An alternative name for the *English process for extracting zinc from its ores. The Champion family was active in several industries in Bristol in the 18th century. In 1738, William Champion was granted a patent for reducing sulfide ores, but the minimal wording of his claim made it ambiguous. His brother John was granted a related patent in 1758, which described in more detail the reduction of blende (zinc sulfide ore) with charcoal. The sulfide must first be converted to the oxide before it can be reduced to the metal. The process was first operated commercially at Warmley, near Bristol, around 1740, using calamine (zinc carbonate ore). It was the only process used for extracting zinc until the end of the 18th century when the *Belgian process was introduced.

British Patents 564 (1738); 726 (1758).

Cocks, E. J. and Walters, B., *A History of the Zinc Smelting Industry in Britain,* George G. Harrap, London, 1968, 7.

Morgan, S. W. K., *Zinc and Its Alloys and Compounds,* Ellis Horwood, Chichester, England, 1985, 17.

Chance   Also called Chance-Claus. A process for recovering sulfur from the calcium sulfide residues from the *Leblanc process. Treatment of a suspension of the residues with carbon dioxide generates hydrogen sulfide, which is converted to sulfur dioxide by the *Claus proces. The sulfur dioxide is converted to sulfuric acid. Developed by A. M. and J. F. Chance 1882 to 1887 and widely used until the Leblanc process was superseded by the *Solvay process.

British Patent 8, 666 (1887).

Hardie, D. W. F., *A History of the Chemical Industry in Widnes,* Imperial Chemical Industries, Widnes, England, 1950, 126.

Chandelon   Also spelled Chandelen. A process for making mercuric cyanate, (mercury fulminate) by dissolving mercury in nitric acid and pouring the solution into aqueous ethanol. Developed by Chandelon in Belgium in 1848.

Urbanski, T., *Chemistry and Technology of Explosives,* translated by M. Jurecki and S. Laverton, Pergamon Press, Oxford, 1964, 150.

Channel Black   Also called Gas Black. One of the processes used to make carbon black; the others are the *Acetylene Black, *Furnace Black, *Thermal Black, and *Thermatomic processes. In the Channel Black process, natural gas was incompletely burnt in small flames, which impinged on cooled channel irons that were continuously moved and scraped. Invented by L. J. McNutt in 1892 and commercialized that year in Gallagher, PA. The last United States plant was closed in 1976.

U.S. Patent 481,240.

Ellis, C., *The Chemistry of Petroleum Derivatives,* Chemical Catalog, New York, 1934, 237.

Kühner, G. and Voll, M., in *Carbon Black Science and Technology,* Donnet, J. B., Bansai, R. C., and Wang, M.-J., Eds., Marcel Dekker, New York, 1993, 57.

Claasen, E. J., in *Inorganic Chemicals Handbook,* Vol. 1., McKetta, J. J., Ed., Marcel Dekker, New York, 1993, 498.

Chardonnet   A process for making "artificial silk" by nitrating cellulose and injecting the nitrate solution into water, thereupon regenerating the cellulose:

$$[C_6H_7O_2(OH)_3]_n \rightleftharpoons [C_6H_7O_2(ONO_2)]_n$$

In the original process the cellulose nitrate itself was used as the fiber (hence its satirical description as "mother-in-law silk"). The regenerating agent is ammonium hydrosulfide. The basic process was first demonstrated by J. W. Swan in London in 1885 but commercialized by Count L. M. H. B. de Chardonnet ("Father of the rayon industry") in France in 1891 and operated there until 1934. The last working factory, that in Brazil, was burnt down in 1949. The other processes for making rayon fibers by regenerating cellulose (*viscose, *cuprammonium) gave superior products. *See also* Rayon.

French Patent 165,349.

J. Soc. Dyers Color., 1914, **30,** 199.

Moncrief, R.W., *Man-Made Fibres,* 6th Ed., Butterworth Scientific, London, 1975, 157.

Chem-Char   A process for destroying organic wastes by pyrolysis on devolatilized coal char in a reducing atmosphere, followed by secondary combustion of the product gases. Developed at the University of Missouri-Columbia.

> Kinner, L. L., McGowin, A., Manahan, S., and Larsen, D. W., *Environ. Sci. Technol.*, 1992, **27**(3), 482.
> *Chem. Br.*, 1993, **29**(6), 460.

ChemCoal   A *coal liquifaction process, using a phenolic solvent, aqueous alkali, and carbon monoxide.

> Porter, C. R. and Rindt, J. R., in *Processing and Utilization of High-Sulfur Coals, IV*, Dugan, P. R., Quigley, D. R., and Attia, Y. A., Eds., Elsevier, Amsterdam, 1991, 651.

Chemetals   A process for reducing manganese oxide to the metal by heating with methane and air. Developed by the Manganese Chemicals Corporation, Baltimore.

> U.S. Patent 3,375,097.

Chemfix   A process for solidifying aqueous wastes, converting them to a solid form suitable for landfill. Silicates and a proprietary setting agent are used. Invented by J. R. Connor in 1970 and offered by Chemfix, Kenner, LA. In 1979, 100 million gallons of waste had been treated in this way in the United States.

> U.S. Patents 3,837,872; 3,841,102.
> Evans, R. J. and Duvel, W. A., Jr., *Pollut. Eng.* 1974, **6**(10), 44.
> Salas, R. K., in *Toxic and Hazardous Waste Disposal*, Pojasek, R. B., Ed., Ann Arbor Science, Ann Arbor, MI, 1979, Chap. 16.

Chemical Vapor Deposition   *See* CVD.

ChemicoB-Basic   A *flue-gas desulfurization process using magnesium oxide slurry.

> Speight, J. G., *Gas Processing*, Butterworth Heinemann, Oxford, 1993, 281.

Chemithon   A process for making detergents by continuous sulfonation of hydrocarbons with sulfur trioxide or oleum. Offered by Lurgi.

Chemo-Trenn   A process for absorbing ammonia, carbon dioxide, hydrogen sulfide, and other undesirable gases from coal gas by absorption in a warm solution of a salt of a weak organic acid such as sodium cresylate. Invented by H. Bähr at I.G. Farbenindustrie, Germany, in 1938.

> German Patents 728,102; 741,222.
> Kohl, A. L. and Riesenfeld, F. C., *Gas Purification*, 4th ed., Gulf Publishing, Houston, TX, 1985, 164.

Chemsweet   A semi-continuous process for removing hydrogen sulfide and other sulfur-containing gases from natural gas. The gas is passed through a suspension of zinc oxide in aqueous zinc acetate. The zinc sulfide produced is normally buried on land. Licensed by C-E Natco, a division of Combustion Engineering. In 1990, 150 units were operating.

> Manning, W. P., *Oil Gas J.*, 1979, **77**(42), 122.
> *Hydrocarbon Process.*, 1992, **71**(4), 94.

Chenot   An early steelmaking process. Iron ore was first reduced to sponge iron; this was mixed with charcoal, manganese, and resin and compressed into small blocks. These blocks were then melted as in the *Crucible process. Invented by A. Chenot in France in 1846 and operated on a small scale in Sheffield in the 1860s, but abandoned after the introduction of the *Open Hearth process.

> British Patent 11,515 (1846).
> Barraclough, K. C., *Steelmaking Before Bessemer, Vol. 2, Crucible Steel*, The Metals Society, London, 1984, 87,299.

Chesney A process for extracting magnesium from seawater. Developed by British Periclase in 1937.

*Chem. Ind. (London),*1992, (5), 154.

Chevron (1) An obsolete xylene isomerization process that used a silica-alumina catalyst.

Chevron (2) A process for separating *p*-xylene from its isomers by continuous crystallization, using liquid carbon dioxide in direct contact with the xylene as the refrigerant. Developed by the Chevron Research Company in 1966.

U.S. Patent 3,467,724.

Chevron WWT [Waste water treatment] An integrated process for treating "sour water" from oil refineries, particularly for removing ammonia, hydrogen sulfide, and carbon dioxide. Only physical processes are used—volatilization and condensation under various conditions. Developed by Chevron Research Company and used in 14 plants worldwide in 1985.

Martinez, D., in *Chemical Waste Handling and Treatment,* Muller, K. R., Ed., Springer-Verlag, Berlin, 1986, 180.
Leonard, J. P., Haritatos, N. J., and Law, D. V., in *Acid and Sour Gas Treating Processes,* Newman, S. A., Ed., Gulf Publishing, Houston, TX, 1985, 734.

Chiyoda *See* Thoroughbred, ABC.

Chlor-Alkali *See* Castner-Kellner.

Chlorex (1) A process for extracting lubricating oil stocks from petroleum fractions using β,β-dichloro diethyl ether. Chlorex is also the trademark for this compound. Developed by Standard Oil Company (Indiana) in the early 1930s and used until the 1960s.

Page, J. M., Buchler, C. C., and Diggs, S. H., *Ind. Eng. Chem.,* 1933, **25**, 418.
Unzelman, G. H. and Wolf, C. J., in *Petroleum Processing Handbook,* Bland, W. F. and Davidson, R. L., Eds., McGraw-Hill, New York, 1967, 3-89.

Chlorex (2) A process for removing hydrogen chloride from hydrocarbons by aqueous alkaline extraction, using a bundle of hollow fibers. Developed by the Merichem Company, Houston, TX and used in three installations in Japan in 1991.

Hydrocarbon Process., 1996, **75**(4), 126.

Chloride One of the two process used today for making titanium dioxide pigment. Mineral rutile, or another mineral rich in titanium, is chlorinated with coke to produce titanium tetrachloride:

$$TiO_2 + 2Cl_2 + C \rightarrow TiCl_4 + CO_2 (+ CO)$$

This titanium tetrachloride is reacted with oxygen at approximately 1,000°C to yield titanium dioxide crystals around 0.2 μm in diameter, the optimum size for scattering visible light:

$$TiCl_4 + O_2 = TiO_2 + Cl_2$$

The oxidation is conducted in a proprietary burner, with extra heat supplied because, although the reaction is exothermic, it is not self-sustaining. The chlorine is recycled.

The process was developed by Du Pont in the 1940s and its first plant started operating in 1958. It has progressively replaced the older *Sulfate Process because it produces less effluent; in 1998, 56 percent of the world capacity for titanium dioxide production used the Chloride Process. *See also* ICON.

Barksdale, J., *Titanium; Its Occurrence, Chemistry, and Technology,* 2nd. ed., Ronald Press, New York, 1966, Chap. 21.
Egerton, T. A., and Tetlow, A., in *Industrial Inorganic Chemicals: Production and Use,* Thompson, R., Ed., Royal Society of Chemistry, Cambridge, 1995, 363.
Chem. Eng. (Rugby, England) 1991, (497), 33.

Chlorine/Hercosett   Also called Hercosett. A process for making wool shrink-resistant by chlorination followed by resin treatment. The resin is made by Hercules, hence the name. Developed by the International Wool Secretariat.

German Patent 2,018,626.
Smith, P. and Mills, J. H., *CHEMTECH*, 1973, **3**, 748.

Chloroff   A process for removing chlorine from organic chlorides such as polychlorinated biphenyls by reaction with hydrogen under pressure, over a proprietary catalyst. Developed by Kinetics Technology International. *See also* Hi-Chloroff.

chlorohydrination   The addion of water and chlorine to an olefin to give a chloroalcohol. Thus ethylene produces ethylene chlorohydrin:

$$CH_2=CH_2 + Cl_2 + H_2O = CH_2OH-CH_2Cl + HCl$$

Chloroalcohols are important intermediates. Propylene chlorohydrin is made similarly and is used for making propylene oxide by hydrolysis with either calcium hydroxide or sodium hydroxide. If calcium hydroxide is used, the byproduct calcium chloride is useless and must be dumped. If sodium hydroxide is used, the byproduct sodium chloride can be recycled to the *Castner-Kellner process.

Chromizing   *See* metal surface treatment.

Chromox   [**Chrom**ium **ox**idation]   A process for destroying organic pollutants in aqueous wastes by oxidation with hydrogen peroxide, catalyzed by Cr^{6+}. Developed by British Nuclear Fuels in 1995, originally for use in nuclear reprocessing.

Chem. Eng. (N.Y.), 1996, **103**(3), 19.

CIL   [**C**arbon **I**n **L**each]   A process for extracting gold from cyanide leach liquors using activated carbon.

Woodhouse, G., in *Trace Metal Removal from Aqueous Solution*, Thompson, R., Ed., Royal Society of Chemistry, London, 1986.

CIP (1)   [**C**arbon **I**n **P**ulp]   A general name for hydrometallurgical extraction processes that use activated carbon in slurries of ground ores. One such process is used for extracting gold from cyanide leach liquors. First operated on a large scale at the Homestake lead mine in South Dakota in 1974. *See also* RIP.

McDougall, G. J. and Fleming, C. A., *Ion Exchange and Sorption Processes in Hydrometallurgy*, Streat M. and Naden D., Eds., John Wiley & Sons, London, 1987, Chap. 2.
Woodhouse, G., in *Trace Metal Removal from Aqueous Solution*, Thompson, R., Ed., Royal Society of Chemistry, London, 1986.

CIP (2)   *See* DR.

Circofer   A *DR process for finely divided ores using reducing gases made from coal in a circulating fluidized bed. Under development by Lurgi in 1995.

Chem. Eng. (N.Y.), 1995, **102**(3), 41.

Circored   A *DR process using hydrogen made by reforming natural gas. Two fluidized beds are used. Developed by Lurgi in 1995 and first installed in Trinidad in 1998.

Chem. Eng. (N.Y.), 1995, **102**(3), 37; 1996, **103**(9), 25.

Citrate   A process for *flue-gas desulfurization by absorption of the sulfur dioxide in aqueous sodium citrate, reacting with hydrogen sulfide to produce elemental sulfur, and recycling the citrate solution. Subsequent modifications involved removing the sulfur dioxide from the citrate solution by steam stripping, or by vacuuming the gas used to make sulfuric acid.

$$Cit^- + HSO_3^- = (HSO_3 \cdot H_2Cit)^{2-}$$

Invented by the U.S. Bureau of Mines in 1968 for use in metal smelters; subsequently piloted at six locations with varying degrees of success. Much of the development work was done by Pfizer, a United States citric acid producer. In 1983 it was operated only at the Saber Refining Company, United States. *See also* Flakt-Boliden.

Chalmers, F. S., *Hydrocarbon Process.,* 1974, **53**(4), 75.

Kohl, A. L. and Riesenfeld, F. C., *Gas Purification,* 4th ed., Gulf Publishing, Houston, TX, 1985, 356.

IEA Coal Research, *The Problems of Sulphur,* Butterworths, London, 1989, 114.

Citrex [**Citr**ic acid **ex**traction] An improved version of the *Citrate process, designed by Peabody Engineereed Systems.

Vasan, S., *Chem. Eng. Prog.,* 1975, **71**(5), 61.

Citrosolv A two-stage process for removing deposits from steam boilers, using citric acid. The first stage uses ammoniated citric acid at pH 3.5 to 4 to remove iron oxide; the second uses a solution containing more ammonia, pH 9.5 to 10, to remove copper oxide, and an oxidant such as sodium nitrite to passivate the surface.

Clandot A process for extracting silver from its ores by extracting with aqueous ferric chloride, precipitating silver iodide by adding zinc iodide, and then reducing the iodide to the metal by heating with zinc. Used in the late 19th century.

Clark The first water-softening process to be operated on a large scale, based on the addition of calculated quantities of calcium hydroxide. Invented in 1841 by T. Clark at Marischal College, Aberdeen. *See also* Porter-Clark.

British Patent 8,875 (1841).

ICI, *Ancestors of an Industry,* Kynoch Press, 1950, 75.

Nordell, E., *Water Treatment for Industrial and Other Uses,* Reinhold Publishing, New York, 1961, 489.

Clanex A solvent extraction process for converting solutions of the nitrates of actinides and lanthanides into their corresponding chlorides. The extractant is a solution of an aliphatic amine in diethylbenzene.

Leuze, R. E. and Lloyd, M. H., *Prog. Nucl. Energy,* 1970, Ser. III, **4**, 596.

Claude (1) Also called Claude-Casale. A high-pressure ammonia synthesis process, developed by G. Claude in the 1920s.

Chem. Eng. Prog., 1952, **48,** 468.

Shearon, W. H., Jr. and Thompson, H. L., in *Modern Chemical Processes,* Vol. 3, Reinhold Publishing, New York, 1954, 16.

Vancini, C. A., *Synthesis of Ammonia,* translated by L. Pirt, Macmillan Press, Basingstoke, England, 1971, 247.

Claude (2) A process for fractionally distilling liquid air, based on the original *Linde process but using two stages. Developed by G. Claude.

Claude, G., *C. R. Acad. Sci.,* 1900, **131,** 447.

Claude-Casale *See* Casale.

Claus A process for removing hydrogen sulfide from gas streams by the catalyzed reaction with sulfur dioxide, producing elementary sulfur. The process has two stages: in the first, one third of the hydrogen sulfide is oxidized with air to produce sulfur dioxide; in the second, this sulfur dioxide stream is blended with the remainder of the hydrogen sulfide stream and passed over an iron oxide catalyst at approximately 300°C. The resulting sulfur vapor is condensed to liquid sulfur.

$$H_2S + \tfrac{3}{2}O_2 = SO_2 + H_2O$$

$$SO_2 + 2H_2S = 3S + 2H_2O$$

The process was invented by C. F. Claus in Germany in 1882 but not introduced into the United States until 1943. It is now a major world source of sulfur for conversion to sulfuric acid. The literature is very extensive.

British Patent 5,958 (1883).

Estep, J. W., McBride, G. T., and West, J. R., *Advances in Petroleum Chemistry and Refining,* Vol. 6., Interscience Publishers, New York, 1962.

Kohl, A. L. and Riesenfeld, F. C., *Gas Purification,* 4th ed., Gulf Publishing, Houston, TX, 1985, 451.

Misra, C., *Industrial Alumina Chemicals,* American Chemical Society, Washington, D.C., 1986, 139.

Downing, J. C. and Goodboy, K. P., in *Alumina Chemicals,* American Ceramic Society, Westerville, OH, 1990, 277.

Nédez, C. and Ray, J.-L., *Catal. Today,* 1996, **27**(1–2), 49.

Clauspol [**Claus** polyethylene gly**col**]   A variation on the *Claus process for removing hydrogen sulfide from gas streams, in which the tail gases are scrubbed with polyethylene glycol to remove residual sulfur dioxide. Clauspol 150 is a modification of this. Developed by the Institut Français du Pétrole.

Bartel, Y., Bistri, Y., Deschamps, A., Renault, P., Simadoux, J. C., and Dutrian, R., *Hydrocarbon Process.,* 1971, **50**(5), 89.

Sulphur, 1974, (111), 48.

Hydrocarbon Process., 1996, **75**(14), 40.

Clayton   A continuous soapmaking process, using centrifugation for separating the soap from the oil. Invented by B. Clayton and operated in the United States from the 1930s.

U.S. Patent 2,219,088.

Shearon, W. H., Jr., Seestrom, H. E., and Hughes, J. P., in *Modern Chemical Processes,* Vol. 2, Reinhold Publishing, New York, 1952, 136.

Cleanair   A process for pretreating gas streams before the Claus and *Stretford processes for removing sulfur compounds. Developed by the J. F. Pritchard Company.

Davis, J. C., *Chem. Eng. (N. Y.),* 1972, **79**(11), 66.

Sulphur, 1974, (111), 52.

Oil Gas J., 1978, **76**(35), 160.

Clean Coke   A process for making coke and chemicals from high-sulfur coals. Developed by the United States Steel Corporation with support from the U.S. Department of Energy. *See also* carbonization.

Schowalter, K. A. and Boodman, N. S., *Chem. Eng. Prog.,* 1974, **70**(6), 76.

Iammartino, N. R., *Chem. Eng.(N. Y.),* 1975, **82**(18), 57.

CLEAR [**C**opper **L**each **E**lectrolysis **A**nd **R**egeneration]   A process for leaching copper from sulfide ores by boiling with aqueous cupric chloride:

$$CuFeS_2 + 3CuCl_2 = 4CuCl + FeCl_2 + 2S$$

Elemental sulfur precipitates and is recovered. Copper powder is produced by electrolyzing the resulting solution. Developed by G. E. Atwood and C. H. Curtis of Duval Corporation at Tucson, AZ. *See also* Cymet.

U. S. Patents 3,785,944; 3,879,272; 4,025,400.

Atwood, G. E. and Livingston, R. W., *Erzmetall,* 1980, **33**(5), 251.

Gupta, C. K. and Mukherjee, T. K., *Hydrometallurgy in Extraction Processes,* CRC Press, Boca Raton, FL, 1990, 12,190.

Climax   A process for making sodium sulfate from sulfuric acid and sodium chloride. Sulfuric acid is sprayed onto a hot fluidized bed of sodium chloride. The products are granular sodium sulfate and hydrogen chloride gas. Invented in 1967 by C. K. Curtis; later developed and commercialized by C. W. Cannon at the Climax Chemical Company at Midland, NM, in the 1970s. Midland was a favorable location because of the proximity of mineral salt and sulfur from petroleum and the availability of cheap transport of the product from the site.

French Patent 1,549,938.

CLINSULF   [Carl von **Lin**de **sulfur**]   A variation of the *Claus process in which the heat from the process is used to heat a second catalytic reactor. The process is designed for gases rich in hydrogen sulfide. First commercialized in 1992 and offered by Linde, Munich.

Heisel, M. P. and Marold, F. J., *Oil Gas J.,* 1989, **87**(33), 37.
Hydrocarbon Process., 1996, **75**(14), 110.

CLINSULF DO   [**D**irect **o**xidation]   A process for recovering sulfur from gas streams containing low concentrations of hydrogen sulfide, under conditions where the conventional *Claus process is not applicable. It can recover sulfur over a wider range than direct oxidation scrubbers. Operated in Austria and South Korea.

Chim. Ind. (Milan), 1996, **78**(6), 731.

CLINSULFᴿSDP   [**S**ub-**d**ew-**p**oint]   A modified *Clinsulf process using an internally cooled reactor which maximizes the conversion rate.

Kunkel. J., *Reports on Science and Technology,* Linde, Munich, 1997, (59), 46.
Chem. Eng. (N. Y.), 1997, **104**(11), 126.
Sulphur, 1996, (243), 45; (245), 13.

Clintox   A process for removing sulfur dioxide from the tail gases from the *Claus process by washing with a proprietary organic scrubbing agent. Offered by Linde, Munich. Four units were operating in 1996.

Heisel, M. P. and Marold, F. J., *Gas Sep. Purif.,* 1987, **1**, 107.
Hydrocarbon Process., 1996, **75**(14), 111.

Clusius   A process for separating isotopes by a combination of thermal diffusion and thermal siphoning. Invented in 1938 by K. Clusius and G. Dickel.

Clusius, K. and Dickel, G. *Naturwiss.,* 1938, **2**, 546.
Clusius, K. and Dickel, G. *Nature (London),* 1939, **144**, 8.

CNA   [**C**oncentrated **n**itric **a**cid]   A general name for processes that make nitric acid more concentrated than the 70 percent made in conventional plants. They include the *CONIA and *SABAR processes. *See also* DSN.

Büchner, W., Schliebs, R., Winter, G., and Büchel, K. H., *Industrial Inorganic Chemistry,* VCH Publishers, Weinheim, Germany, 1989, 62.

CNC   A chemical/biological process for oxidizing cyanide ion in wastewater. The wastewater is mixed with a proprietary soluble reagent and passed through a biological reactor in which cyanide is oxidized to carbon dioxide and nitrogen. Developed and offered by Radian Corporation, Austin, TX.

Chem. Eng. (N.Y.), 1989, **96**(3), 17.

CNG   [**C**onsolidated **N**atural **G**as]   A process for removing acid gases from natural gas and *syngas, using supercritical carbon dioxide. Under development since 1973 by the Consolidated Natural Gas Research Company with assistance from the U.S. Department of Energy and Helipump Corporation. Liquid carbon dioxide is first used to extract the sulfur compounds. Crystallization at the triple point separates these sulfur compounds from the carbon dioxide. The process is claimed to be more effective than other acid gas processes in removing a wide range of trace impurities as well. In 1986 it was still being developed.

U.S. Patent 4,270,937.

Kohl, A. L. and Riesenfeld, F. C., *Gas Purification,* 4th ed., Gulf Publishing, Houston, TX, 1985, 818.

Auyang, L., Liu, Y. C., Petrik, M., and Siwajek, L., in *Acid and Sour Gas Treating Processes,* Newman, S. A., Ed., Gulf Publishing, Houston, TX, 1985, 497.

Massey, L. G. and Brown, W. R., in *Recent Developments in Separation Science,* Li, N. N., and Calo, J. M., Eds., CRC Press, Boca Raton, FL, 1986, 15.

CO₂ Acceptor   A fluidized bed coal gasification process in which the heat is provided by the exothermic reaction of carbon dioxide with calcium oxide. Developed by the Conoco Coal Development Company in the 1970s.

Hydrocarbon Process., 1975, **54**(4), 120.

Hebden, D. and Stroud, H. J. F., in *Chemistry of Coal Utilization,* 2nd. Suppl. Vol., Elliott, M. A., Ed., John Wiley & Sons, New York, 1981, 1642.

coal cleaning   A general name for processes for removing sulfur from coal. The sulfur is present mostly as iron pyrites (FeS). Such processes with special names that are described in this dictionary are: Ames (2), Gravimelt, JPL Chlorinolysis, Ledgemont, Magnex, PETC, SABA, TRW Gravichem, TRW Meyers.

Coalcon   A coal gasification process using a fluidized bed operated with hydrogen. Developed by Union Carbide Corporation and the Chemical Construction Company, based on work on liquid-phase hydrogenation completed by Union Carbide in the 1950s. A 20-ton per day pilot plant was operated in the 1960s, but a planned larger demonstration plant was abandoned because of cost.

Ferretti, E. J., *Chem. Eng. Prog.,* 1976, **72**(8), 62.

Morgan, W. D., *Chem. Eng. Prog.,* 1976, **72**(8), 64.

coal gasification   A general name for processes for converting coal to gaseous products. They are to be distinguished from *carbonization processes in which the production of carbonaceous products is the main aim, although some gas may also be produced. Those with special names that are described in this dictionary are: Atgas, Babcock and Wilcox, Bi-Gas, Bubiag, CAFB, CGCC, Clean coke, Coalcon, CO₂ Acceptor, COED, COGAS, Dynagas, H-COAL, HTW, Hydrane, Hyflex, HYGAS, IG-Hydrogenation, KBW, Kerpely, KilnGas, Koppers Kontalyt, Koppers-Totzek, KR, K-T, LC, Lurgi, Lurgi-Ruhrgas, Mond gas, Morgas, ORC, Otto-Rummel, Panindco, PEATGAS, Pintsch Hillebrand, PRENFLO, Riley-Morgan, Ruhrgas, Saarburg-OTTO, Shell Coal Gasification, Shell-Koppers, Winkler, Stoic, Synthane, Texaco, Thyssen-Galoczy, U-COAL, U-GAS, Viad, WD-IGI, Wellman, Wellman-Galusha, Winkler, Woodhall-Duckham (2).

See also carbonization.

coal liquifaction   A general name for processes for converting coal to liquid products, which are usually further converted to liquid fuels. Those with special names that are described in this dictionary are: Bergius, Bergius-Pier, Burgess, ChemCoal, COSTEAM, CSD, CSF, Dynagas, Exxon Donor Solvent, IG-NUE, LSE, Pott-Broche, Ruhrkohle/VEBA, SRC, Still, Synthoil.

CO-C-Iron   *See* DR.

Codir   A direct reduction ironmaking process. It uses coal as the reductant in a rotary kiln. Developed by Krupp, Germany. First operated at the Dunswart plant in South Africa in 1973; two plants were operating in India in 1994. *See also* DR, SL/RN.

Ullmann's Encyclopedia of Industrial Chemistry, 5th ed., Vol. A14, VCH Publishers, Weinheim, Germany, 1989, 563.

COED [Char-Oil-Energy-Development] A coal gasification process, based on carbonization in successive fluidized beds operated with non-oxygen-containing gases. The fluidizing gases were generated by partial combustion of part of the product char. The process was developed by FMC Corporation, under contract to the Office of Coal Research. It was replaced by *COGAS.

Dainton, A. D., in *Coal and Modern Coal Processing,* Pitt, G. J. and Millward, G. R., Eds., Academic Press, London, 1979, 141.
Probstein, R. F. and Hicks, R. E., *Synthetic Fuels,* McGraw-Hill, New York, 1982, 280.

COGAS [Coal gasification] A multi-stage coal gasification process. In the three initial stages, coal is pyrolyzed in fluidized beds to produce oils and gas. In the subsequent stages the char is gasified with steam. The heat for the pyrolysis is provided by transferring some of the hot gas produced in the gasifiers. The pyrolysis section is similar to that in the *COED process. Developed by the COGAS Development Company, UK, in the 1970s. A large pilot plant was operated by the British Coal Utilization Research Association, Leatherhead, in 1974.

Dainton, A. D., in *Coal and Modern Coal Processing,* Pitt, G. J. and Millward, G. R., Eds., Academic Press, London, 1979, 141.
Hebden, D. and Stroud, H. J. F., in *Chemistry of Coal Utilization,* 2nd. Suppl. Vol., Elliott, M. A., Ed., John Wiley & Sons, New York, 1981, 1701.
Probstein, R. F. and Hicks, R. E., *Synthetic Fuels,* McGraw-Hill, New York, 1982, 282.

co-generation A general name for processes for making fuel gas and electric power from coal. Those with special names that are described in this dictionary are: APAC, GEGas, ICG-GUD, IGCC, STEAG.

COIL [Concurrent oil] A process for concurrently hydrogenating coal and heavy oil feedstocks. Developed by Hydrocarbon Research.

coking *See* carbonization.

Cold Acid A process for polymerizing isobutene, mainly into dimers and trimers, for making high-octane gasoline blending components. It is catalyzed by 60 to 70 percent sulfuric acid at 25 to 35°C. Developed by the Shell Companies. *See also* Hot Acid.

The Petroleum Handbook, 3rd ed., Shell Petroleum, London, 1948, 228.

Cold Hydrogenation A process for selectively hydrogenating petroleum fractions made by *steam-reforming, in order to produce gasoline. Developed by Bayer and now in use in 70 refineries and chemical complexes worldwide.

Krönig, W., *Hydrocarbon Process.,* 1970, **49**(3), 121.
Lauer, H., *Erdoel Kohle Erdgas Petrochem.,* 1983, **36**(6), 249.

Cold lime-soda *See* Porter-Clark.

Coldstream A process for recycling cemented carbides by heating them to 1,700°C, cooling in nitrogen gas, and blowing them at high speed against a carbide plate.

Collin A process for removing hydrogen sulfide from coal gas by absorption in aqueous ammonia. The hydrogen sulfide was regenerated and used for making sulfuric acid. Invented by F. Collin in 1940 and operated in England and Europe in the 1950s.

German Patent 743,088.
T. H. Williams, *Gas World,* 1955, **142**(509), 7.

Claxton, G., *Benzoles, Production and Uses,* National Benzole & Allied Products Assoc., London, 1961, 212.

Kohl, A. L. and Riesenfeld, F. C., *Gas Purification,* 4th ed., Gulf Publishing, Houston, TX, 1985, 167.

Colox   An aerobic biological treatment system for municipal and industrial wastes. The biomass is fixed as a film on granules in a fixed bed. Developed in Texas by Tetra Technologies.

Chem. Mark. Rep., 1991, 2 Dec., 5,27.

Stephenson, T., Mann, A., and Upton, J., *Chem. Ind. (London),* 1993, (14), 533.

Combifining   A petroleum refining process which removes asphaltenes, sulfur, and metals from residues, before further treatment. The catalyst is an activated petroleum coke in a fluidized bed, operated under hydrogen pressure at 380 to 420°C.

Weisser, O. and Landa, S., *Sulphide Catalysts, their Properties and Applications,* Pergamon Press, Oxford, 1973, 352.

COMBISULF   A process for removing sulfur compounds from the gases from combined-cycle power plants with integrated coal gasification. Carbonyl sulfide is removed by catalytic hydrolysis; hydrogen sulfide is removed by selective absorbtion in aqueous MEDA; and the sulfur is recovered from both processes by a modified *Claus unit. Developed by Krupp Koppers and first operated in Spain in 1996. *See* PRENFLO.

Comex   An ion-exchange process for removing heavy metals from wastewater by extraction into water-insoluble acids.

Boyadzhiev, L. and Khadzhiev, D., *Khim. Ind. (Sofia),* 1981, **9**, 404 (*Chem. Abstr.,* **97**, 60331).

Cominco   [**Co**nsolidated **Min**ing & Smelting **Co**mpany]   A process for absorbing sulfur dioxide from smelting operations. The sulfur dioxide is absorbed in an aqueous solution of ammonium sulfite; regeneration is by acidification with sulfuric acid. The ammonium sulfate byproduct is sold. Operated at the Cominco smelter at Trail, Canada, and at other smelters and sulfuric acid plants in the United States. Licensed by the Olin Mathieson Corporation. The name has been applied also to a lead extraction process.

Kohl, A. L. and Riesenfeld, F. C., *Gas Purification,* 4th ed., Gulf Publishing, Houston, TX, 1985, 364.

Comofining   [**Co**balt **Mo**lybdenum **Refining**]   A hydrorefining process for making lubricating oils. Developed by Lurgi and Wintershall and operated in Salzbergen, Germany.

Conrad, C. and Hermann, R., *Erdoel Kohle,* 1964, **17**, 897.

Compagnie AFC   A variation of the *Hall-Héroult process for making aluminum metal, in which the electrolyte is a mixture of aluminum fluoride, sodium fluoride, and barium chloride. The process is operated at a lower temperature than the Hall-Héroult process, and the product has a purity of 99.99 percent instead of 99.5 percent. Developed in 1934 by the Cie. des Produits Chimiques et Electrometallurgiques Alais, Froges et Carmarque, in France, and still operated by that company.

Dennis, W. H., *A Hundred Years of Metallurgy,* Gerald Duckworth, London, 1963, 155.

Comprex   A process for treating sulfide ores by high-temperature pressure leaching.

Nogueira, E. D., Regife, J. M., Redondo, A. L., and Zaplana, M., *Complex Sulfide Ores* (conf. proc.), 1980 (*Chem. Abstr.,* **94**, 69161).

Comprimo   A version of the *Claus process offered by Comprimo Engineers & Contractors, The Netherlands. In 1983, plants using this process were being installed in Italy, Kuwait, France, and Japan. *See also* Superclaus.

Eur. Chem. News, 1983, **41**(1104), 22.

Comurhex A process for making uranium dioxide by reducing ammonium di-uranate by heating it in hydrogen. Operated by the French company of that name.

Concat A process for removing residual sulfur-containing gases from the off-gases from the *Claus process, by oxidation to sulfur trioxide and hot condensation to sulfuric acid. Developed by Lurgi and first operated at Port Sulfur, LA, in 1974.

CONIA A process for making nitric acid simultaneously at two concentrations. *See also* CNA, DSN.

> Hellmer, L., *Chem. Eng. Prog.,* 1972, **68**(4), 67.
> Büchner, W., Schliebs, R., Winter, G., and Büchel, K. H., *Industrial Inorganic Chemistry,* VCH Publishers, Weinheim, Germany, 1989, 64.

CONOSOX A complex *flue-gas desulfurization process using potassium carbonate solution as the wet scrubbing medium. The product potassium bisulfite is converted to potassium thiosulfate and then reduced with carbon monoxide to potassium carbonate for re-use. The sulfur is recovered as hydrogen sulfide, which is converted to elemental sulfur by the *Claus process. Developed by the Conoco Coal Development Company and piloted in 1986.

Conox A process for beneficiating sulfide ores by selective flotation. Developed and offered by Lurgi.

Consortium The Consortium für Elektrochemische Industrie, founded by A. Wacker in Germany in 1903, is the corporate research laboratory of Wacker-Chemie. Many processes have been developed in this laboratory, but the one for which it is best known is the *Wacker process for making acetaldehyde; this has also been called the Consortium process.

Contact [From the German, Kontaktverfahren, meaning catalytic process] The process now universally used for making sulfuric acid. Sulfur dioxide, made either by burning sulfur or by roasting sulfide ores, is oxidized to sulfur trioxide over a heterogeneous catalyst, typically containing vanadia, and then absorbed in sulfuric acid in several stages. The process has a long, complex history, involving various methods for overcoming catalyst poisoning. Invented in 1831 by P. Phillips, Jr., a vinegar manufacturer in Bristol. Initially a platinum catalyst was used, but this was replaced by vanadia in 1895. It operated for many years in competition with the *Chamber process, but eventually displacing it because, on a large scale, it was cheaper and because it yielded oleum, which is necessary for some sulfonation processes. *See also* Mannheim (2), Schröder-Grillo, Tenteleff.

> British Patent 6,096 (1831).
> Miles, F. D., *The Manufacture of Sulfuric Acid by the Contact Process,* Gurney & Jackson, London, 1925.
> Trickett, A. A., in *Industrial Inorganic Chemicals: Production and Use,* Thompson, R., Ed., Royal Society of Chemistry, Cambridge, 1995, 93.

ConvEx A process for converting an HF-catalyzed *alkylation plant to one using sulfuric acid. Developed by Stratco.

> *Chem. Eng. (N.Y.),* 1994, **101**(5), 23.

Cooledge A process for forming tungsten powder into wire, important in the development of the electric light bulb. Developed in 1908.

Copaux A method for extracting beryllium from beryl. The ore is heated with sodium fluorosilicate at 850°C. Leaching with water dissolves the beryllium fluoride, leaving the silica and most of the aluminum fluoride as an insoluble residue. Addition of sodium hydroxide precipitates beryllium as the hydroxide. The process was invented by H. Copaux and has been in use in France since 1915 and in the United Kingdom since World War II.

Copaux, H., *C. R. Acad. Sci.,* 1919, **168,** 610.

Bryant, P. S., in *Extraction and Refining of the Rarer Metals,* The Institute of Mining and Metallurgy, London, 1957, 310.

Everest, D. A., *The Chemistry of Beryllium,* Elsevier, Amsterdam, 1964, 104.

Copaux-Kawecki   An improved version of the *Copaux process for extracting beryllium from beryl, which permits recovery of the fluorine. Addition of ferric sulfate to the dilute sodium fluoride solution remaining after the separation of the beryllium hydroxide precipitates sodium tetrafluoroferrate, which is then used in place of sodium fluorosilicate.

COPE   [Claus Oxygen-based Process Expansion]   A modification of the *Claus process, which improves the recovery of the sulfur. The combustion stage uses oxygen instead of air. Introduced in 1985 and now licensed by Air Products & Chemicals and Goar, Allison & Associates. In 1990, six units were operating in the United States.

Hydrocarbon Process., 1992, **71**(4), 95.

Schendel, R. L., *Oil Gas J.,* 1993, **91**(39), 63.

Hydrocarbon Process., 1996, **75**(4), 112.

Copeland   A process for oxidizing organic wastes in a fluidized bed of inert particles. The wastes may be solid, liquid, or gaseous, and the oxidant is air. Inorganic residues are collected as granular solids and the heat generated is normally utilized. Developed and marketed by Copeland Systems, Oak Brook, IL, United States, and used in a wide variety of industries.

COPISA   [CO pressure induced selective adsorption]   A process for separating carbon monoxide from the effluent gases from steel mills by a two-stage *PSA unit. Developed jointly by Kawasaki Steel Corporation and Osaka Oxygen Industry. In the first stage, carbon dioxide is removed by activated carbon. In the second stage, carbon monoxide is removed by sodium mordenite.

Suzuki, M., in *Adsorption and Ion Exchange: Fundamentals and Applications,* LeVan, M. D., Ed., American Institute of Chemical Engineers, New York, 1988, 123.

Coppee   *See* carbonization.

COPSA   [CO pressure swing adsorption]   A process similar to *COPISA but using activated charcoal impregnated with cuprous chloride as the adsorbent for carbon monoxide. Developed by Mitsubishi Kakoki Kaisha.

Suzuki, M., in *Adsorption and Ion Exchange: Fundamentals and Applications,* LeVan, M. D., Ed., American Institute of Chemical Engineers, New York, 1988, 123.

COREX   A two-stage ironmaking process. Iron ore is reduced in a vertical shaft furnace and then melted in a melter gasifier, which also generates reducing gases for the shaft furnace. Developed by DVAI, Düsseldorf, and first used in 1989 at the Pretoria works of Iscor, South Africa. Ten plants were being planned in 1997.

Chem. Eng. Int. Ed., 1991, **98**(5), 32.

Downie, N. A., *Industrial Gases,* Blackie Academic, London, 1997, 302.

Corinthian   A misspelling of *Carinthian.

Cosden   A process for polymerizing isobutene to a polyisobutene having a molecular weight between 300 and 2,700.

Weissermel, K. and Arpe, H.-J., *Industrial Organic Chemistry,* 3rd ed., VCH Publishers, Weinheim, Germany, 1997, 73.

Coslettizing   *See* metal surface treatment.

COSMOS [Cracking oil by steam and molten salts]   A catalytic process for cracking petroleum or heavy oils. The catalyst is a molten mixture of the carbonates of lithium, sodium, and potassium. Developed by Mitsui and piloted in 1977.

Hu, Y. C., in *Chemical Processing Handbook,* Marcel Dekker, New York, 1993, 783.

Cosorb [CO absorbtion]   A process for recovering carbon monoxide by absorption in a solution of cuprous aluminum chloride in toluene. Three stages are involved—absorption, desorption, and washing. Invented by Esso Research and Engineering Company and then developed by Tenneco Chemicals in the early 1970s. Piloted in 1976, after which many large-scale plants were built worldwide. The solution has also been proposed for transporting acetylene. A variation, referred to as a PSA system (pressure shifting absorption), but not to be confused with *pressure swing adsorption, was developed in Japan jointly by Mitsubishi Kakoki Kaisha and Nichimen Corporation.

U.S. Patents 3,651,159; 3,767,725.
Walker, D. G., *CHEMTECH,* 1975, **5,** A62, 308.
Haase, D. J. and Walker, D. G., *Chem. Eng. Prog.,* 1974, **70**(5), 74.
Haase, D. J., Duke, P. M., and Cates, J. W., *Hydrocarbon Process.,* 1982, **61**(3), 103.
Keller, G. E., Marcinowsky, A. E., Verma, S. K., and Williamson, K. D., in *Separation and Purification Technology,* Li, N. N. and Calo, J. M., Eds., Marcel Dekker, New York, 1992, 64.
Weissermel, K. and Arpe, H.-J., *Industrial Organic Chemistry,* 3rd ed., VCH Publishers, Weinheim, Germany, 1997, 23.

COSTEAM   A process for obtaining both gas and electric power from coal. The coal is first liquified by a process which is catalyzed by modifying iron compounds naturally present in some coals. Developed on a laboratory scale by the Pittsburgh Energy Technology Center in the 1970s.

Dainton, A. D., in *Coal and Modern Coal Processing,* Pitt, G. J. and Millward, G. R., Eds., Academic Press, London, 1979, 172.
Alpert, S. B. and Wolk, R. H., in *Chemistry of Coal Utilization,* 2nd. Suppl. Vol., Elliott, M. A., Ed., John Wiley & Sons, New York, 1981, 1934.

Cowles   An electrothermal process for making aluminum alloys. A mixture of bauxite, charcoal, and the metal forming the alloy (usually copper), was heated in an electric furnace and the molten alloy tapped from the base. The process cannot be used for making aluminum alone because in the absence of the other metal the product would be aluminum carbide. Invented by the Cowles brothers and operated in Cleveland, OH in 1884 and later in Stoke-on-Trent, England. The electrical efficiency was poor and the process was superseded by the Hall-Héroult process.

Dennis, W. H., *A Hundred Years of Metallurgy,* Gerald Duckworth, London, 1963, 145.

CP [Continuous polymerization]   A continuous process for making high-density polyethylene, based on the *Ziegler process but using a much more active catalyst so that de-ashing (catalyst removal) is not required. Developed by Mitsui Petrochemical Industries and upgraded into its *CX process, which was first licensed in 1976.

C$_5$ Pentesom [Pentene isomerization]   A process for isomerizing linear pentenes to isopentenes, catalyzed by a zeolite. The isopentenes are intended for use as intermediates in the production of ethers for use as fuel additives. Developed by UOP in 1992 and offered for license in 1993. *See also* C$_4$ Butesom.

CPG [Clean Power Generation]   A version of the *IGCC process which incorporates the *Texaco coal gasifier. Developed and offered by Humphreys and Glasgow.

cracking   The pyrolysis of petroleum fractions to produce lower molecular weight materials. *See* catalytic cracking, thermal cracking.

Creighton   An electrolytic process for reducing sugars to their corresponding polyols. Glucose is thus reduced to sorbitol, mannose to mannitol, and xylose to xylitol. The electrodes are made of amalgamated lead or zinc; the electrolyte is sodium sulfate. Invented in 1926 by H. J. Creighton.

> U.S. Patents 1,612,361; 1,653,004; 1,712,951; 1,712,952; 2,458,895.
> Creighton, H. J., *Trans. Electrochem. Soc.,* 1939, **75**, 289.
> Pigmann, W. W. and Goepp, R. M., *Chemistry of the Carbohydrates,* Academic Press, New York, 1948, 238.
> Creighton, H. J., *J. Electrochem. Soc.,* 1952, **99**, 127C.

Cresex   [**Cres**ol **ex**traction]   One of the *Sorbex processes. This one extracts *p*- or *m*-cresol from mixed cresols, and cresols as a class from higher alkyl phenols. By 1990, one plant had been licensed.

CRG   [**C**atalytic **R**ich **G**as]   A process for making town gas and rich gas from light petroleum distillate (naphtha). The naphtha is reacted with steam over a nickel-alumina catalyst yielding a gas mixture rich in methane. Developed by British Gas and used in the United Kingdom in the 1960s, but abandoned there after the discovery of North Sea gas. In 1977, 13 plants were operating in the United States.

> *Gas Making and Natural Gas,* British Petroleum, London, 1972, 126.

CRIJ   *See* carbonization.

CRISFER   A glassmaking process developed by Rhône-Poulenc around 1985.

Cros   A process for making diammonium phosphate, offered by Davy Corporation.

Crosfield   A continuous soapmaking process developed by J. Crosfield & Sons, Warrington, England and used in the manufacture of "Persil" from 1962.

> Musson, A. E., *Enterprise in Soap and Chemicals,* Manchester University Press, Manchester, 1965, 364.
> *Kirk-Othmer's Encyclopedia of Chemical Technology,* 3rd ed., Vol. 21, John Wiley & Sons, New York, 1983, 174.

Cross   A high-pressure, mixed-phase, thermal process for cracking petroleum, introduced in the United States in 1924 by the Cross brothers, further developed by the MW Kellogg Company, and widely used in the 1920s and 1930s. Eventually, 130 units were built in the United States and abroad.

> Spitz, P. H., *Petrochemicals, the Rise of an Industry,* John Wiley & Sons, New York, 1988, 168.

Cross-Bevan-Beadle   *See* Viscose.

Crucible   *See* Huntsman.

Cryoplus   A cryogenic process for separating hydrocarbon gases. Developed and licensed by Technip. In 1992, more than ten units were operating.

> *Hydrocarbon Process.,* 1992, **71**(4), 96.

CrystalSulf   A process which uses a nonaqueous solvent/catalyst system to remove sulfuric acid from high-pressure natural gas. This project, part of the GRI Basic Research programme, has been conducted by Radian Corporation.

> *Sulphur,* 1996, (246), 58,61.
> *Oil & Gas J.,* 1997, **95**(29), 54.

CSA [Catalytic solvent abatement] A process for removing chlorinated solvents from waste gases by catalytic oxidation. Two catalysts are used in series and the products are carbon dioxide, water, and hydrogen chloride. Developed in Germany by Hoechst and Degussa and licensed by Tebodin in The Netherlands.

Hydrocarbon Process., 1993, **72**(8), 77.

CSD [Critical solvent deashing] A process for removing insoluble material from coal before liquifaction, using toluene as a solvent under subcritical conditions. Used in the Kerr-McGee and National Coal Board processes.

CSF [Consol Synthetic Fuels] A two-stage *coal liquifaction process. In the first stage, the coal is extracted with process-derived oil and the ash removed. In the second, the extract is catalytically hydrogenated. Piloted by the Consolidation Coal Company, Cresap, WV, from 1963 to 1972. *See also* H-Coal, SRC, Synthoil.

Alpert, S. B. and Wolk, R. H., in *Chemistry of Coal Utilization,* 2nd. Suppl. Vol., Elliott, M. A., Ed., John Wiley & Sons, New York, 1981, 1954.

CSMP [Cupola surface melting process] A process for vitrifying residues from the incineration of municipal wastes. Offered by ML Entsorgungs und Energieanlagen.

CT-121 [Chiyoda Thoroughbred] *See* Thoroughbred.

CT-BISA [Chiyoda Thoroughbred bisphenol-A] A catalytic process for making Bisphenol-A from phenol and acetone. The catalyst is an acidic ion-exchange resin. The product is used for making polycarbonate resins. Developed and offered by Chiyoda Corporation, Japan. The first plant was operated in Tobata, Japan, in 1997.

CTX A *BAF process.

Stephenson, T., Mann, A., and Upton, J., *Chem. Ind. (London),* 1993, (14), 533.

Cumene *See* Hock.

Cumox [Cumene oxidation] A process for making acetone and phenol by oxidizing cumene, based on the *Hock process. This version was further developed and licensed by UOP. Three plants were operating in 1986. UOP now licenses the Allied-UOP Phenol process, which combines the best features of Cumox and a related process developed by the Allied Chemical Corporation.

Pujado, P. R., Salazar, J. R., and Berger, C. V., *Hydrocarbon Process.,* 1976, **55**(3), 91.
Hydrocarbon Process., 1981, **60**(11), 198.
Pujado, P. R., in *Handbook of Chemicals Production Processes,* Meyers, R. A., Ed., McGraw-Hill, New York, 1986, 1.16-1.

Cupellation An ancient metallurgical operation, still in use, for removing lead and other base metals from silver by blowing air over the surface of the molten metal. The lead oxidizes to lead monoxide (litharge), which floats on the molten silver and is separated off. The molten litharge dissolves the other base metal oxides present. A cupel is the shallow refractory dish in which the operation is conducted.

Old Testament, Jer., **vi,** 29.
Dennis, W. H., *A Hundred Years of Metallurgy,* Gerald Duckworth, London, 1963, 193.

Cuprammonium A process for making regenerated cellulose fibers. Cellulose, from cotton or wood, is dissolved in ammoniacal copper sulfate solution (Schweizer's reagent, also called cuprammonium sulfate). Injection of this solution into a bath of dilute sulfuric acid regenerates the cellulose as a fiber. The process was invented in 1891 by M. Fremery and J. Urban at the Glanzstoff-Fabriken, Germany, developed there, and subsequently widely adopted worldwide. *See also* Bemberg, Chardonnet, Viscose.

> Moncrieff, R. W., *Man-made Fibres,* 6th ed., Newnes-Butterworths, London, 1975, 224.

Cuprasol   Also called EIC. A process for removing hydrogen sulfide and ammonia from geothermal steam by scrubbing with an aqueous solution of copper sulfate. The resulting copper sulfide slurry is oxidized with air, and the copper sulfate re-used. The sulfur is re-covered as ammonium sulfate. Developed by the EIC Corporation, MA, and demonstrated by the Pacific Gas & Electric Company at Geysers, CA, in 1979.

> U.S. Patent 4,192,854.
> Kohl, A. L. and Riesenfeld, F. C., *Gas Purification,* 4th ed., Gulf Publishing, Houston, TX, 1985, 547.
> *Chem. Eng. News,* 1979, **57**(49), 29.

Cuprex   [**Copper** extraction]   A process for extracting copper from sulfide ores, combining chloride leaching, electro-winning, and solvent extraction. Piloted by a consortium consisting of ICI, Nerco Minerals Company, and Tecnicas Reunidas.

> Dalton, R. F., Price, R., Ermana, H., and Hoffman, B., in *Separation Processes in Hydrometallurgy,* Davies, G. A., Ellis Horwood, Chichester, England, 1987, 466.

CVD   [**Chemical Vapor Deposition**]   A general term for any process for depositing a solid on a solid surface by a chemical reaction from reactants in the gas phase. To be distinguished from Physical Vapor Deposition (PVD), in which no chemical reaction takes place. (For the ten international conferences held on this between 1967 and 1987 see the reference by Stinton *et al.* below.)

> Stinton, D. P., Besman, T. W., and Lowden, R. A., *Am. Ceram. Soc. Bull.,* 1988, **67,** 350.
> Hocking, M. G., Vasantasree, V., and Sidky, P. S., *Metallic and Ceramic Coatings,* Longman, Harlow, 1989.
> *Journal of Chemical Vapor Deposition,* Technomic Publishing, Lancaster, PA, July, from 1992.

CVI   [**Chemical Vapor Infiltration**]   A ceramic manufacturing process in which the pores of a matrix are filed by *CVD.

> Chiang, Y.-M., Haggerty, J. S., Messner, R. P., and Demetry, C., *Am. Ceram. Soc. Bull.,* 1989, **68**, 420.

CX   A continuous process for making high-density polyethylene, based on the *Ziegler process, but using a much more active catalyst so that de-ashing (catalyst removal) is not required. Developed by Mitsui Petrochemical Industries from its *CP process. First licensed in 1976; by 1990, 20 licenses had been granted worldwide.

CyAM   [**Cyanide ammonia**]   A process for reducing the cyanide concentration in the ammonia liquor from coke manufacture, so that the liquor may be fed to an activated sludge effluent treatment plant. Developed by the United States Steel Corporation and used by that company in two of its coking plants.

> Glassman, D., *Ironmaking Conf. Proc.,* 1976, **35**, 121 (*Chem. Abs.,* **89**, 168444).

Cyan-cat   A process for destroying gaseous hydrogen cyanide by catalytic oxidation over platinum metal at approximately 300°C.

> Martinez, D., in *Chemical Waste Handling and Treatment,* Muller, K. R., Ed., Springer-Verlag, Berlin, 1986, 133.

Cyanamide   *See* Frank-Caro.

cyanidation   *See* cyanide.

cyanide Also called cyanidation. A process for extracting gold from crushed rock by contacting it with aqueous sodium or potassium cyanide in the presence of air. The gold is converted to an aurocyanide:

$$4Au + 8KCN + 2H_2O + O_2 = 4KAu(CN)_2 + 4KOH$$

The solution is then reduced with metallic zinc:

$$2KAu(CN)_2 + Zn = K_2Zn(CN)_4 + 2Au$$

First patented by J. W. Simpson in 1884 and developed by J. S. MacArthur and R. W. and W. Forrest in Glasgow in 1887. The first commercial application was at the Crown Mine, New Zealand in 1889, followed shortly by mines in South Africa and the United States. If zinc dust is used, the process is known as the *Merrill-Crowe process.

Dennis, W. H., *A Hundred Years of Metallurgy,* Gerald Duckworth, London, 1963, 269.

Yannopoulos, J. C., *The Extractive Metallurgy of Gold,* Van Nostrand Reinhold, New York, 1991, 67,143.

Cyclar [**Cycl**ization of light hydrocarbons to **ar**omatics] A catalytic process for converting light hydrocarbons, typically mixtures of C_3 and C_4 acyclic aliphatic hydrocarbons, to aromatic liquid products. Also called dehydrocyclodimerization (DHCD) because the mechanism is believed to be dehydrogenation, followed by dimerization, followed by cyclization. Developed jointly by UOP and British Petroleum. The catalyst, which is zeolite ZSM-5 containing gallium, is continuously regenerated in a separate unit. A large demonstration plant was started up in Grangemouth, Scotland, in 1989. The first industrial plant was built at Sabic's plant at Yanbu, Saudi Arabia, in the late 1990s.

U.S. Patents 4,175,057; 4,180,689.

Franck, H.-G. and Stadelhofer, J. W., *Industrial Aromatic Chemistry,* Springer-Verlag, Berlin, 1988, 88.

Doolan, P. C. and Pujado, P. R., *Hydrocarbon Process.,* 1989, **68**(9), 72.

Chem. Eng. (Rugby, England), 1991, (491), 12.

Gosling, C. D., Wilcher, F. P., Sullivan, L., and Mountford, R. A., *Hydrocarbon Process.,* 1991, **70**(12), 69.

Jenneret, J. J., in *Handbook of Petroleum Refining Processes,* Meyers, R. A., Ed., McGraw-Hill, New York, 1997, 2.27.

Cyclopol A process for making cyclohexanone from benzene, the intermediates being cyclohexane and cyclohexanol. Developed and licensed by Polimex-Cekop. In 1997, 20 percent of world demand for cyclohexanone was made by this process.

Haber, J. and Borowiak, M., *Appl. Catal. A: General,* 1997, **155**(2), 292.

Cycloversion A petroleum treatment process which combined *catalytic reforming with *hydrodesulfurization. The catalyst was bauxite. The process differed from the *Houdry process in that the catalyst bed temperature was controlled by injecting an inert gas. Developed by the Phillips Petroleum Company and used in the United States in the 1940s.

Pet. Refin., 1960, **39**(9), 205.

Unzelman, G. H. and Wolf, C. J., in *Petroleum Processing Handbook,* Bland, W. F. and Davidson, R. L., Eds., McGraw-Hill, New York, 1967, 61.

Little, D. M., *Catalytic Reforming,* PennWell Publishing, Tulsa, OK, 1985, xiv.

Cycom A combustion process for treating solid industrial wastes. The solids are fed into the top of a vertical gas-fired cylinder. The ash melts and flows down the walls for recovery. Developed by IGT, Chicago, and piloted with several industrial wastes in the 1990s.

Chem. Eng. (N.Y.), 1996, **103**(4), 25.

Cymet (1) [**Cy**prus **Met**allurgical] A process for extracting copper from sulfide ores. It involves leaching, solvent extraction, and electrowinning. Developed by the Cyprus Metallurgical Processes Corporation, Golden, CO. Superseded by *Cymet (2).

Cymet (2)   [**Cy**prus **Met**allurgical]   A process for extracting copper from sulfide ores. Copper is leached from the ore using aqueous ferric and cupric chloride solution:

$$CuFeS_2 + 3FeCl_3 = CuCl + 4FeCl_2 + 2S$$

The copper is crystallized out as cuprous chloride, which is then reduced with hydrogen in a fluidized bed reactor. Developed by Cyprus Mines Corporation, successor to the Cyprus Metallurgical Processes Corporation, which developed Cymet (1). *See also* CLEAR.

> U.S. Patent 3,972,711.
> Kruesi, P. R., Allen, E. S., and Lake, J. L., *CIM Bull.*, 1973, **66**, 81. *Eng. Min. J.*, 1977, **178**(11), 27.
> Gupta, C. K. and Mukherjee, T. K., *Hydrometallurgy in Extraction Processes,* Vol. 1, CRC Press, Boca Raton, FL, 1990, 12,188.

Cymex   [**Cym**ene **ex**traction]   One of the *Sorbex processes, for extracting *p*-cymene or *m*-cymene from cymene isomers. By 1990, one plant had been licensed.

CYTOX   An aerobic sewage treatment process using pure oxygen, similar to *UNOX.

CZD   [**C**ombined **z**one **d**ispersion]   Also called Bechtel CZD. A *flue-gas desulfurization process in which a slurry of a chemically reactive form of lime (pressure-hydrated dolomitic lime) is injected into the duct. The water in the slurry evaporates quickly and the solid product is collected downstream in an electrostatic precipitator. Developed by Bechtel International in the late 1980s, intended for retrofitting to an existing plant.

> PCT Patent Appl. 80/1377.

Czrochralski   A process for growing large single crystals. The bulk of the material is melted in a crucible. A single crystal of the same material is lowered onto the surface of the melt and then slowly pulled upward, producing a cylindrical single crystal known as a boule. Invented by J. Czochralski as a method for determining the velocity of crystallization of molten metals.

> Czochralski, *Z. Phys. Chem.*, 1917, **92**, 219.
> Vere, A. W., *Crystal Growth; Principles and Progress,* Plenum Press, New York, 1987, 67.

D

3D   [**D**iscriminatory **d**estructive **d**istillation]   A thermal deasphalting process which uses the same short contact time concept as the *MSCC process and a circulating solid for heat transfer between reactor and generator. It is claimed to be most effective on heavy contaminated whole crude oils or residues. Developed by Bar-Co and now offered by UOP.

> *Eur. Chem. News,* 1995, **64**(1682), 28.
> *Hydrocarbon Process.,* 1996, **75**(11), 96.

Daicel   A process for making propylene oxide by oxidizing propylene with peroxyacetic acid. The peroxyacetic acid is made by reacting together ethyl acetate, acetaldehyde, a metal ion catalyst, and air. Developed by Daicel Chemical Industries, Japan from 1966 and commercialized from 1969 to 1980. *See also* Propylox.

U.S. Patents 3,654,094; 3,663,574.
Yamagishi, K., Kageyama, O., Haruki, H., and Numa, Y., *Hydrocarbon Process.,* 1976, **55**(11), 102.
Weissermel, K. and Arpe, H.-J. *Industrial Organic Chemistry,* 3rd ed., VCH Publishers, Weinheim, Germany, 1997, 270.

Daniell A process for making gas from rosin, the residue from the distillation of turpentine. Used in the United States in the 19th century for making gas for lighting. Invented by J. F. Daniell, better known for his invention of an electrical cell.

Davies, D. I., *Chem. Br.,* 1990, **26**(10), 946.

Dapex [Di-alkylphosphoric acid extraction] A process for the solvent extraction of uranium from sulfuric acid solutions using di-(2-ethylhexyl) phosphoric acid (HDEHP). The HDEHP is dissolved in kerosene containing 4 percent of tributyl phosphate. The uranium is stripped from the organic phase by aqueous sodium carbonate and precipitated as uranyl peroxide (yellow cake). The process was no longer in use in 1988. *See also* Amex.

Chem. Eng. News, 1956, **34**(21), 2590.
Blake, C. A., Baes, C. F., Jr., and Brown, K. B., *Ind. Eng. Chem.,* 1958, **50,** 1763.
Danesi, P. R., in *Developments in Solvent Extraction,* Alegret, S., Ed., Ellis Horwood, Chichester, England, 1988, 206.
Johnston, B. E., *Chem. Ind., (London),* 1988, (20), 658.

DAP-Mn Also called the Manganese Dioxide Process. A *flue-gas desulfurization process using a fluidized bed of manganese dioxide, which becomes converted to manganous sulfate. The adsorbent is regenerated with ammonia.

Speight, J. G., *Gas Processing,* Butterworth Heinemann, Oxford, 1993, 303.

DAV A *DR ironmaking process, using coal as the reductant. Operated since 1985 at the Davsteel plant in Vanderbijlpark, South Africa. *See* DR.

Davey A modification of the *Parkes process for removing silver from lead. A water-cooled tray is floated on the molten lead. Invented by T. R. A. Davies in 1970 and operated by Penarroya in Brazil, France, Greece, and Spain.

U.S. Patent 4,356,033.
Davey, T. R. A. and Bied-Charreton, B., *J. Met.,* 1983, **35**(8), 37.

Davison A process for making phosphate fertilizer.

Day-Kesting *See* Kesting.

DCC [Deep catalytic cracking] A general term for processes which convert heavy petroleum feedstocks and residues to light hydrocarbons. One such process, for making C_3-C_5 olefins, was developed by the Research Institute of Petroleum Processing, China, and licensed through Stone & Webster. Five units were operating in China in 1997.

Hunt, D. A., in *Handbook of Petroleum Refining Processes,* Meyers, R. A., Ed., McGraw-Hill, New York, 1997, 3.101.

DCH (1) [Direct contact hydrogenation] A process which uses catalytic *hydroprocessing to re-refine spent lubricating oils to produce clean fuel, lubricating oil base stocks, or hydrocarbon feedstock for cracking. Developed by UOP but not commercialized as of 1991.

DCH (2) An early name for *HyChlor.

Deacon Also called Deacon-Hurter. A process for oxidizing hydrogen chloride to chlorine, using atmospheric oxygen and a heterogeneous catalyst:

$$4HCl + O_2 = 2H_2O + 2Cl_2$$

The catalyst is cupric chloride, supported on calcined clay. The resulting chlorine is mixed with nitrogen and steam, but that is not disadvantageous when the product is used to make bleaching powder. Invented by H. Deacon in Widnes, England in 1868, in order to utilize the hydrochloric acid byproduct from the *Leblanc process. W. Hurter improved on Deacon's original invention by passing the gases downward through the catalyst column. Used in competition with the *Weldon proces, until both processes were made obsolete by the invention of the electrolytic process for making chlorine from brine. More modern variations, using superior catalysts, have been operated in The Netherlands and India.

British Patent 1403, 1868.

U.S. Patent 85,370, 1868.

Hardie, D. W. F., *A History of the Chemical Industry in Widnes,* Imperial Chemical Industries, Widnes, England, 1950, 67.

Redniss, A., in *Chlorine, Its Manufacture, Properties and Uses* Sconce, J. S., Ed., Reinhold, New York, 1962, 251.

Weissermel, K. and Arpe, H.-J. *Industrial Organic Chemistry,* 3rd ed., VCH Publishers, Weinheim, Germany, 1997, 219.

dealkylation   In organic chemistry, this word can be used to describe any reaction in which one or more alkyl groups is removed from a molecule. In process chemistry, the word has a more restricted meaning—the conversion of toluene or xylene to benzene. Since the process is operated in an atmosphere of hydrogen, it is also called hydrodealkylation. Named processes which achieve this are *Detol, *Hydeal, *Litol.

Dean   A process for extracting manganese from low-grade ores by extraction with aqueous ammonium carbamate. Invented by R. S. Dean and used since 1950 by the Manganese Chemical Company, now part of Diamond Shamrock Chemical Company.

Weiss, S. A., *Manganese, the Other Uses,* Garden City Press, Letchworth, Hertfordshire, 1977, 152,155.

Debatox   A rotary kiln system for recycling consumer battery materials developed by Sulzer Chemtech. The system first shreds the batteries and then incinerates them. Carbon, plastics, and paper are burnt. Dioxins are destroyed in an afterburner, and mercury is condensed in a scrubber. The residual solids, containing zinc, manganese, and iron, can be recycled by standard smelters.

Chem. Eng. (N.Y.), 1996, **103**(4), 19.

Decompozon   [**Decomp**ose **ozon**e]   A process for destroying ozone in gas streams by passage through a fixed bed of a proprietary catalyst containing nickel. Developed by Ultrox International, Santa Ana, CA.

DeDiox   A process for destroying polychlorinated dioxins and furans in flue-gases by catalytic oxidation with hydrogen peroxide. The catalyst is based on silica and the process is operated at 80 to 100°C. Developed by Degussa from 1994. The business was offered for sale in 1998.

Eur. Chem. News, 1998, **69**(1801), 33.

Deep Shaft   A high-intensity biological treatment process for purifying domestic sewage and biodegradable industrial effluents. The waste is circulated very rapidly through a vertical loop reactor, 30 to 150 m in height, usually installed underground. Compressed air, introduced at the middle of both legs, drives the circulation. Because of its rapid movement, the sludge operates at a much higher density than is normal in *Activated Sludge plants. Developed by ICI in the 1970s as a spin-off of its *Pruteen process. By 1994 more than 60 plants had been installed worldwide. The process was acquired by Davy International in 1995. *See also* VerTech.

British Patent 1,473,665.

Dunlop, E. H., in *New Processes of Waste Water Treatment and Recovery,* Ellis Horwood, Chichester, England, 1978, 177.

Horan, N. J., *Biological Wastewater Treatment Systems,* John Wiley & Sons, Chichester, England, 1990, 69.

Gray, N. F., *Activated Sludge: Theory and Practice,* Oxford University Press, Oxford, 1990, 113.

Water Sewage Int., 1991, **12,** 30.

Chem. Eng. (Rugby, England), 1994, (569), 23.

DeFine [**Di-ol**e**fine** saturation]   A process for converting di-olefins to mono-olefins by selective dehydrogenation. Developed by UOP for use with its *Pacol process. First commercialized in 1986 and now incorporated in all new Pacol plants. Six units were operating in 1990.

U.S. Patents 4,523,045; 4,523,048; 4,761,509.

Vora, B., Pujado, P., Imai, T., and Fritsch, T., *Chem. Ind. (London),* 1990, (6), 187.

de Florez cracking   An early gas-phase thermal process for cracking petroleum. Developed by L. de Florez.

U.S. Patent 1,437,045.

Ellis, C., *The Chemistry of Petroleum Derivatives,* Chemical Catalog, New York, 1934, 112.

Deglor [**De**toxification and **gl**assification **of** **r**esidues]   A process for vitrifying wastes, such as fly ash, by heating to 1,400°C in an electric furnace. Some of the heat is provided by radiant heaters, some by passage of electricity through the melt. Developed and piloted by ABB in Switzerland from the 1980s, commercialized in Japan in 1996.

Environ. Bus. Mag., 1995, (15), 6.

DegOX [**Deg**ussa **ox**idation]   A pulp-bleaching process developed by Degussa. The active species is peroxomonosulfuric acid (Caro's acid). The first full-scale commercial trial was held in 1994.

Eur. Chem. News, 1994, **61**(1619), 38.

Chem. Eng., 1997, **104**(1), 21.

Chem. Eng., 1997, **104**(4), 10.

DeGuide   A sugar extraction and purification process in which barium is recycled. Addition of barium hydroxide to molasses precipitates barium saccharate. A slurry of this is treated with carbon dioxide, forming barium carbonate and releasing the sugar. The barium carbonate is reconverted to the hydroxide by a two-stage process involving monobarium and tribarium silicates:

$$2BaCO_3 + BaSiO_3 = Ba_3SiO_5 + 2CO_2$$

$$Ba_3SiO_5 + 2H_2O = BaSiO_3 + 2Ba(OH)_2$$

Dahlberg, H. W. and Brown, R. J., in *Beet Sugar Technology,* McGinnis, R. A., Ed., The Beet Sugar Development Foundation, Fort Collins, CO, 1971, 573.

Degussa   Also called BMA. The process by which this large German company is best known is its version of the *Andrussov process for making hydrogen cyanide. Methane and ammonia are reacted in the absence of air, at approximately 1,400°C, over a platinum metal catalyst:

$$NH_3 + CH_4 = HCN + 3H_2$$

Heat is provided by passing the gases through externally heated ceramic tubes. The tubes are made of corundum, and lined with catalytic amounts of platinum metals. Developed by Degussa and operated by that company in Mobile, AL; Wesseling, Germany; and Antwerp, Belgium.

German Patents 882,985; 959,364.

Endter, F., *Chem. Ing. Tech.,* 1958, **30,** 305.

Hydrocarbon Process., 1967, **46**(11), 189.

Dowell, A. M., III, Tucker, D. H., Merritt, R. F., and Teich, C. I., in *Encyclopedia of Chemical Processing and Design,* McKetta, J. J. and Cunningham, W. A., Eds., Marcel Dekker, New York, 1988, **27,** 11.

Wittcoff, H. A. and Reuben, B. G., *Industrial Organic Chemicals,* John Wiley & Sons, New York, 1996, 245.

Weissermel, K. and Arpe, H.-J. *Industrial Organic Chemistry,* 3rd ed., VCH Publishers, Weinheim, Germany, 1997, 45.

dehydrocyclodimerization   *See* Cyclar.

dehydrogenation   A generic name for catalytic processes that remove hydrogen from hydrocarbons. Alkanes are thus converted to alkenes; di-alkenes to mono-alkenes. Those with special names which are described in this dictionary are: Catadiene, Catofin, DeFine, Oleflex, Pacol, Phillips (3), Styro Plus.

De la Breteque   An electrochemical method for extracting gallium from sodium aluminate solution, developed by Schweizerische Aluminium (Alusuisse) in 1955.

Palmear, I. J., in *The Chemistry of Aluminium, Gallium, Indium, and Thallium,* Downs, A. J., Ed., Blackie, London, 1993, 88.

Wilder, J., Loreth, M. J., Katrack, F. E., and Agarwal, J. C., in *Inorganic Chemicals Handbook,* Vol. 2., McKetta, J. J., Ed., Marcel Dekker, New York, 1993, 946.

De Laval Centripure   A continuous soapmaking process. Fat and aqueous sodium hydroxide are fed countercurrently to a vertical reactor through which much of the product is continuously recirculated, thus emulsifying the reactants.

U.S. Patent 2,727,915.

Palmqvist, F. T. E. and Sullivan, F. E., *J. Am. Oil Chem. Soc.,* 1959, **36,** 173.

Delayed coking   *See* carbonization.

Delay for Decay   A process for trapping radioactive gases (e.g., xenon, krypton, iodine) from nuclear power plants until their radioactivities have decayed to acceptable levels. Activated carbon is the usual adsorbent, and the gases are first dried with a zeolite.

Sherman, J. D. and Yon, C. M., in *Kirk-Othmer's Encyclopedia of Chemical Technology,* 4th ed., Vol. 9, John Wiley & Sons, New York, 1991–1998, 561.

delignification   A general name for processes which dissolve lignin from wood pulp. Those with special names which are described in this dictionary are Alcell, ASAM.

Demet (1)   A catalytic process for cracking petroleum.

Unzelman, G. H. and Wolf, C. J., in *Petroleum Processing Handbook,* Bland, W. F. and Davidson, R. L., Eds., McGraw-Hill, New York, 1967, 3-14.

DEMET (2)   A process for removing metal contaminants from spent *FCC catalysts by a series of pyrometallurgical and hydrometallurgical procedures. These typicaly include calcination, chlorination, and sulfiding. The demetallized catalyst can be re-used. Developed by ChemCat Corporation, LA, in 1988; first operated commercially in New Jersey in 1989. Now operated by Coastal Catalyst Technology.

U.S. Patent 4,686,197.

Elvin, F. J., *Hydrocarbon Process.,* 1989, **68**(10), 71.

Pavel, S. K. and Elvin, F. J., in *Extraction and Processing for the Treatment and Minimization of Wastes,* Hager, J., *et al.,* Eds., The Minerals, Metals & Materials Society, 1993, 1015.

Hydrocarbon Process., 1994, **73**(11), 102.

Demex [**Dem**etallization by **ex**traction] A process for removing metal compounds from heavy petroleum fractions, after vacuum distillation, by solvent extraction and supercritical solvent recovery. The solvent is typically a mixture of octanes and pentanes. Developed jointly by UOP and the Instituto Mexicano del Petroleo; seven units were operating in 1988.

Hydrocarbon Process., 1988, **67**(9), 66.

Houde, E. J., in *Handbook of Petroleum Refining Processes,* Meyers, R. A., Ed., McGraw-Hill, New York, 1997, 10.53.

Den A batch process for making the fertilizer "superphosphate." The **den** is the vat into which the mixture of phosphate rock and sulfuric acid is dumped after mixing. There is also a continuous-den process. Not to be confused with **DEN** [Deutsch - Englisch - Norwegische Gruppe], a fertilizer cartel operated in the three countries from which the name is derived from 1929 to 1930. *See also* Davison, Oberphos.

Demmerle, R. L. and Sackett, W. J., in *Modern Chemical Processes,* Vol. 1, Reinhold Publishing, New York, 1950, 33.

Dennis-Bull A process for making phenol by first sulfonating benzene, the benzenesulfonic acid being extracted into water. Invented in 1917 by H. Bull and L. M. Dennis. *See also* Tyrer.

U.S. Patents 1,211,923; 1,247,499.

Kenyon, R. L. and Boehmer, N., in *Modern Chemical Processes,* Vol. 2, Reinhold Publishing, New York, 1952, 35.

De Nora An electrolytic process for making chlorine and sodium hydroxide solution from brine. The cell has a mercury cathode and graphite anodes. It was developed in the 1950s by the Italian company Oronzio De Nora, Impianti Elettrochimici, Milan, based on work by I. G. Farbenindustrie in Germany during World War II. In 1958 the Monsanto Chemical Company introduced it into the United States in its plant at Anniston, AL. *See also* Mercury cell.

Kenyon, R. L. and Gallone, P., in *Modern Chemical Processes,* Vol. 3, Reinhold Publishing, New York, 1954, 207.

Berkey, F. M., in *Chlorine, Its Manufacture, Properties and Uses,* Sconce, J. S., Ed., Am. Chem. Soc. Monogr. No. 154, Reinhold Publishing, New York, 1962, 203.

Denox A generic name for processes for removing nitrogen oxides from flue-gases by catalyzed reaction with ammonia.

DeNOx (1) A *Denox process for removing nitrogen oxides from the gaseous effluents from nitric acid plants. The oxides are reduced with ammonia, over a catalyst containing potassium chromate and ferric oxide. Developed by Didier Werke in the 1980s.

Chauvel, A., Delmon, B., and Hölderich, W. F., *Appl. Catal. A: Gen.,* 1994, **115,** 179.

DeNOx (2) A *Denox process for removing nitrogen oxides from the gaseous effluents from nitric acid plants. The oxides are reduced by ammonia, over a catalyst containing iron and chromium. Developed by La Grande Paroise in the 1970s and used in eight plants in the 1980s.

Chauvel, A., Delmon, B., and Hölderich, W. F., *Appl. Catal. A: Gen.,* 1994, **115,** 179.

DeNOx (3)   A *Denox process for removing nitrogen oxides from the gaseous effluents from chemical plants. The catalyst is vanadia on alumina. Developed by Rhône-Poulenc in the 1970s and used in 25 plants by 1994.

French Patent 2,450,784.

Chauvel, A., Delmon, B., and Hölderich, W. F., *Appl. Catal. A: Gen.,* 1994, **115,** 179.

DENOX   (Shell DENOX) A low-temperature, add-on *SCR system that operates at between 120 and 350°C. The honeycomb catalyst contains vanadium and titanium.

van der Grift, C. J. G., Woldhuis, A. F., and Maaskant, O. L., *Catal. Today,* 1996, **27**(1-2), 23.

Deoxo   A family of catalytic process using a precious metal supported on alumina. The basic reaction is that of oxygen with hydrogen to produce water. Thus oxygen is removed from hydrogen, or gases containing hydrogen, and hydrogen is removed from oxygen. Another version is used for destroying ozone in the atmosphere of aircraft cabins. Developed and supplied by Engelhard Industries.

Deoxy   A process for removing small concentrations of oxygen from natural gas. The gas is passed over a hot catalyst, which converts the oxygen to carbon dioxide.

Hydrocarbon Process., 1996, **75**(4), 114.

DEPA-TOPO   [**di** (2-**e**thylhexyl) **p**hosphoric **a**cid and **t**ri**o**ctyl**p**hosphine **o**xide]   A process for recovering uranium from *wet-process phosphoric acid, by solvent extraction with a mixture of the two named reagents. Developed at Oak Ridge National Laboratory and first commercialized in 1978 by Freeport Minerals Corporation and Wyoming Mineral Corporation.

Chem. Eng. Int. Ed., 1979, **86**(26), 87.

DESAL   [**Desal**ination]   A process for de-ionizing brackish waters by the use of two weak-electrolyte ion-exchange resins.

Kunin, R., in *Ion Exchange for Pollution Control,* Vol. 1, Calomon, C. and Gold, H., Eds., CRC Press, Boca Raton, FL, 1979, 111.

DESONOX   [Degussa **SOx NOx**]   A process for removing SO_x and NO_x from flue-gases. The NO_x is reacted with ammonia by the *SCR process, using a zeolite catalyst to give nitrogen, and the SO_2 is catalytically oxidized to SO_3 and converted to sulfuric acid. Developed by Degussa, in association with Lurgi and Lentjes, and demonstrated at the Hafen heat and power station in Munster from 1985 to 1988.

Chem. Eng. (N.Y.), 1993, **100**(12), 97.

Chauvel, A., Delmon, B., and Hölderich, W. F., *Appl. Catal. A: Gen.,* 1994, **115,** 186.

Wieckowska, J., *Catal. Today,* 1995, **24**(4), 452.

Desorex   A process for removing impurities from industrial gases by adsorption on activated carbon. Offered by Lurgi.

Desox   A *flue-gas desulfurization process in which limestone slurry absorbs the sulfur dioxide, forming calcium sulfite. This is then oxidized to saleable gypsum:

$$CaCO_3 + SO_2 = CaSO_3 + CO_2$$

$$2CaSO_3 + O_2 = 2CaSO_4$$

Developed in Japan.

Finan. Times (London), 1984, 15 Aug., 10.

Destrugas   A process for destroying organic wastes by pyrolysis in an indirectly heated vertical retort. The comminuted waste is fed from the top, and raw gas and coke are withdrawn from the bottom.

Martinez, D., in *Chemical Waste Handling and Treatment,* Muller, K. R., Ed., Springer-Verlag, Berlin, 1985, 148.

DEsulf A process for removing hydrogen sulfide from coke-oven gas by scrubbing with aqueous ammonia. Developed by Didier Engineering.

Hydrocarbon Process., 1986, **65**(4), 81.

DESUS A *hydrotreating process developed and offered by VEB Petrolchemisches Kombinat Schwedt. Operated in the Schwedt oil refinery, Germany, since 1988.

De Sy *See* DR.

Detal [Detergent alkylation] A process for making "detergent alkylate," i.e., alkyl aromatic hydrocarbons such as linear alkyl benzenes, as intermediates for the manufacture of detergents, by reacting C_{10}–C_{13} olefins with benzene in a fixed bed of an acid catalyst. Developed by UOP and CEPSA as a replacement for their *Detergent Alkylate process, which uses liquid hydrogen fluoride as the catalyst. Demonstrated in a pilot plant in 1991 and first commercialized in Canada in 1996. Offered by UOP.

Vora, B., Pujado, P., Imai, T., and Fritsch, T., *Chem. Ind. (London),* 1990, 19 March, 187.
Hydrocarbon Process., 1991, **70**(3), 130.

Detergent Alkylate A process for making "detergent alkylate," i.e., alkyl aromatic hydrocarbons such as linear alkyl benzenes, as intermediates for the manufacture of detergents, by reacting C_{10}–C_{13} olefines with benzene using liquid hydrogen fluoride as the catalyst. This technology has been commercialized since 1968 and, as of 1990, 28 units were operating. *See* Detal.

U.S. Patents 4,467,128; 4,523,048.
Vora, B., Pujado, P., Imai, T., and Fritsch, T., *Chem. Ind. (London),* 1990, (6), 187.

DETOL [De-alkylation of toluene] A process for making benzene by de-alkylating toluene and other aromatic hydrocarbons. Developed by the Houdry Process and Chemical Company, and generally similar to its *Litol process for the same purpose. The catalyst is chromia/alumina. Licensed by Air Products & Chemicals. In 1987, 12 plants had been licensed.

Hydrocarbon Process., 1963, **42**(11), 161.
Lorz, W., Craig, R. G., and Cross, W. J., *Erdoel Kohle Erdgas Petrochem.,* 1968, **21,** 610.
Hydrocarbon Process., 1987, **66**(11), 66.

deTOX A process for vitrifying the ash from municipal solid waste incineration. The wastes are melted in a furnace heated with a submerged electric arc, operated with a deep bed and a cold top. Either sand or flyash are added to ensure the formation of a vitreous phase. Volatile heavy metals are trapped beneath the cold top. Developed in the late 1980s by Dunston Ceramics and Cookson Group but not yet commercialized.

Deville (1) The first commercial process for making aluminum metal. Molten sodium aluminum chloride was reduced by heating with metallic sodium. Invented by H. É. St-Claire Deville and operated around 1854 to 1890. Superseded by the Hall-Héroult process. *See also* Cowles.

Deville, H. É. St.-C., *C. R. Acad. Sci.,* 1854, **38,** 279.
Deville, H. É. St.-C., *C. R. Acad. Sci.,* 1854, **39,** 321.
Chem. Ind. (London), 1992, (11), 403.

Deville (2) An early process for making sodium by reducing sodium carbonate with carbon at or above 1,100°C. Developed in 1886 and used until it was superseded by electrolytic processes. *See* Downs and Castner (4).

Fleck, A., *Chem. Ind. (London),* 1947, (66), 515.

Deville and Debray   A process for extracting the platinum metals from their ores. The ore is heated with galena (lead sulfide ore) and litharge (lead oxide) in a reverberatory furnace. The platinum forms a fusible alloy with the metallic lead, which is also formed. Invented by H. É. St-Claire Deville and H. J. Debray.

Partington, H. R., *A History of Chemistry,* Vol. 4, Macmillan, London, 1964, 499.

Deville-Pechiney   An obsolete process for making alumina from bauxite. Bauxite was roasted with sodium carbonate, yielding sodium aluminate:

$$Al_2O_3 + Na_2CO_3 = 2NaAlO_2 + CO_2$$

The product was digested with warm water, which dissolved the sodium aluminate and left the other materials as an insoluble residue. Passage of carbon dioxide through the clarified liquor precipitated aluminum as hydroxide:

$$2NaAlO_2 + CO_2 + 3H_2O = 2Al(OH)_3 + Na_2CO_3$$

All the silica present in the bauxite was converted to insoluble sodium aluminosilicate, which represented a loss of sodium and aluminum. The aluminum hydroxide was calcined to the oxide, and the sodium carbonate solution was concentrated for re-use. The process was developed by H. É. St-Claire Deville in the 1860s; the carbon dioxide stage had been invented earlier by H. L. Le Chatelier. It was superseded by the *Bayer process.

Dennis, W. H., *A Hundred Years of Metallurgy,* Gerald Duckworth, London, 1963, 147.

DeVOx   A catalytic oxidation process for destroying volatile organic compounds in effluent gases. The catalyst contains a non-noble metal and can easily be regenerated. Typical operating temperatures for 95 percent VOC conversion are 175 to 225°C for oxygenates, and 350°C for toluene. Developed in 1995 by Shell, Stork Comprimo, and CRI Catalysts. First installed in 1996 at Shell Nederland Chemie's styrene butadiene rubber facility at Pernis.

Eur. Chem. News, 1995, **64**(1686), 26.
Eur. Chem. News, 1996, **66**(1714), 24.
Environ. Bus. Mag., 1996, (17), 7.

dewaxing   A general term to describe processes for removing waxes from petroleum streams. These waxes are an ill-defined group of saturated hydrocarbons in the molecular-weight range of 225 to 1,000, mostly with unbranched molecules. Those dewaxing processes with special names which are described in this dictionary are Bari-Sol, Catalytic Dewaxing, Endewax, isodewaxing, MLDW, MSDW.

DEZ   [Diethyl zinc]   A process for preserving books and documents by treatment with diethyl zinc vapor, which neutralizes any residual acidity from the papermaking process. The articles are placed in a low-pressure chamber and suffused by diethyl zinc vapor. This vapor reacts with the moisture in the paper to yield zinc oxide, which neutralizes the acid. The process was developed by Texas Alkyls (a joint company of Akzo and Hercules) and the U.S. Library of Congress. A pilot plant was set up in 1988, capable of treating batches of 300 books, and plans to build two larger plants were announced in 1989.

Chem. Br., 1989, **25**(10), 975.
Wedinger, R., *Chem. Br.,* 1992, **28**(10), 898.

DF   [light distillate fraction]   A process for oxidizing light naphtha (a mixture of C_4–C_8 hydrocarbons) to acetic and other carboxylic acids. It operates in the liquid phase at 150 to 200°C. Developed by Distillers Company (now BP Chemicals) in England in the 1960s; it was still operated there by that company in 1992.

U.S. Patent 2,825,740.

Pennington, J., in *An Introduction to Industrial Chemistry,* Heaton, C. A., Ed., Leonard Hill, London, 1984, 376.

Weissermel, K. and Arpe, H.-J. *Industrial Organic Chemistry,* 3rd ed., VCH Publishers, Weinheim, Germany, 1997, 174.

DH *See* Dortmund-Hoerder.

DHC [Druck-Hydrogenium-Cracken, German, meaning pressure hydrocracking] A vapor-phase *hydrocracking process for making transport fuels from heavy petroleum fractions. Developed and operated in Germany in the 1950s and 60s.

Höring, M., Öttinger, W., and Reitz, O., *Erdoel Kohle Erdgas Petrochemie,* 1963, **16**, 361.

Öttinger, W. and Reitz, O., *Erdoel Kohle Erdgas Petrochemie,* 1965, **18**, 267.

DHCD *See* Cyclar.

DHD A petroleum reforming process operated in Germany from 1940. The catalyst was molybdena on alumina.

Nonnenmacher, H., *Brennst.-Chem.,* 1950, **31**, 138 (*Chem. Abs.,* **44**, 8098).

DHDS [Diesel deep hydrodesulfurization] A petroleum refining process developed by the Instituto Mexicano del Petroleo (IMP) with plans for it to be in operation at the Pemex refinery at Cadereyka, Mexico, in 1999.

DHR [Druck-Hydrogenium-Raffination; German, meaning hydrorefining] A petroleum refining process developed by BASF in Germany. *See* DHC.

Öhttinger, W., *Erdoel Kohle,* 1953, **6**, 693.

DIAMOX A process for removing hydrogen sulfide and hydrogen cyanide from coke oven gas by absorption in aqueous ammonia. Developed by Mitsubishi Chemical Industries and Mitsubishi Kakoki Kaisha and operated in Japan.

Kohl, A. L. and Riesenfeld, F. C., *Gas Purification,* 4th ed., Gulf Publishing, Houston, TX, 1985, 168.

Hydrocarbon Process., 1986, **65**(4), 81.

Dianor [Diamond Alkali, Oronozio de Nora] A process for cracking naphtha to ethylene, adapted for small-scale operation so that polyvinyl chloride could be made by developing nations. Developed in the 1970s by Chem Systems and the two named companies, but not commercialized.

Spitz, P. H., *Petrochemicals, the Rise of an Industry,* John Wiley & Sons, New York, 1988, 411.

Diaphragm cell A family of electrochemical *chlor-alkali processes using cells with semipermiable membranes which minimize diffusion between the electrodes. The overall reaction is:

$$2NaCl + 2H_2O = 2NaOH + H_2 + Cl_2$$

Chlorine is liberated at the anode and hydrogen at the cathode, where the sodium hydroxide is formed. Development started in the mid-19th Century. In 1962, it was stated that several hundred cell designs had been patented and more than 30 types of diaphragm cells were in use in the United States. The diaphragms of the cells used in the 1960s had membranes of asbestos deposited on steel. In the 1970s, diaphragms made of cation-exchange membranes began to be used and the cells were called *Membrane cells. *See also* Castner-Kellner.

Hardie, D. W. F., *Electrolytic Manufacture of Chemicals from Salt,* Oxford University Press, London, 1959, 24.

Kircher, M. S., in *Chlorine, Its Manufacture, Properties and Uses,* Sconce, J. S., Ed., Reinhold, New York, 1962, 81.

Purcell, R. W., in *The Modern Inorganic Chemicals Industry,* Thompson, R., Ed., The Chemical Society, London, 1977, 106.

Diazo   A family of reprographic processes (including **Diazotype**), based on the coupling of diazonium compounds with dye couplers to form colored compounds. Exposure of the diazonium compounds to near-ultraviolet radiation destroys them, so illuminated areas do not develop color.

Dieselmax   A petroleum cracking process which combines mild *hydrocracking with *thermal cracking to maximize the production of middle distillate without using more hydrogen than hydrocraking alone. Developed by UOP.

Diesulforming   A *hydrodesulfurization process which used a molybdenum-containing catalyst. Developed by the Husky Oil Company and first operated in Wyoming in 1953.

Oil Gas J., 1956, **54**(46), 165.
Unzelman, G. H. and Wolf, C. J., in *Petroleum Processing Handbook,* Bland, W. F. and Davidson, R. L., Eds., McGraw-Hill, New York, 1967, 3-42.

DIFEX   [Dimethyl formamide **extraction**]   A process for separating butadiene from cracked petroleum fractions by extracting with dimethyl formamide.

Smeykal, K. and Luetgert, H., *Chem. Tech. (Berlin),* 1962, **14**, 202.

Dilchill   [Dilute, chill]   A process for dewaxing petroleum by controlled crystallization, with cooling accomplished by the incremental addition of a cold solvent. Developed by Exxon Research & Engineering Company.

Sequeira, A., Jr., in *Petroleum Processing Handbook,* McKetta, J. J., Ed., Marcel Dekker, New York, 1992, 658.

Dimersol   A family of processes for dimerizing single or mixed olefines, catalyzed by mixtures of trialkyl aluminum compounds and nickel salts. Developed by IFP and first commercialized in 1977; by 1997 it was used in 26 plants.

Weissermel, K. and Arpe, H.-J. *Industrial Organic Chemistry,* 3rd ed., VCH Publishers, Weinheim, Germany, 1997, 84.

Dimersol E   A process for making gasoline from ethylene. The catalyst is a soluble *Ziegler-type catalyst containing nickel. Developed by IFP in the 1980s and operated at an undisclosed location since 1988.

Chauvel, A., Delmon, B., and Hölderich, W. F., *Appl. Catal. A: Gen.,* 1994, **115**, 173.

Dimersol G   A process for dimerizing propylene to a mixture of isohexenes, suitable for blending into high-octane gasoline. The process is operated in the liquid phase with a dissolved homogeneous catalyst. Developed by IFP and first operated at Alma, MI, in 1977.

Andrews, J. W. Bonnifay, P. L., Cha, B. J., Barbier, J. C., Douillet, D., and Raimbault, J., *Hydrocarbon Process.,* 1976, **55**(4), 105.
Andrews, J. W. and Bonnifay, P. L., *Hydrocarbon Process.,* 1977, **56**(4), 161.
Weismantel, G. E., *Chem. Eng. (N.Y.),* 1980, **87**(12), 77.

Dimersol X   A process for dimerizing mixed butenes to mixed octenes. Selective hydrogenation, catalyzed by a soluble Ziegler catalyst, is used. The spent catalyst is discarded. The process was developed by IFP and first operated at Kashima, Japan, in 1980. BASF has used the process in Ludwigshafen since 1985.

Leonard, J. and Gaillard, J. F., *Hydrocarbon Process.,* 1981, **60**(3), 99.
Boucher, J. F., Follain, G., Fulop, D., and Gaillard, J., *Oil Gas J.,* 1982, **80**(13), 84.

Dimox   [Directed **metal** oxidation]   *See* Lanxide.

Dipen [**Di**nitrogen **pen**toxide]   A process for making dinitrogen pentoxide by oxidizing dinitrogen tetroxide with ozone. The product, dissolved in nitric acid or an organic solvent, is used as a nitrating agent. Developed in the 1990s by QVF Process Systems and the Defence Research Agency, UK.

   Eur. Chem. News, 1995, **64**(1694), 20.

Direct   *See* American.

Direct reduction   *See* DR.

disproportionation   In industrial chemistry, this term usually means the catalytic conversion of toluene to a mixture of xylene isomers and benzene, but other reactions also are known by this name. Those with special names that are described in this dictionary are: MSTDP, MTDP, Raecke, Triolefine, Xylenes-plus.

DISTAPEX   A process for removing aromatic hydrocarbons from pyrolysis gasoline or coke-oven benzole by extractive distillation with added N-methyl pyrrolidone. The operating temperature is at least 170°C. Developed by Lurgi. First announced in 1961; by 1993, 22 plants had been built.

Distex   A family of extractive distillation processes used in the petroleum industry from 1940. In one such process, furfural is used as the extracting agent for separating butadiene from other C_4 hydrocarbons.

   Buell, C. K. and Boatright, G. R., *Ind. Eng. Chem.,* 1947, **39,** 695.

DKS   *See* carbonization.

D-LM   [**D**wight-**L**loyd **M**cWane]   A process for prereducing iron ore. A mixture of the ore, noncoking coal, and limestone is pelletized and carbonized. The reduced pellets are then fed to an electric furnace. Commercialized in Mobile, AL. *See also* DR.

   Ban, T. E. and Worthington, B. W., *J. Met.,* 1960, **12,** 937.

DMO   [**D**irect **m**ethane **o**xidation]   A process for converting methane to methanol or synthetic liquid fuels. Under development by Catalytica in 1997.

   Oil Gas J., 1997, **95**(25), 16.

Dobanol   A process for making linear fatty alcohols from *syngas. Developed by Deutsche Shell Chemie.

DOC   [**d**irect **o**xy**c**hlorination]   A one-stage process for making vinyl chloride from ethylene and chlorine. Piloted by Hoechst in Germany in 1989.

Doctor   *See* Doktor.

DODD   A process for the deep desulfurization of middle petroleum distillates. Introduced by Exxon in 1989.

   Absci-Halabi, M., Stanislaus, A., and Qabazard, H., *Hydrocarbon Process.,* 1997, **76**(2), 45.
   Zaczepinski, S., in *Handbook of Petroleum Refining Processes,* Meyers, R. A., Ed., McGraw-Hill, New York, 1997, 8.63.

Doktor   Also spelled Doctor. A process for deodorizing oils by converting trace mercaptans to disulfides by the use of alkaline sodium plumbite and sulfur. The reactions are:

$$2RSH + Na_2PbO_2 = Pb(RS)_2 + 2NaOH$$

$$Pb(RS)_2 + S = R_2S_2 + PbS$$

The spent solution is regenerated by blowing air through it:

$$PbS + 2O_2 + 4NaOH = Na_2PbO_2 + Na_2SO_4 + 2H_2O$$

No longer used as a manufacturing process, but still used as a qualitative test for mercaptans in petroleum fractions. The presence of mercaptans is indicated by a black precipitate of lead sulfide.

Unzelman, G. H. and Wolf, C. J., in *Petroleum Processing Handbook,* Bland, W. F. and Davidson, R. L., Eds., McGraw-Hill, New York, 1967, 3-125.

Donau Chemie   A process for making saleable gypsum as a byproduct of phosphoric acid manufacture. The product is used to make partition panels for buildings.

Becker, P., *Phosphates and Phosphoric Acid,* 2nd. ed., Marcel Dekker, New York, 1989, 560.

Dored   *See* DR.

Dorr   One of the two wet processes for making phosphoric acid by the acidulation of phosphate rock; the other is the *Haifa process. The Dorr process uses sulfuric acid. Phosphate rock is primarily apatite, $Ca_5(PO_4)_3F$. The calcium phosphate portion generates orthophosphoric acid and calcium sulfate:

$$Ca_3(PO_4)_2 + 3H_2SO_4 = 3CaSO_4 + 2H_3PO_4$$

The fluoride portion is either removed as gaseous silicon tetrafluoride, if silica is added, or as the sparingly soluble sodium hexafluorosilicate, which remains with the calcium sulfate.

There are several variants of the Dorr process which differ according to the treatment of the calcium sulfate. Some variants produce the dihydrate, gypsum, others produce the hemihydrate. The variants also differ in the concentration of the phosphoric acid produced, but it is never more than 43 percent. The basic process was patented by Lawes in England in 1842 but the presently used variant was developed by the Dorr-Oliver company in the 1930s.

Stevens, H. M., *Phosphorus and Its Compounds,* Interscience, New York, 1961.
Beveridge, G. S. G. and Hill, R. G., *Chem. Process. Eng. (London),* 1968, **49**(7), 61.
Childs, A. F., in *The Modern Inorganic Chemicals Industry,* Thompson, R., Ed., The Royal Society of Chemistry, London, 1977, 375.

Dorr-Oliver   One of the wet processes for making phosphoric acid. *See* Dorr, Jacobs-Dorr.

Dortmund-Hoerder   Also called DH. A steelmaking process in which the molten metal is vacuum degassed.

Double Loop   A process for removing hydrogen sulfide from natural gas. In the first loop an organic solvent absorbs the hydrogen sulfide from the gas. In the second, an aqueous solution of an iron chelate converts this to elemental sulfur. Developed by Radian International in the 1990s.

Quinlan, M. P., Echterhof, L. W., Leppin, D., and Meyer, H. S., *Oil Gas J.,* 1997, **95**(29), 54.

Double steeping   *See* SINI.

Dow bromine   An electrolytic process for extracting bromine from brines. Ferric bromide was an intermediate. Developed by H. H. Dow, founder of the Dow Chemical Company.

Chem. Eng. News, 1997, **75**(27), 43.

Dow-Phenol   A process for making phenol by oxidizing molten benzoic acid with atmospheric oxygen. The catalyst is cuprous benzoate.

Buijs, W., *Catal. Today,* 1996, **27**(1-2), 159.

Dowa   A *flue-gas desulfurization process in which the sulfur dioxide is absorbed in a basic aluminum sulfate solution. The product solution is oxidized with air and reacted with limestone to produce gypsum. Developed by the Dowa Mining Company, Japan, in the early 1980s. In 1986 it was in use in nine plants in Japan and the United States.

Kohl, A. L. and Riesenfeld, F. C., *Gas Purification,* 4th ed., Gulf Publishing, Houston, 1985, 371.

Downs   A process for making sodium by electrolyzing a molten eutectic mixture of sodium chloride and calcium chloride at 580°C. Invented in 1922 by J. C. Downs at Du Pont, and widely used ever since. In the United Kingdom it was first used in 1937.

U.S. Patent 1,501,756.
Sittig, M., *Sodium, Its Manufacture, Properties and Uses,* Reinhold Publishing, New York, 1956, 21.
Hardie, D. W. F., *Electrolytic Manufacture of Chemicals from Salt,* Oxford University Press, London, 1959, 28.

Dow seawater   A process for extracting magnesium from seawater. Calcined dolomite (CaO·MgO, dololime), or calcined oyster shell, is added to seawater, precipitating magnesium hydroxide. This is flocculated, sedimented, and filtered off. For use as a refractory it is calcined; for the manufacture of magnesium chloride for the manufacture of magnesium metal, it is dissolved in hydrochloric acid. Developed by the Dow Chemical Company and later operated by the Steetly Company in West Hartlepool, UK, and Sardinia.

Mantell, C. L., in *Chlorine, Its Manufacture, Properties and Uses,* Sconce, J. S., Ed., American Chemical Society Monogr. No. 154. Reinhold Publishing, New York, 1962, 578.

DPG hydrotreating   A process for converting petroleum pyrolysates into high-octane gasoline. Two stages of selective hydrogenation are used. Developed by Lummus Crest. In 1986, 28 systems were operating.

Hydrocarbon Process., 1986, **65**(9), 85.

DR   [**D**irect **R**eduction]   A general name for processes for making iron or steel by reducing iron oxide ores at temperatures below the melting points of any of the ingredients. The product is known as DRI [**d**irect **r**educed **i**ron). The reducing agent may be carbon in some form, or carbon monoxide, or hydrogen, or a mixture of these.

The Commission of the European Communities commissioned a literature study of all the DR processes. It was published by the Verlag Stahleisen mbH, Düsseldorf, updated several times, and then translated into English and published in 1979 by The Metals Society, London. The Report classified the processes into three groups: those that were of great industrial significance, those of limited industrial significance, and those without industrial application. The names of all the processes, in their three groups, are listed as follows. Most of those in the first two categories can be found in their respective alphabetical locations in this dictionary.

Processes of great industrial significance: Allis-Chalmers, Armco, Esso Fior, HIB, HyL, Kawasaki, Kinglor-Metor, Koho, Krupp sponge iron, MIDREX, NSC, SDR, SL/RN, SPM.

Processes with limited industrial significance: Carbotherm, CIP (2), Echeverria, Elkem, H-D, Hoganas, ICEM, Krupp-Renn, Novalfer, Rotored, Wiberg-Soder.

Processes without industrial application: Aachen rotary furnace reduction, Arthur D. Little, Basset, Bouchet-Imphy, CO-C-iron, De Sy, DLM, Dored, Electric fluidized bed, Elkem, Finsider, Flame-Smelting, Freeman, Heat-Fast, H-iron, Jet Smelting, Kalling (Avesta), Kalling (Domnarvets), Madaras, Nu-Iron, Orcarb, ORF, Purofer, Republic Steel, Stelling, Strategic-Udy, Sturzelberg, UGINE, VOEST.

The following additional DR processes have been operated since 1979 and are described in separate entries: Accar, Circored, CODIR, DAV, DRC, OSIL, Plasmared, Purofer, SIIL, SPIREX, Tisco, USKO.

Kirk-Othmer's Encyclopedia of Chemical Technology, 4th ed., Vol. 14, John Wiley & Sons, New York, 1991–1998, 855.

Dr. C. Otto   *See* carbonization.

Dravo-Ruthner   A development of the *Ruthner process for recovering hydrochloric acid and iron oxide from steel pickling liquors. Developed by Ruthner Industrieplanungs, Vienna, with Dravo Corporation, Pittsburgh. *See also* Ruthner.

DRC   [Direct reduction by coal]   A direct reduction ironmaking process, using coal as the reductant. First operated in Rockwood, United States, in 1978, and now operating in two plants in South Africa and one in China. *See* DR.

Drizo   A variation of the glycol process for removing water vapor from natural gas, in which the water is removed from the glycol by stripping with a hydrocarbon solvent, typically a mixture of pentanes and heavier aliphatic hydrocarbons. The process also removes aromatic hydrocarbons. Last traces of water are removed from the triethylene glycol by stripping with toluene in a separate, closed loop. Invented in 1966 by J. C. Arnold, R. L. Pearce, and H. G. Scholten at the Dow Chemical Company. Twenty units were operating in 1990.

    U.S. Patent 3,349,544.
    Hydrocarbon Process., 1975, **54**(4), 81.
    Isalski, W. H., *Separation of Gases,* Clarendon Press, Oxford, 1989, 218.
    Smith, R. S., *Hydrocarbon Process.,* 1990, **69**(2), 75.
    Hydrocarbon Process., 1992, **71**(4), 98.
    Chem. Eng. (Rugby, England), 1993, (549), 17.

Dry Box   *See* Iron Sponge.

DRYPAC   A *flue-gas desulfurization process in which an aqueous lime suspension is injected into a spray drier. Developed by Flakt Industri, Sweden. In 1986 it was in use in 17 plants in Sweden, Denmark, Finland, Germany, and the United States.

DS (1)   [Dual sludge]   A two-stage sewage treatment process, combining aerobic and anaerobic processes, which yields methane for use as fuel. Offered by Linde, Munich.

DS (2)   A general name for a process for making iron or steel, which give a liquid iron product, similar to the pig iron produced in a blast furnace. Examples are SKF, KR (2), Midrex.

DSN   [Direct Strong Nitric]   A general name for processes for concentrating nitric acid from 50–70 percent to 98 percent by adding dinitrogen tetroxide.

    G. Fauser, *Chem. Met. Eng.,* 1932, **39**, 430 (*Chem, Abstr.,* **26**, 5388).
    Newman, D. J. and Klein, L. A., *Chem. Eng. Prog.,* 1972, **68**(4), 62.
    Hellmer, L., *Chem. Eng. Prog.,* 1972, **68**(4), 67.
    Ohroi, T., Okubo, M., and Imai, O., *Hydrocarbon Process.,* 1978, **57**(11), 163.
    Marzo, L. M. and Marzo, J. M., *Chem. Eng. (N.Y.),* 1980, **87**(22), 54.

D-TOX   An oxidative process for destroying unsaturated chlorinated compounds such as vinyl chloride in gas streams. Developed by Ultrox International, Santa Ana, CA.

Dual Alkali   A *flue-gas desulfurization process. The sulfur dioxide is absorbed in aqueous sodium hydroxide and partially oxidized, and this liquor is then treated with calcium hydroxide to regenerate the scrubbing solution and precipitate calcium sulfate. Developed by Combustion Equipment Associates and Arthur D. Little.

    LaMantia, C. R., Lunt, R. R., and Shah, I. S., *Chem. Eng. Prog.,* 1974, **70**(6), 66.

Dualayer Distillate   A process for extracting organic acids from petroleum fractions, using an aqueous solution of sodium cresylate. Developed by the Mobil Oil Company in the 1950s.

Unzelman, G. H. and Wolf, C. J., in *Petroleum Processing Handbook,* Bland, W. F. and Davidson, R. L., Eds., McGraw-Hill, New York, 1967, 3-113.

Dualayer Gasoline A process for extracting mercaptans and other organic sulfur compounds from petroleum distillates by solvent extraction with aqueous sodium or potassium hydroxide and a proprietary solubilizer.

Unzelman, G. H. and Wolf, C. J., in *Petroleum Processing Handbook,* Bland, W. F. and Davidson, R. L., Eds., McGraw-Hill, New York, 1967, 3-113.

Dual-Spectrum Also called Workman. A thermographic copying process. A transparent film base, coated with 4-methoxy-1-naphthol and a photo-reducible dye such as erythrosine, receives the image, which is then transferred to a paper sensitized with sodium behenate. Invented in 1961 by W. R. Workman at the Minnesota Mining and Manufacturing Company, MN, and commercialized for office copying, but later superseded by various electrophotographic processes.

U.S. Patent 3,094,417.

Dubbs One of a number of thermal (*i.e.,* noncatalytic) processes for cracking petroleum, widely used in the 1920s and 30s. Invented by J. A. Dubbs and C. P. Dubbs and promoted by The Universal Oil Products Company (now UOP). One of the original patents (U.S. 1,123,502), filed in 1909 but not granted until 1915, was mainly concerned with methods for breaking an emulsion by subjecting it to heat and pressure, but subsequent amendments and filings extended it to cover continuous distillation and condensation under pressure. Widely adopted from 1923 and used until the introduction of *catalytic cracking.

U.S. Patents 1,100,717; 1,123,502.
Enos, J. L., *Petroleum Progress and Profits,* MIT Press, Cambridge, MA, 1962, Chap. 2.
Stanley, H. M., in *Propylene and Its Industrial Derivatives,* Hancock, E. G., Ed., Ernest Benn, London, 1973, 14.
Achillades, B., *Chem. Ind. (London),* 1975, (8), 337.

Dubrovai A petroleum cracking process, deriving its heat from partial combustion. Operated in the USSR.

Asinger, F., *Paraffins, Chemistry and Technology,* translated by B. J. Hazzard, Pergamon Press, Oxford, 1968, 588.

Duftschmid A variation of the *Fischer-Tropsch process in which synthesis gas and an oil are circulated over a fixed bed of iron catalyst in order to increase the yield of olefins from the gas.

Asinger, F., *Paraffins, Chemistry and Technology,* translated by B. J. Hazzard, Pergamon Press, Oxford, 1968, 151.

Dunlop A process for making foam rubber which uses sodium fluorosilicate to coagulate the rubber particles and deactivate the surfactants. *See also* Talalay.

Madge, E. W., *Latex Foam Rubber,* John Wiley & Sons, New York, 1962.

Dunn *See* Wendell Dunn.

Duo-Sol A process for separating aromatic from aliphatic hydrocarbons by partition between two solvents. The first solvent (Selecto or Selectox) is a mixture of phenol and cresylic acids; the second is liquid propane. Developed by the Max B. Miller Company and licensed by Milwhite Company.

Petrol. Refin., 1960, **39**(9), 234.
Unzelman, G. H. and Wolf, C. J., in *Petroleum Processing Handbook,* Bland, W. F. and Davidson, R. L., Eds., McGraw-Hill, New York, 1967, 3-90.

Duplex (1)   An integrated steelmaking process in which iron is converted to steel in a *Bessemer furnace with a basic lining, and the molten product is transferred to a basic lined arc furnace in which the remaining impurities are oxidized. Developed in Germany and widely used there and elsewhere around 1900.

Dennis, W. H., *A Hundred Years of Metallurgy,* Gerald Duckworth, London, 1963, 111.

Duplex (2)   A process for making sodium perborate from both sodium peroxide and hydrogen peroxide. Developed in the United States from the 1920s and used until the development of the present process, which uses hydrogen peroxide as the only source of the peroxygen. *See also* Acid.

DuraTherm   A process for treating waste products from oil refineries by thermal desorption. The wastes are passed through a rotating, externally heated metal drum containing a rotating helix. Volatile waste products are swept from the drum in a stream of nitrogen and condensed.

Oil Gas J., 1996, **94**(38), 49.

Dutch   An obsolete process for making basic lead carbonate pigment (white lead) by exposing metallic lead to vinegar and carbon dioxide. Reportedly first described by Theophrastos around 300 BC. Also known as the Stack process because the metal ingots were arranged in vertical stacks. *See also* Thompson-Stewart.

Sherwood Taylor, F., *A History of Industrial Chemistry,* Heinemann, London, 1957, 83.

Duval   *See* CLEAR.

Dwight-Lloyd   A pyrometallurgical process for simultaneously roasting and smelting a ground ore. The ore is contained in a series of shallow iron boxes with perforated bases which are drawn continuously through a furnace having a downward draft of air. Developed by A. S. Dwight and R. L. Lloyd at the Greene Consolidated Smelter in Cananea, Brazil, in 1906; acquired by Lurgi in 1909, and now widely used for ferrous and nonferrous ores.

Dennis, W. H., *A Hundred Years of Metallurgy,* Gerald Duckworth, London, 1963, 60.

DWN   [Druckwechsel nitrogen]   A proprietary *PSA system for separating nitrogen from air. Developed and offered by Linde. *See also* DWO.

DWO   [Druckwechsel oxygen]   A proprietary *PSA system for separating oxygen from air. Developed and offered by Linde. *See also* DWN.

Dynacracking   A petroleum cracking process which combines the best features of the *catalytic cracking and *thermal cracking processes. It converts heavy oil feedstocks to fuel gas, gasoline, and fuel oil. No catalyst is used. Developed in the 1950s by Hydrocarbon Research, but not commercialized.

Rakow, M. S. and Calderon, M., *Chem. Eng. Prog.,* 1981, **77**(2), 31.

Dynagas   A noncatalytic process for hydrogenating coal to produce gas or light oil fuels. Developed by Hydrocarbon Research.

Dynaphen   A process for converting mixed alkyl phenols (from coal liquids or lignin) to benzene, phenol, and fuel gas, by noncatalytic hydrogenation at high temperature. Developed and offered by Hydrocarbon Research.

Dynatol   A continuous process for making sorbitol. Developed by Hydrocarbon Research before 1982.

E

EARS [Enhanced acid regeneration system]   A process for recovering hydrochloric acid from the *ERMS ilmenite beneficiation process. It may be used also for recovering waste pickle liquor. The acid liquor containing ferrous chloride is evaporated at low temperature to form iron chloride pellets, which are fed to a pyrohydrolysis reactor. This generates hydrochloric acid and iron oxide pellets, which can be used for steel production or disposed of as inert landfill. Developed by E. A. Walpole at the University of Newcastle, Australia, from the early 1990s and piloted by Austpac Gold (now Austpac Resources).

Eastman   Also known as the Tennessee Eastman Challenge Process. Not an actual chemical process but a theoretical chemical engineering exercise, based on a proprietary process.

> Downs, J. J. and Vogel, E. F., *Comput. Chem. Eng.,* 1993, **17**, 245.
> Luyben, W. L., *Ind. Eng. Chem. Res.,* 1996, **35**(10), 3280.
> Wu, K.-L. and Yu, C.-C, *Ind. Chem. Eng. Res.,* 1997, **36**(6), 2239 (23 refs.).

Eastman-Halcon   A process for making acetic anhydride from *syngas. The basic process is the carbonylation of methyl acetate. Methanol is made directly from the carbon monoxide and hydrogen of syngas. Acetic acid is a byproduct of the cellulose acetate manufacture for which the acetic anhydride is needed. The carbonylation is catalyzed by rhodium chloride and chromium hexacarbonyl.

> Wittcoff, H. A. and Reuben, B. G., *Industrial Organic Chemicals,* John Wiley & Sons, New York, 1996, 323.

EB   *See* Ethylbenzene.

Ebara [Electron beam ammonia reaction]   A dry process for removing sulfur and nitrogen oxides from flue-gas. A beam of high energy electrons is injected into the gas, to which a stoichiometric quantity of ammonia has been added. The product, a mixture of ammonium sulfate and ammonium nitrate, is collected downstream by an electrostatic precipitator or a bag filter. Developed by Ebara Corporation, Japan, and piloted in Indianapolis in 1986.

> IEA Coal Research, *Flue-Gas Desulfurization Handbook,* Butterworths, London, 1987.

Ebex [Ethylbenzene extraction]   A version of the *Sorbex process, for extracting ethylbenzene from mixtures of aromatic C_8 isomers. The adsorbent is a zeolite. It had not been commercialized as of 1984.

> de Rossett, A. J., Neuzil, R. W., Tajbl, D. G., and Braband, J. M., *Sep. Sci. Technol.,* 1980, **15**, 637.
> Ruthven, D. M., *Principles of Adsorption and Adsorption Processes,* John Wiley & Sons, New York, 1984, 400.

EBMax   A continuous, liquid-phase process for making ethylbenzene from ethylene and benzene, using a zeolite catalyst. Developed by Raytheon Engineers and Constructors and Mobil Oil Corporation and first installed at Chiba Styrene Monomer in Japan in 1995. Generally similar to the *Mobil/Badger process, but the improved catalyst permits the reactor size to be reduced by two thirds.

> *Chem. Eng. (N.Y.),* 1995, **109**(9), 21.

EBV   *See* carbonization.

ECF [Elemental chlorine free]   A generic term for pulp-bleaching processes which use chlorine dioxide and other oxidants in place of elemental chlorine. *See also* TCF.

Chem. Eng. (N.Y.), 1997, **104**(4), 33.

Nelson, P. J., in *Environmentally Friendly Technologies for the Pulp and Paper Industries,* Young, R.A. and Akhar, M., Eds., John Wiley & Sons, New York, 1998, 215.

Echeverria   *See* DR.

Ecoclear   A continuous process for destroying organic contaminants in wastewater by treatment with ozone in the presence of a proprietary catalyst. Developed in 1995 by Eco Purification Systems, The Netherlands.

Econ-Abator   A process for oxidizing hydrogen sulfide and other sulfur compounds in waste gases by fluid bed combustion in the presence of an oxide catalyst. Licensed by ARI Technologies. In 1992 there were 90 installations.

Hydrocarbon Process., 1992, **71**(4), 99.

Econamine   A process for removing acid gases from natural gas by selective absorption in diglycolamine (also called [2-(2-aminoethoxy) ethanol], and DGA). Developed by the Fluor Corporation, the El Paso Natural Gas Company, and the Jefferson Chemical Company and widely used. Later versions, developed by Fluor Daniel International, include the Fluor Daniel Econamine and the *Econamine FG processes. More than 30 units were operating in 1996. *See also* Aromex.

Dingman, J. C. and Moore, T. F., *Hydrocarbon Process.,* 1968, **47**(7), 138.

Hydrocarbon Process., 1971, **50**(4), 101.

Kohl, A. L. and Riesenfeld, F. C., *Gas Purification,* 4th ed., Gulf Publishing, Houston, TX, 1985, 39.

Hydrocarbon Process., 1996, **75**(4), 116.

Econamine FG   [Flue gas]   A process for removing carbon dioxide from flue-gases by dissolution in an aqueous solution of monoethanolamine and a proprietary corrosion inhibitor. Originally developed by the Dow Chemical Company under the designation "Gas/Spec FT," the process was acquired in 1989 by the Fluor Corporation and is now licensed by that company.

Econocat   A process for deodorizing gases by catalytic oxidation over manganese dioxide activated with other metals. Developed and sold by Cortaulds Engineering.

Chem. Eng. (N.Y.), 1984, **91**(13), 156.

Edeleanu (1)   A process for extracting aromatic hydrocarbons from kerosene using liquid sulfur dioxide. Developed in Romania in 1908 by L. Edeleanu to improve the burning properties of lamp kerosene; in 1990 the original process was still in use in commercial plants. The company founded by the inventor in 1910, Edeleanu Gesellschaft mbH, now a part of the RWE Group, Germany, is today better known for its range of processes for making lubricating oils, waxes, and certain organic chemicals.

British Patent 11,140 (1908.)

Edeleanu, L., *J. Inst. Petr. Technol.,* 1932, **18,** 900.

The Petroleum Handbook, 3rd ed., Shell Petroleum, London, 1948, 183.

Edeleanu (2)   A process for extracting higher aliphatic hydrocarbons from hydrocarbon mixtures by extraction with aqueous urea.

Weissermel, K. and Arpe, H.-J. *Industrial Organic Chemistry,* 3rd ed., VCH Publishers, Weinheim, Germany, 1997, 78.

EDS   *See* Exxon Donor Solvent.

Efflox   A process for destroying cyanide wastes in hydrometallurgical effluents by the use of Caro's acid. The cyanide ion is oxidized to cyanate ion.

EFFOL   A process for making nylon salt (hexamethylenediamine di-adipate). Developed by Rhône-Poulenc in the 1980s and now used in its plant at Chalampé, France. It produces much less effluent than the earlier process.

EHD   [Electrohydrodimerization]   Also known as Electrodimerization. An electrolytic process for converting acrylonitrile to adiponitrile. *See* Monsanto.

The name has been used also for the electrodimerization of formaldehyde to ethylene glycol, being developed by Electrosynthesis, Amherst, NY, in 1984.

EIC   *See* Cuprasol.

Eichner   An early thermographic copying process.

ELCOX   A *flue-gas desulfurization process in which the sulfur dioxide is oxidized electrochemically to sulfuric acid, using an organometallic catalyst (*e.g.*, cobalt phthalocyanine) adsorbed on activated carbon. Developed by the Central Laboratory of Electrochemical Power Sources, Sofia, Bulgaria.

> European Patent Appl. 302,224.
> Vitanov, T., Budevski, E., Nikolov, I., Petrov, V., Naidenov, V., and Christov, Ch., in *Effluent Treatment and Waste Disposal,* Institution of Chemical Engineers, Rugby, England, 1990, 251.

Electrodimerization   *See* EHD.

Electrofining   A process for purifying petroleum fractions by extracting them with various liquid reagents and then assisting their separation by means of an electric field. Developed by the Petreco Division of Petrolite Corporation, and first operated in California in 1951.

> Unzelman, G. H. and Wolf, C. J., in *Petroleum Processing Handbook,* Bland, W. F. and Davidson, R. L., Eds., McGraw-Hill, New York, 1967, 3-137.

Electrohydrodimerization   *See* EHD.

Electropox   [Electrochemical partial oxidation]   Also called Pox. An electrochemical process for oxidizing methane to *syngas. It combines the partial oxidation and steam reforming of methane with oxygen separation in a single stage. Invented in 1988 by T. J. Mazanec at BP Chemicals. An industrial-academic consortium to develop the process was formed in 1997.

> U.S. Patents 4,802,958; 4,793,904.
> European Patent Appl. 399,833.
> Mazanec, T. J., Cable, T. L., and Frye, J. G., Jr., *Solid State Ionics,* 1992, **53-56,** 111.
> Mazanec, T. J., Cable, T. L., and Frye, J. G., Jr., in *The Role of Oxygen in Improving Chemical Processes,* Felizon, M. and Thomas, W. J., Eds., Royal Society of Chemistry, Cambridge, 1993, 212.
> Mazanec, T. J., in *The Activation of Dioxygen and Homogeneous Catalytic Oxidation,* Barton, D. H. R., Martell, A. E., and Sawyer, D. T., Eds., Plenum Press, New York, 1993, 85.
> Mazenac, T. J., *Interface,* 1996, **5**(4), 46.
> *Chem. Ind. (London),* 1997, (22), 911.

Electroslag   Also called ESR. A general term for any electrolytic metal extraction process in which the metal is produced in the molten state beneath a layer of molten slag. Used mainly for ferrous alloys.

> Duckworth, W. E. and Hoyle, G., *Electroslag Refining,* Chapman & Hall, London, 1969.

ElectroSlurry   An electrolytic process for extracting copper from sulfide ores, liberating elemental sulfur. Developed by the Envirotech Research Center, Salt Lake City, UT.

> *Chem. Eng. (N.Y.),* 1980, **87**(25), 35.

Elektrion Also called Volto. A method of increasing the molecular weight, and thus the viscosity, of a mixture of light mineral oil and a fatty oil, by subjecting it to an electric discharge in a hydrogen atmosphere.

Elektron A family of processes, operated by Magnesium Elektron, UK, for making magnesium, magnesium-zirconium alloys, and zirconium chemicals. In the 1920s and 30s, the names "elektron" and "elektronmetall" were used colloquially in Germany for magnesium metal.

Elf-SRTI An industrial chromatographic process developed by Elf Aquitane Dévelopment and the Société de Recherches Techniques et Industrièlles. Multiple beds are used. Used for separating perfume ingredients and proposed for separating aliphatic hydrocarbons.

> Bonmati, R. G., Chapelet-Letourneux, G., and Margulis, J. R., *Chem. Eng. (N.Y.)*, 1980, **87**(6), 70. *Anal. Chem.*, 1980, **52**(4), 481A.
> Yang, R. T., *Gas Separation by Adsorption Processes*, Butterworths, Guildford, 1987, 212.

Elkem *See* DR.

Elkington An early electrolytic process for plating gold and silver from cyanide solutions. Invented by G. R. and H. Elkington in 1842.

Ellis A process for making isopropyl alcohol from light olefin mixtures by treatment with concentrated sulfuric acid. Operated in World War I by the Melco Chemical Company, as an intermediate for the production of acetone for airplane "dope."

> Ellis, C., *The Chemistry of Petroleum Derivatives*, Chemical Catalog, New York, 1934, 349.

Eloxal [Electrolytic oxidation of aluminium] An electrolytic process for applying an oxide film to the surface of aluminum in order to protect it from further oxidation or to make it capable of adsorbing a dyestuff.

> Institut Fresenius, *Waste Water Technology*, Springer-Verlag, Berlin, 1989, 70.

ELSE [Extremely low sulphur emission] A *flue-gas desulfurization process in which the sulfur dioxide is absorbed by zinc oxide. Developed by Amoco, United States.

> Wieckowska, J., *Catal. Today*, 1995, **24**(4), 445.

Elsorb A process for recovering sulfur from the sulfur dioxide in flue-gases. The gases are scrubbed with aqueous sodium phosphate and the sulfur dioxide is recovered by heating the scrubbing liquor. Developed in the 1980s by the Norwegian Institute of Technology, Trondheim, and piloted at the Vitkovice Steel Works, Czechoslovakia. The first commercial plant was built for Esso Norway.

Eluex An early process for extracting uranium from its ores, using both ion-exchange and solvent extraction. Developed by the National Lead Company, United States.

> Merritt, R. C., *The Extractive Metallurgy of Uranium*, U.S. Atomic Energy Authority, 1971, 209. *Eng. Min. J.*, 1978, **179**(12), 84.
> Eccles, H. and Naylor, A. *Chem. Ind. (London)*, 1987, (6), 174.

Eluxyl A process for separating *p*-xylene from its isomers, using an adsorbent-solvent technique. The process is based on simulated countercurrent adsorption where the selective adsorbent is held stationary in the adsorption column. The feed mixture to be separated is introduced at various levels in the middle of the column, as in the *Sorbex process. The *p*-xylene product can be more than 99.9 percent pure. Developed by IFP and Chevron Chemical. A large pilot plant was built in Chevron's site at Pascacougla, MS, in 1994 and a commercial plant on the site was announced in 1996, Since then, the process has been widely licensed.

> *Eur. Chem. News*, 1994, **62**(1648), 18.

Emersol A process for separating stearic and palmitic acids from oleic acid by fractional crystallization from aqueous methanol.

Emert Also known as the Gulf process, the University of Arkansas process, and SSF. A process for converting cellulose to ethanol by simultaneous saccharification and fermentation. Invented by G. H. Emert.

> Worthy, W., *Chem. Eng. News,* 1981, **59**(49), 36.

Empty cell *See* Rueping.

Ensaco A process for making carbon black by the incomplete combustion of used car tires. Developed and commercialized by IMM, Belgium.

Enciforming [National Chemical re**forming**] A petroleum *reforming process that converts pyrolysis gasoline to mixtures of propane, butane, and aromatic hydrocarbons, thereby obviating the usual hydrogenation and solvent extraction processes. The catalyst is a ZSM-5–type zeolite containing both iron and a platinum metal. Developed by the National Chemical Laboratory, Pune, India, since 1988, but not yet commercialized.

> Indian Patent Appl. 526/DEL/88.

Endewax [National Chemical **dewax**ing] A process for dewaxing heavy petroleum fractions by treatment with a catalyst which converts the long-chain hydrocarbons to shorter ones. The catalyst is a ZSM-5–type zeolite in which some of the aluminum has been replaced by iron. Developed by the National Chemical Laboratory, Pune, India, and piloted in 1991.

> Indian Patent Appl. 904/DEL/89; 905/DEL/89.

Enersludge An *OFS process for making fuel oil from sewage sludge. The dried sludge is heated to 450°C in the absence of oxygen, thereby vaporizing about half of it. The vapors are then contacted with hot char from the reaction zone, where catalyzed reactions and thermal cracking convert the lipids and proteins to hydrocarbons. The oil yield is variable and typically 25 percent of the sludge mass. Developed in the 1980s at the Wastewater Technology Center, Burlington, Canada, based on earlier work by E. Bayer at Tübingen University, Germany, and piloted in Australia and Canada. Marketed by Enersludge, Australia.

> U.S. Patents 4,618,735 (process); 4,781,796 (equipment).
> European Patent 52,334.
> Canadian Patent 1,225,062.
> Fernandes, X., *Water Waste Treat.,* 1991, **34**(9), 114.

Engeclor A process for making ammonium chloride by passing gaseous ammonia and hydrogen chloride into an aqueous suspension of the product. Developed by the Brazilian company of the same name.

> Bamforth, A. W. and Sastry, S. R. S., *Chem. Process Eng. (London),* 1972, **53**(2), 72.
> *Ullmann's Encyclopedia of Industrial Chemistry,* 5th ed., Vol. A2, VCH Publishers, Weinheim, Germany, 1989, 259.

Engel A process for making potassium carbonate from potassium chloride obtained from the salt deposits at Stassfurt, Germany. The basis of the process is the formation of the sparingly soluble double salt $MgKH(CO_3)_2 \cdot 4H_2O$ when carbon dioxide is passed into a suspension of magnesium carbonate in aqueous potassium chloride:

$$3MgCO_3 + 2KCl + CO_2 + 5H_2O = 2MgKH(CO_3)_2 \cdot 4H_2O + MgCl_2$$

The double salt is decomposed by hot water and magnesia, forming potassium carbonate and insoluble hydrated magnesium carbonate:

$$2MgKH(CO_3)_2 \cdot 4H_2O + MgO = 3MgCO_3 \cdot 3H_2O + K_2CO_3$$

Invented by C. R. Engel in France in 1881.

German Patent 15,218.
Mellor, J. W., *Comprehensive Treatise on Inorganic and Theoretical Chemistry,* Vol. 2., Longmans Green, London, 1922, 716.

Engel-Precht   An improved version of the *Engel process, developed in 1889 by H. Precht at the Salzbergwerk Neu Stassfurt, Germany, and operated until 1938.

German Patent 50,786.
Thorpe's Dictionary of Applied Chemistry, 4th ed., Vol. 10, Longmans Green, London, 1950, 150.

English   A process for extracting zinc from its ores by reduction with charcoal, invented by the *Champion brothers in Bristol in the 18th Century.

Cocks, E. J. and Walters, B., *A History of the Zinc Smelting Industry in Britain,* George G. Harrap, London, 1968, 8.

Ensio-Fenox   A process for removing chlorinated phenols from pulp-bleaching effluents. It combines anaerobic and aerobic fermentation processes.

Hakulinen, R. and Salkinoja-Salonen, M., *Proc. Biochem.,* 1982, **17,** 18.
Speece, R. E., in *Toxicity Reduction in Industrial Effluents,* Lankford, P. W., and Eckenfelder, W. W., Jr., Eds., Van Nostrand Reinhold, New York, 1990, 146.

ENSOL   A combined process for converting *syngas to methanol and then to ethanol. Acetic acid is an intermediate. Developed by Humphries & Glasgow, in conjunction with BASF and Monsanto.

Chem. Ind. (London), 1985, (8), 240.
Winter, C. L., *Hydrocarbon Process.,* 1986, **65**(4), 71.

ENSORB   [ExxoN adSORBtion]   A process for separating linear from branched hydro-carbons, using a zeolite molecular sieve. The adsorbed gases are desorbed using ammonia. It operates in a cyclic, not a continuous, mode. Developed by Exxon Research & Engineering Company, and used by that company on a large scale at the Exxon refinery in Baytown, TX.

Asher, W. J., Campbell, M. L., Epperly, W. R., and Robertson, J. L., *Hydrocarbon Process.,* 1969, **48**(1), 134.
Yang, R. T., *Gas Separation by Adsorption Processes,* Butterworths, Guildford, England, 1987, 262.

EnZone   [**En**zyme o**zone**]   A pulp bleaching process using the enzyme xylanase in com-bination with oxygen, ozone, and hydrogen peroxide. Invented by K.-E. L. Eriksson at the University of Georgia and piloted in 1998.

Chem. Eng. News, 1998, **76**(12), 42.

Enzink   [**En**zyme de**ink**ing]   A paper deinking process using cellulase enzymes. Invented by K.-E. L. Eriksson and J. L. Yang at the University of Georgia and commercialized in 1994.

Chem. Eng. News, 1998, **76**(12), 42.

EOF   [**E**nergy **o**ptimizing **f**urnace]   An oxygen steelmaking process in which part of the heat is provided by the combustion of carbon powder blown beneath the surface of the molten iron. Developed by the KORF group and being considered for use in India in 1987.

Bose, P. K., *Gas Sep. Purif.,* 1987, **1**, 30.

Eolys   A process for removing 80 to 90 percent of the particulate carbon from diesel ex-haust gases. It uses a catalytic fuel additive containing cerium. Developed by Rhône-Poulenc in 1995 and licensed to Sumitomo Metal Mining Company in Japan in 1996.

EP2   A process for polymerizing olefins in the slurry phase. Developed by Borealis.

Epal A process for making linear aliphatic alcohols by reacting ethylene with triethyl aluminum and oxidizing the products. Similar to *Alfol, but incorporating a trans-alkylation stage that permits a wider range of products to be made. Developed by Ethyl Corporation (now Albermarle Corporation) and operated in the United States since 1964.

Ullmann's Encyclopedia of Industrial Chemistry, 5th ed., Vol. A10, VCH Publishers, Weinheim, Germany, 1989, 284.

Erco A group of processes for making chlorine dioxide by reducing sodium chlorate with a chloride:

$$NaClO_3 + NaCl + H_2SO_4 = ClO_2 + 1/2Cl_2 + Na_2SO_4 + H_2O$$

Developed by W. H. Rapson at the Electric Reduction Company of Canada, now Erco Industries, United States. Erco R-3 uses chloride in sulfuric acid as the reductant; Erco R-3H uses mixed hydrochloric and sulfuric acids; and Erco R-5 uses chloride in hydrochloric acid.

Canadian Patents 825,084; 826,577.

Erdmann *See* Normann.

Erdölchemie A process for treating the waste from the *ammoxidation process for making acrylonitrile, yielding ammonium sulfate. Developed by the eponymous German company, a joint venture of Bayer and BP Chemicals.

Erifon A process for making cellulose textiles flame-resistant. Titanium and antimony oxychlorides are applied from acid solution and the cloth is then neutralized with sodium carbonate solution. Invented by Du Pont in 1947. *See also* Titanox FR.

U.S. Patent 2,570,566.
Gulledge, H. C. and Seidel, G. R., *Ind. Eng. Chem.,* 1950, **42**, 440.

Erlenmeyer An early process for making potassium cyanide from potassium ferrocyanide by heating it with sodium:

$$K_4Fe(CN)_6 + 2Na = 4KCN + 2NaCN + Fe$$

First commercialized in 1876. *See also* Rodgers.

ERMS [Enhanced roasting magnetic separation—also known as Ernie's reductive magnetic separation] A process for beneficiating ilmenite, yielding an off-white product containing more than 99 percent of TiO_2. The ilmenite is roasted in a fluid bed, leached with hydrochloric acid, and beneficiated by dry magnetic separation. The acid is recycled by the *EARS process. The product can be used as a *chloride-process feed material, or as a low-grade pigment. Developed by E. A. Walpole at the University of Newcastle, Australia, from the early 1990s and piloted by Austpac Gold (now Austpac Resources).

World Patent WO 92/04121.
Ind. Miner. (London), 1996, (349), 10.

Escambia (1) A process for oxidizing isobutene to *a*-hydroxy-isobutyric acid (HIBA), for use as an intermediate in the manufacture of methacrylates. The oxidant was dinitrogen tetroxide, N_2O_4. Operated by the Escambia Chemical Corporation, FL, in 1965 before its destruction by an explosion in 1967. It has not been used again.

British Patent 954,035.

Escambia (2) A process for oxidizing propylene to propylene oxide. A transition metal catalyst is used in an organic solvent at high temperature and pressure. Developed by the Escambia Corporation.

Dumas, T., and Bulani, W., *Oxidation of Petrochemicals: Chemistry and Technology,* Applied Science Publishers, London, 1974, 34.

E-SOX   A *flue-gas desulfurization process. The gas is contacted with calcium hydroxide or sodium hydroxide solution in a spray drier, and the resulting particles are trapped in an electostatic precipitator. Developed and piloted by the U.S. Environmental Protection Agency.

Espig   A process for making synthetic emeralds by the flux reaction technique. Beryllia and alumina are dissolved in molten lithium molybdate, and silica is floated on the melt. The emerald crystals form at the base of the melt, but because they tend to float and mix with the silica, a platinum screen is suspended in the middle of the melt. Invented by H. Espig.

> Espig, H., *Chem. Technol. (Berlin)*, 1960, **12**, 327.
> Elwell, D., *Man-made Gemstones*, Ellis Horwood, Chichester, England, 1979, 59.

ESR   *See* Electroslag.

Esso Fior   *See* Fior.

Estasolvan   A process for removing acid gases from liquified petroleum gases by absorption in tributyl phosphate and separation by fractional distillation. Developed by the Institut Français du Pétrole and Friedrich Uhde. No commercial plants were operating in 1985.

> Franckowiak, S. and Nitschke, E., *Hydrocarbon Process.*, 1970, **49**(5), 145.
> Kohl, A. L. and Riesenfeld, F. C., *Gas Purification*, 4th ed., Gulf Publishing, Houston, TX, 1985, 861.

ESTER   A batch process for immobilizing nuclear waste in a borosilicate glass for long-term disposal. Developed in Italy in the 1970s and installed at the Euratom Research Centre, Ispra, in 1981. Intended for use in the radioactive pilot plant (Impianto Vetrificatione Eurex, IVEX) at the European Extraction Plant (UREX) at Saluggia, Italy.

> Lutze, W., in *Radioactive Waste Forms for the Future*, Lutze, W. and Ewing, R. C., Eds., North-Holland, Amsterdam, 1988, 12.

ESTEREX   A process for extracting neutral and acid esters from the products of sulfuric acid-calysed alkylation processes, using sulfuric acid passed through a bundle of hollow fibers immersed in them. Developed by the Merichem Company, Houston, TX.

> *Hydrocarbon Process.*, 1996, **75**(4), 126.

Esterfip   A process for transesterifying vegetable oils to their methyl esters, for use as diesel fuels. Developed by IFP and Sofiproteol, France.

> *Eur. Chem. News*, 1994, **62**(1648), 18.

Ethermax   A process for making ethers (*e.g.*, methyl *t*-butyl ether) by reacting tertiary olefins with alcohols. Conversion levels are increased by using Reaction with Distillation technology (also called *RWD) in which the reaction takes place in a distillation column containing the catalyst—a sulfonic ion-exchange resin. Equilibrium limitations are overcome by continuously removing the products in the RWD section as the reaction occurs. Developed jointly by Hüls, UOP, and Koch Engineering Company, and licensed by UOP. Eleven units were operating in 1996.

> *Chem. Eng. Int. Ed.*, 1991, **98**(7), 44.
> DeGarmo, J. L., Parulekar, V. N., and Pinjala, V., *Chem. Eng. Prog.*, 1992, **88**(3), 43.
> *Hydrocarbon Process.*, 1996, **75**(11), 113.
> Davis, S., in *Handbook of Petroleum Refining Processes*, Meyers, R. A., Ed., McGraw-Hill, New York, 1997, 13.9.

Etherol   A process for making oxygenated fuels (*e.g.*, methyl *t*-butyl ether) from C_4–C_6 hydrocarbons by reacting them with methanol over an acid resin catalyst in a fixed bed reactor under mild conditions. Developed by BP with Erdoel Chemie and first used in a refinery at Vohburg, Germany, in 1986. Four units were operating and one was under construction in 1988.

> *Hydrocarbon Process.*, 1988, **67**(9), 83.

Ethoxene A catalytic process for dehydrogenating ethane to ethylene. Developed by Union Carbide. Acetic acid is a minor by product.

Hu, Y. C., in *Chemical Processing Handbook,* McKetta, J. J., Ed., Marcel Dekker, New York, 1993, 810.

Ethylbenzene Also called UOP Ethylbenzene. A liquid-phase process for making ethylbenzene by reacting ethylene with benzene, catalyzed by a zeolite. The process is usually coupled with one for converting ethylbenzene to styrene. Developed by Unocal Corporation and now licensed by UOP and ABB Lummus Crest.

Eureka A process for upgrading bitumen and heavy oils by delayed coking, which yields a heavy pitch rather than a coke. Developed by the Kureha Chemical Industry Company and operated in Japan since 1976 and in China since 1988.

Takahashi, R. and Washimi, K. *Hydrocarbon Process.,* 1976, **55**(11), 93.
Aiba, T., Kaji, H., Suzuki, T., and Wakamatsu, T., *Chem. Eng. Prog.,* 1981, **77**(2), 37.

EVA-ADAM [Einzelrohr Verzugsanlage/ADA Methanator] Also called ADAM-EVA. A cyclic process for transporting energy, produced in a nuclear reactor, by a gas pipeline. Heat from the reactor is used for *steam reforming methane. The gaseous products are passed along a pipe, at the end of which the reaction is catalytically reversed. Developed by Kernforschungsanlage Jülich, Germany, and piloted there from 1979 to 1981.

Parmon, V. N., *Catal. Today,* 1997, **35**(1-2), 153.
Weissermel, K. and Arpe, H.-J. *Industrial Organic Chemistry,* 3rd ed., VCH Publishers, Weinheim, Germany, 1997, 21.

Excer A process for making uranium tetrafluoride by electrolytic reduction of a uranyl fluoride solution, precipitation of a uranium tetrafluoride hydrate, and ignition of this.

EXOL N Also called EXOL N Extraction. A solvent extraction process for purifying feedstocks for making lubricating oil. The solvent is N-methyl pyrrolidone. Developed and licensed by Exxon Research & Engineering Company. Seventeen units had been installed by 1994.

Bushnell, J. D. and Fiocco, R. J., *Hydrocarbon Process.,* 1980, **59**(5), 119.
Hydrocarbon Process., 1994, **73**(11), 140.

Extafax An early thermographic copying process.

Exxon Donor Solvent Also known as EDS. A coal liquifaction process in which coal in solution in tetrahydronaphthalene is hydrogenated, using a cobalt/molybdenum/alumina catalyst. So-called because the hydrogen is "donated" by the tetrahydronaphthalene to the coal. Developed from the *Pott-Broche process. Piloted by Exxon Research & Engineering Company in the 1970s and operated at 250 ton/day in the Exxon refinery in Baytown, TX, from 1980 to 1982.

Furlong, L. E., Effron, E., Vernon, L. W., and Wilson, E. L., *Chem. Eng. Prog.,* 1976, **72**(8), 69.
Eur. Chem. News, 1982, **39**(1047), 11.
Maa, P. S., Trachte, K. L., and Williams, R. D., *The Chemistry of Coal Conversion,* Schlosberg, R. H., Ed., Plenum Publishing, New York, 1985.

Exxpol [Exxon **pol**ymerization] A gas-phase process for making polyethylene from ethylene. The process uses single-site catalysis (SSC), based on a zirconium metallocene catalyst. Developed by Exxon Chemical Company in 1990 with plans to be commercialized in 1994.

Eur. Chem. News, 1992, **57**(1514), 24,27.

F

Falconbridge   Also called the matte leach process. A process for extracting copper and nickel from matte (a sulfide ore that has been roasted to remove most of the sulfur). Most of the nickel is leached out with hydrochloric acid and recovered as nickel chloride crystals. The leach residue is roasted and leached with sulfuric acid to dissolve the copper. The process has been operated in Canada and Norway since 1970.

> Thornhill, P. G., Wigstol, E., and Van Weert, G., *J. Met.,* 1971, **23**(7), 13.
> Burkin, A. R., *Extractive Metallurgy of Nickel,* John Wiley & Sons, Chichester, England, 1987, 121.
> Gupta, C. K. and Mukherjee, T. K., *Hydrometallurgy in Extraction Processes,* Vol. 1, CRC Press, Boca Raton, FL, 1990, 15,117.
> Hill, J., in *Insights into Speciality Inorganic Chemicals,* Thompson, D., Ed., Royal Society of Chemistry, Cambridge, 1995, 18.

Fan steel   A process for extracting tungsten from wolframite, $FeWO_4$. The ore is mixed with sodium carbonate and heated to 800°C, forming sodium tungstate. This is leached out and treated with calcium chloride, precipitating calcium tungstate. The metal is produced via tungstic acid, ammonium tungstate and tungstic oxide.

FAST   A *BAF process.

> Stephenson, T., Mann, A., and Upton, J., *Chem. Ind. (London),* 1993, (14), 533.

FASTMELT   *See* FASTMET.

FASTMET   A *DR process, using pulverized coal and iron ore fines, heated in a rotary hearth furnace. Under development by Midrex Corporation and Kobe Steel from 1991; a pilot plant was operated by Kobe Steel in Japan in 1996.

A variation, known as FASTMELT, conveys the hot iron powder to an adjacent melter. *See also* Midrex.

Fauser   An early process for making ammonia. Developed by G. Fauser in Italy in 1924.

> Vancini, C. A., *Synthesis of Ammonia,* translated by L. Pirt, Macmillan Press, Basingstoke, England, 1971, 230.

Fauser-Montecatini   A *reforming process for making *syngas from heavy hydrocarbons by gasifying with preheated steam and oxygen. Widely operated in Europe and Asia in the 1960s.

FBA   [Fixed bed alkylation]   An *alkylation process developed by Amoco Corporation and Haldor Topsoe. To be demonstrated at Amoco's Yorktown, VA, refinery in 1997 to 1998.

> *Oil & Gas J.,* 1996, **94**(14), 69.

FCC   [Fluid catalytic cracking]   A process for converting various heavy liquid petroleum fractions into high-octane gasoline and other fuels. Developed by Universal Oil Products (now UOP) and several oil companies in the 1930s and first commercialized by Standard Oil of New Jersey (now Exxon) at Baton Rouge, LA, in 1942. Continuously improved since then, especially in the mid-1960s with the replacement of the original silica-alumina catalyst by a zeolite. The catalyst is now typically a zeolite Y, bound in a clay matrix. The feed is vaporized and contacted in a pipeline reactor with co-currently flowing microspheroidal catalyst particles. The catalyst is then separated from the hydrocarbon products and is continuously regenerated by burning off the coke in a fluidized bed. The process is licensed by UOP and several hundred units are in operation worldwide.

Enos, J. L., *Petroleum Progress and Profits,* MIT Press, Cambridge, MA, 1962, Chap. 6.
Venuto, P. B. and Habib, E. T., Jr., *Fluid Catalytic Cracking with Zeolite Catalysts,* Marcel Dekker, New York, 1979.
Tajbl, D. G., in *Handbook of Petroleum Refining Processes,* Meyers, R. A., Ed., McGraw-Hill, New York, 1986, 2-9.
Magee, J. S. and Mitchell, M. M., Jr., *Fluid Catalytic Cracking: Science and Technology,* Elsevier, Amsterdam, 1993.
Dwyer, J. and Rawlence, D. J., *Catal. Today,* 1993, **18**(4), 487.
Sadeghbeigi, R., *Fluid Catalytic Cracking Handbook,* Gulf Publishing, Houston, TX, 1995.

FEAST [Further exploitation of advanced Shell technology]   Not a single process but a range of processes for converting cyclic di-olefins into alpha-omega dienes. The catalyst is based on rhenium on alumina. Operated in France since 1986.

Chem. Eng. (N.Y.), 1987, **94**(11), 22.
Chaumont, P. and John, C. S., *J. Mol. Catal.,* 1988, **46**, 317.

Feld   Also called Thionite. An early process proposed for removing hydrogen sulfide and ammonia from coal gas by absorption in an aqueous solution of ammonium thionates. Investigated by W. Feld in Germany in the early 1900s; operated at Königsberg, but never fully developed.

German Patent 237,607.
Kohl, A. L. and Riesenfeld, F. C., *Gas Purification,* 4th ed., Gulf Publishing, Houston, TX, 1985, 484.

Fernbach–Strange–Weizmann   *See* Weizmann.

Ferrite   *See* Löwig.

Ferrofining   A mild *hydrotreating process for purifying lubricating oils. The catalyst contained cobalt, molybdenum, and iron (hence the name). Developed by the British Petroleum Company and first operated in Dunkirk, England in 1961.

Dare, H. F. and Demeester, J., *Pet. Refin.,* 1960, **39**(11), 251.
Hydrocarbon Process., 1964, **43**(9), 187.
Unzelman, G. H. and Wolf, C. J., in *Petroleum Processing Handbook,* Bland, W. F. and Davidson, R. L., Eds., McGraw-Hill, New York, 1967, 3-45.

FERROSEP   A magnetic process for removing iron contaminants from petroleum residues before desulfurization. Developed by Nippon Oil Company and Nippon Petroleum Refining Company and operated in Japan since 1992.

Japanese Patent H5 35754.

Ferrox   A process for removing hydrogen sulfide from petroleum refining streams by absorption in an aqueous solution of sodium carbonate containing suspended ferric hydroxide. The absorbent is regenerated by blowing air through it, producing elementary sulfur. The process was invented in 1921 by F. D. Mann, Jr., at the Standard Development Company, and subsequently developed by the Koppers Company and widely used; later it was replaced mainly by the *Thylox process. *See also* Gluud.

U.S. Patents 1,525,140; 1,841,419.
Kohl, A. L. and Riesenfeld, F. C., *Chem. Eng. (N.Y.),* 1959, **66**, 152.
Kohl, A. L. and Riesenfeld, F. C., *Gas Purification,* 4th ed., Gulf Publishing, Houston, TX, 1985, 489.

Fersona   A process for stabilizing the calcium sulfite/sulfate waste from *FGD processes, so that it may be used for landfill. The waste is mixed with ferric sulfate waste from another process (*e.g.*, metallurgical leaching) to form sparingly soluble basic sodium ferric sulfates. Developed in the 1970s at the Battelle Columbus Laboratories, OH, under contract with Industrial Resources. *See also* Sinterna.

> U.S. Patents 3,876,537; 3,984,312; 4,034,063.
> Dulin, J. M., in *Toxic and Hazardous Waste Disposal*, Pojasek, R. J., Ed., Ann Arbor Science, Ann Arbor, MI, 1979, Chap. 18.

FGD   [Flue gas desulfurization]   *See* *flue-gas desulfurization.

Fina/Badger   A process for making styrene by dehydrogenating ethylbenzene. The reaction takes place at high temperature, low pressure, and in the presence of steam and a proprietary heterogeneous catalyst. The hydrogen produced is used to provide process heat. Developed by The Badger Company in the 1960s, first operated on a large scale by Union Carbide Corporation at Seadrift, TX, and now widely used.

FINGAL   [Fixation in Glass of Active Liquors]   A batch process for immobilizing nuclear waste in a borosilicate glass. Developed by the United Kingdom Atomic Energy Authority from 1958 and piloted at its Windscale Works 1962–1966. After a lapse of several years, the project was resumed in 1972 under the acronym *HARVEST.

> Lutze, W., in *Radioactive Waste Forms for the Future*, Lutze, W. and Ewing, R. C., Eds., North-Holland, Amsterdam, 1988, 14.

FINMET   An ironmaking process now used by BHP in Australia and under construction in Venezuela.

Finsider   *See* DR.

Fior   Also called Esso Fior. A direct reduction ironmaking process, using natural gas as the reductant, in a fluidized bed. Operated in Venezuela since 1976. Licensed by Davy Corporation. *See* DR.

FIPS   [Fission Product Solidification]   A process for immobilizing the radioactive waste products from the thorium fuel cycle in a borosilicate glass for long-term storage. Developed at Kernforschungsanlage Jülich, Germany, from 1968, until abandoned in favor of *PAMELA in 1977.

> Lutze, W., in *Radioactive Waste Forms for the Future*, Lutze, W. and Ewing, R. C., Eds., North-Holland, Amsterdam, 1988, 8.

Fischer   A process for removing hydrogen sulfide from coal gas by absorption in an aqueous solution of potassium ferrocyanide and bicarbonate; the solution is regenerated electrochemically with the production of elemental sulfur. Operated at the Hamburg gasworks in the 1930s.

> Müller, H., *Gas Wasserfach.*, 1931, **74**, 653.
> Kohl, A. L. and Riesenfeld, F. C., *Gas Purification*, 4th ed., Gulf Publishing, Houston, TX, 1985, 511.

Fischer-Tropsch   A process for converting synthesis gas (a mixture of carbon monoxide and hydrogen) to liquid fuels. Modified versions were known as the *Synol and *Synthol processes. The process is operated under pressure at 200 to 350°C, over a catalyst. Several different catalyst systems have been used at different periods, notably iron-zinc oxide, nickel-thoria on kieselgühr, cobalt-thoria on kieselgühr, and cemented iron oxide. The main products are C_5-C_{11} aliphatic hydrocarbons; the aromatics content can be varied by varying the process conditions. The basic reaction was discovered in 1923 by F. Fischer and

H. Tropsch, working at the Kaiser Wilhelm Institute for Coal Research in Mülheim, Germany. In 1984, Mako and Samuel wrote "The quantity of patents and literature that has appeared on the subject in the past 60 years makes it virtually impossible to retrace stepwise the developments of the Fischer-Tropsch synthesis." The first full-scale plant was built by Ruhr Chemie at Holten, 1934–1936, and by 1939 eight more plants had been built in Germany. All these plants were disabled by air attacks in 1944. Used in the *SASOL coal gasification plant in South Africa. The first modern F-T plant outside Africa was built by Shell in Malaysia in 1993, based on natural gas. In the 1990s, many companies developed related processes for making liquid fuels from *syngas. *See also* Synthine.

German Patent 484,337.
Fischer, F. and Tropsch, H., *Ber. Dtsch. Chem. Ges.*, 1923, **56**, 2428.
Fischer, F. and Tropsch, H., *Brennstoff-Chem.*, 1923, **4**, 193.
Fischer, F. and Tropsch, H., *Brennstoff-Chem.*, 1932, **13**, 62.
Fischer, F. and Tropsch, H., *Brennstoff-Chem.*, 1935, **16**, 1.
Storch, H. H., Golumbic, N., and Anderson, R. B., *The Fischer-Tropsch and Related Syntheses*, John Wiley & Sons, New York, 1951.
Dry, M. E., *Hydrocarbon Process.*, 1982, **61**(8), 121.
Anderson, R. B., *The Fischer-Tropsch Synthesis*, Academic Press, Orlando, FL, 1984.
Mako, P. F. and Samuel, W. A., *Handbook of Synfuels Technology*, Meyers, R. A., Ed., McGraw-Hill, New York, 1984, Chap. 2.
Dry, M. E., in *Chemicals from Coal: New Processes*, Payne, K. R., Ed., John Wiley & Sons, Chichester, England 1987.
Bartholomew, C. H., in *New Trends in CO Activation*, Guczi, L., Ed., Elsevier, Amsterdam, 1991, 159.
Weissermel, K. and Arpe, H.-J. *Industrial Organic Chemistry*, 3rd ed., VCH Publishers, Weinheim, Germany, 1997, 21.
Chem. Eng. (N.Y.), 1997, **104**(4), 39.

Flakt-Boliden   A variation on the *Citrate process for *flue-gas desulfurization in which the sulfur dioxide is removed from the citrate solution by vacuum. Developed by Flakt, United States, and piloted in 1980 at the TVA Electric Power Research Institute, Muscle Shoals, AL.

Fläkt-Hydro   A *flue-gas desulfurization process which uses seawater as the scrubbing liquor. The pH of the effluent is raised with calcium hydroxide before it is discharged to sea. Developed by Norsk Viftefabrikk. Now owned by ABB Fläkt Industri. As of 1996, 16 plants had been installed worldwide.

Bafy, R., Coughlan, J., and Reynolds, S. K., *Env. Protect. Bull.*, 1991, **12**, 21.
Radojevic, M., *Chem. Br.*, 1996, **32**(11), 47.

Flame-Smelting   *See* DR.

FLC   [Flame chamber]   A high-temperature process for pyrolyzing solid waste. The granulated waste is passed down the axial space of a vertical, co-axial reactor, and heated by a central combustion chamber.

Martinez, D., in *Chemical Waste Handling and Treatment*, Muller, K. R., Ed., Springer-Verlag, Berlin, 1986, 150.

Fleming   An early liquid-phase, thermal process for cracking petroleum. *See also* Dubbs.

Flesch-Winkler   *See* Winkler.

Fletcher   A new process proposed for making titanium dioxide pigment from ilmenite, based on its dissolution in hydrochloric acid. Developed by Fletcher Titanium Products, a subsidiary of Fletcher Challenge, a large paper company in New Zealand, based on original work performed in 1979 at the New Zealand Department of Scientific and Industrial Research, and the Victoria University of Wellington. The hydrochloric acid was available in the paper mills, and the intended product was anatase, suitable for use in pigmenting paper. In 1989 a pilot unit at Gracefield on the North Island was said to be producing at the rate of 3 tons per year, and a unit for producing 3,000 tons had been designed. The process was sold to Sherwin-Williams Company in 1989 and development was continuing in 1992.

> *Eur. Chem. News,* 1988, **51**(1354), 24.
> *Ind. Miner. (London),* 1989, (257), 13.
> *Chem. Week,* 1989, **114**(16), 9.

FLEXICOKING   A process for thermally cracking the residues from petroleum distillation. It combines fluidized-bed (or fixed-bed) coking of residiuum with coke gasification. The products are mainly C_4 hydrocarbons. Developed from the *FLUID COKING process by Esso Research & Engineering Company in the early 1970s and licensed by that company and the Union Oil Company of California. Five units were operating in 1988.

> *Oil Gas J.,* 1975, **73**(10), 53.
> Allan, D. E., Metrailer, W. J., and King, R. C., *Chem. Eng. Prog.,* 1981, **77**(12), 40.
> *Hydrocarbon Process.,* 1994, **73**(11), 96.
> Roundtree, E. M., in *Handbook of Petroleum Refining Processes,* Meyers, R. A., Ed., McGraw-Hill, New York, 1997, 122.3.

Flexicracking   A version of the *FCC process developed by Exxon Research & Engineering Company. Seventeen units were operating in 1996.

> *Hydrocarbon Process.,* 1996, **75**(11), 121.

Flexomer   A gas-phase process for making ethylene-propylene co-polymers. Developed by Union Carbide Corporation and first commercialized in 1989.

FLEXSORB   A group of gas-treating processes using proprietary hindered amines, developed by Exxon Research & Engineering Company and announced in 1983. FLEXSORB SE removes hydrogen sulfide; FLEXSORB HP removes carbon dioxide; and FLEXSORB PS removes both gases. Twenty-nine units were operating in 1992.

> Goldstein, A. M., Edelman, A. M., and Ruziska, P. A., in *Acid and Sour Gas Treating Processes,* Newman, S. A., Ed., Gulf Publishing, Houston, TX, 1985, 319.
> *Hydrocarbon Process.,* 1992, **71**(4), 100.
> *Hydrocarbon Process.,* 1996, **75**(4), 118.

Flintshire   An early lead-smelting process in which galena was roasted in a reverberatory furnace.

flue-gas desulfurization   Often abbreviated to FGD. A general term for the removal of sulfur dioxide from the off-gases from power stations and smelters. Many processes for accomplishing this have been developed; those with special names that are described in this dictionary are: Abgas-Turbo-Wascher von Kroll, ADVACATE, ASARCO, Battersea, BF/Uhde, Bischoff, CEC, Citrate, Citrex, CONOSOX, CT-121, CZD, DAP-Mn, Desonox, Desox, Dowa, DRYPAC, Dual Alkali, Ebara, ELSE, E-SOX, Flakt-Boliden, Flakt-Hydro, Formate (2), FW-BF, HALT, Holter, Howden, Ispra Mark 13A, Kranz MWS, LIFAC, LIMB, Molten carbonate, Nahcolite, Neutrec, NOXSO, Pox-O-Tec, RCE, Reinluft, Saarburg-Holter, SDA, SGFD, SHU, SNOX, SOLINOX, Stackpol, Stone & Webster/Ionics, Sulfidine, SULF-X, Sultrol, Thoroughbred, Walther, Wellman-Lord, WSA, WSA-SNOX, WUK.

> Ashley, M. J. and Greaves, R. A., *Chem. Ind. (London),* 1989, (3), 60.
> Wieckowska, J., *Catal. Today,* 1995, **24**(4), 405.

FLUID COKING A noncatalytic, thermal process for converting bitumen and coal liquids to lighter hydrocarbon fluids and gases. Developed by the Exxon Research & Engineering Company and used commercially since 1954. *See also* FLEXICOKING.

> Massenzio, S. F., in *Handbook of Synfuels Technology,* Meyers, R. A., Ed., McGraw-Hill, New York, 1984, 6-5.
> Allan, D. E., Metrailer, W. J., and King, R. C., *Chem. Eng. Prog.,* 1981, **77**(12), 40.

Fluid Hydroforming An early catalytic *reforming process in which the catalyst was used in a continuously regenerated fluidized bed. Developed by the MW Kellogg Company.

> Ciapetta, C. F., Dobres, R. M., and Baker, R. W., in *Catalysis,* Vol. **6,** Emmett, P. H., Ed., Reinhold Publishing, New York, 1958, 495.

Fluohmic Also called Shawinigan. A process for making hydrogen cyanide by passing a mixture of ammonia and methane through a fluidized bed of coke. The bed is heated by passing an electric current through it, hence the name. The process is economic only where cheap electricity is available. Operated by Shawinigan Chemicals at Shawinigan Falls, Quebec, from 1960 to 1968.

> Dowell, A. M., III, Tucker, D. H., Merritt, R. F., and Teich, C. I., in *Encyclopedia of Chemical Processing and Design,* McKetta, J. J. and Cunningham, W. A., Eds., Marcel Dekker, New York, 1988, **27,** 10.

Fluor Econamine *See* Econamine.

Fluorodec [**Fluor**ine on **de**mand by electrolysis] An electrolytic process and apparatus for generating fluorine. The electrolyte is molten KHF_2, the fluorine is liberated at a nickel anode. Offered by Fluorogas, UK.

Fluor Solvent A process for removing carbon dioxide from natural gas and various industrial gas streams by dissolution in propylene carbonate. Carbon dioxide is much more soluble than other common gases in this solvent at low temperatures. The process cannot be used when hydrogen sulfide is present. The process was invented in 1958 by A. L. Kohl and F. E. Miller at the Fluor Corporation, Los Angeles. It is now licensed by Fluor Daniel. The first plant was built for the Terrell County Treating plant, El Paso, TX in 1960; by 1985, 13 plants were operating.

> U.S. Patents 2,926,751; 2,926,752; 2,926,753.
> Kohl, A. L. and Buckingham, P. A., *Pet. Refin.,* 1960, **39,** 193.
> Kohl, A. L. and Riesenfeld, F. C., *Gas Purification,* 4th ed., Gulf Publishing, Houston, TX, 1985, 845.

FMC *See* carbonization.

Folkins A process for making carbon disulfide from methane and sulfur at elevated temperature and pressure. A complex separation system removes the hydrogen sulfide from the products so that this sulfur can be re-used. The process can be operated catalytically or non-catalytically. Developed in 1948 by H .O. Folkins and others at the Pure Oil Company, Chicago.

> U.S. Patent 2,568,121.
> Folkins, H. O., Miller, E., and Hennig, H., *Ind. Eng. Chem.,* 1950, **42,** 2202.

Fondon *See* Cazo.

Footing A process for displacing platinum metals from their chloride solutions by adding metallic iron. *See* cementation.

Footner   *See* metal surface treatment.

Formate (1)   A process for making sodium dithionite by reacting sodium formate with sulfur dioxide in aqueous methanol.

> Bostian, L. C., in *Speciality Inorganic Chemicals,* Thompson, R., Ed., Royal Society of Chemistry, London, 1981, 65.

Formate (2)   A *flue-gas desulfurization process. Potassium formate solution reduces the sulfur dioxide to thiosulfate, and then to hydrosulfide.

> Buckingham, P. A. and Homan, H. R., *Hydrocarbon Process.,* 1971, **50**(8), 121.
> Speight, J. G., *Gas Processing,* Butterworth Heinemann, Oxford, 1993, 289.

Formcoke   *See* carbonization.

Formex   A process for extracting aromatic hydrocarbons from petroleum reformate, using N-formyl morpholine at 40°C. Developed by SNAM-Progetti.

> Cinelli, E., Noe, S., and Paret, G., *Hydrocarbon Process.,* 1971, **51**(4), 141.
> Bailes, P. J., in *Handbook of Solvent Extraction,* Lo, C. C., Baird, M. H. I., and Hanson, C., Eds., John Wiley & Sons, Chichester, England, 1983, 18.2.4.

Formox   [**Form**aldehyde by **ox**idation]   A process for oxidizing methanol to formaldehyde, using a ferric molybdate catalyst. Based on the *Adkins-Peterson reaction, developed by Reichold Chemicals, and licensed by that company and Perstorp, Sweden. Acquired by Dyno Industries in 1989. The process uses formaldehyde produced in this way to make formaldehyde-urea resin continuously. A plant using this process was to be built in Ghent by 1991, owned jointly by Dyno and AHB-Chemie. Licensed to 35 sites worldwide. Several other companies operate similar processes.

> *Chem. Week,* 1990, 7 Mar, 7.
> Weissermel, K. and Arpe, H.-J. *Industrial Organic Chemistry,* 3rd ed., VCH Publishers, Weinheim, Germany, 1997, 37.

Fosbel   [**Fos**eco **Bel**ret]   A ceramic welding process for repairing refractories. A dry mixture of a refractory and a metallic powder is projected, in a stream of oxygen, onto the surface to be repaired. Oxidation of the metal produces the necessary high temperature. Developed by Foseco, UK, and Belret, Belgium, hence the name.

Foster Wheeler-Stoic   *See* Stoic.

Foulis-Holmes   A process for removing hydrogen cyanide from coal gas by scrubbing with a suspension of freshly precipitated ferrous carbonate. Invented by W. Foulis and P. Holmes in England in the 1890s.

> British Patents 9,474 (1893); 15,168 (1896).
> Hill, W. H., in *Chemistry of Coal Utilization,* Vol. 2, Lowry, H. H., Ed., John Wiley & Sons, New York, 1945, 1097.

Four-step   *See* Aldol.

FR   *See* Titanox FR.

Frank-Caro   Also called the Cyanamide process. An early process for fixing atmospheric nitrogen. Lime and carbon were heated to produce calcium carbide; this was reacted with nitrogen to give calcium cyanamide, which was hydrolyzed with steam to yield ammonia and calcium carbonate. Developed by A. Frank and N. Caro from 1895 at Dynamit, Germany, and used in Germany, Norway, and Italy until it was replaced by the *Haber-Bosch process after World War I.

> German Patents 88,363; 108,971.
> Vancini, C. A., *Synthesis of Ammonia,* translated by L. Pirt, Macmillan Press, Basingstoke, England, 1971, 2.

Morris, P. J. T., *Chem. Ind. (London),* 1983, (18), 710.

Frasch (1)   A process for extracting sulfur from underground deposits, developed by H. Frasch between 1890 and 1902 at Sulfur Mine, LA. Three concentric pipes are inserted into a hole drilled into the deposit. The outermost pipe carries water superheated to 140 to 165°C, which melts the sulfur; hot air is forced down the central pipe, which forces the molten sulfur up through the intermediate annular space. Only a small proportion of sulfur deposits have the appropriate geology for extraction in this way. Because of this invention, sulfur came to be exported from America to Europe, instead of from Sicily to America. In 1991 the process was operated in the United States, Mexico, Poland, and Iraq.

U.S. Patents 461,429; 461,430; 461,431.
J. Soc. Chem. Ind., 1914, **33,** 539.
Haynes, W., *The Stone that Burns,* D. Van Nostrand, New York, 1942.
Shearon, W. H., Jr. and Pollard, J. H., *Ind. Eng. Chem.,* 1950, **42,** 2188.
Loughbrough, R., *Ind. Miner. (London),* 1991, July, 19.

Frasch (2)   A process for removing sulfur compounds from petroleum fractions by distillation from copper oxide.

Freeman   *See* DR.

French   Also known as the Indirect process. A process for making zinc oxide, in the form of a white pigment, from metallic zinc. The zinc is melted and vaporized in a current of carbon monoxide. The vapor is oxidized with air in a second chamber, forming zinc oxide and carbon dioxide. The fume passes through a settling chamber, where oversized particles settle out, and from there go to a bag house, where the product is collected. Confusingly, the French process has been operated by North American Oxide Company, at Clarksville, TN. *See also* American.

Fresnel   A predecessor of the *Solvay process, proposed by A. J. Fresnel in France in 1811.

Cohen, J. M., *The Life of Ludwig Mond,* Methuen, London, 1956, 267.

Fricker   A process for producing zinc oxide by direct oxidation of zinc vapor. Operated by the Fricker's Metal Company at Luton and Burry Port, UK, in the 1920s and 30s, subsequently acquired by the Imperial Smelting Corporation. Also operated by the Anglo American Corporation, South Africa, after World War II.

Cocks, E. J. and Walters, B., *A History of the Zinc Smelting Industry In Britain,* George G. Harrap, London, 1968, 28,93,95,140.

Fritz Winkler   *See* Winkler.

Fröhler   *See* Munich.

F-S   [Ferrous sulfate]   A process for removing ammonia, hydrogen sulfide, and hydrogen cyanide from coke-oven gas by scrubbing with aqueous ferrous sulfate solution obtained from steel pickling. A complex series of reactions in various parts of the absorption tower yield ammonium sulfate crystals and hydrogen sulfide (for conversion to sulfur or sulfuric acid) as the end products. Developed in Germany by F. J. Collin A.G.

Dixon, T. E., *Iron Age,* 1955, **175**(12), 91.
Kohl, A. L. and Riesenfeld, F. C., *Gas Purification,* 4th ed., Gulf Publishing, Houston, TX, 1985, 563.

F-T   *See* Fischer-Tropsch.

FTC   [Fluid-bed thermal cracking]   A continuous thermal cracking process for making synthesis gas from heavier petroleum fractions.

Full cell   *See* Bethell.

Fumaks   A process for removing hydrogen sulfide from coke-oven gas by oxidation with picric acid. Developed by Sumitomo Metals Industries, and used in 11 units in Japan from 1972 to 1997.

Kohl, A. L. and Riesenfeld, F. C., *Gas Purification,* 4th ed., Gulf Publishing, Houston, TX, 1985, 539.

Fumaks-Rhodaks   A combination of the *Fumaks and *Rhodaks processes for removing both hydrogen sulfide and hydrogen cyanide from coke-oven gas. Developed by the Osaka Gas Company and marketed by Sumitomo Chemical Engineering Company.

Furnace   Also called the Readman process. A process for making elementary phosphorus in an electric furnace. The raw materials are phosphate rock, coke, and silica. Large amounts of electric power are consumed, so the process is economic only where cheap power is available. The overall reaction is:

$$2Ca_3(PO_4)_2 + 6SiO_2 + 10C = 6CaSiO_3 + 4P + 10CO$$

Invented by J. B. Readman in Edinburgh in 1888 and now operated on a very large scale in at least ten countries.

British Patent 14,962 (1888).

Hartlapp, G., in *Phosphoric Acid,* Vol. 1, Part II, Slack, A. N., Ed., Marcel Dekker, New York, 1968, 927.

Childs, A. F., in *The Modern Inorganic Chemicals Industry,* Thompson, R., Ed., The Royal Society of Chemistry, London, 1977, 378.

Hocking, M. B., *Modern Chemical Technology and Emission Control,* Springer-Verlag, Berlin, 1985, 184.

Furnace Black   One of the three principal processes used for making carbon black; the others being the *Thermal Black and the *Channel Black processes. In the Furnace Black process, aromatic fuel oils and residues are injected into a high velocity stream of combustion gases from the complete burning of an auxiliary fuel with an excess of air. Some of the feedstock is burned, but most of it is cracked to yield carbon black and hydrogen. The products are quenched with water.

Franck, H.-G. and Stadelhofer, J. W., *Industrial Aromatic Chemistry,* Springer-Verlag, Berlin, 1988, 384.

Kühner, G. and Voll, M., in *Carbon Black Science and Technology,* Donnet, J.-B., Bansai, R. C., and Wang, M. J., Eds., Marcel Dekker, New York, 1993, 14.

Fuse-quench   Also called the Kjellgren-Sawyer process. A process for extracting beryllium from beryl. The beryl is fused at 1,600°C and then rapidly quenched by pouring through a water jet of high velocity. The glassy product is heated to 900°C to precipitate beryllia from its solid solution in silica, and then extracted with sulfuric acid. Operated by the Brush Beryllium Company, Cleveland, OH.

Schwenzfrier, C. W., Jr., in *The Metal Beryllium,* White, D. W., Jr. and Burke, J. E., Eds., American Society for Metals, Cleveland, OH, 1955.

Everest, D. A., *The Chemistry of Beryllium,* Elsevier, Amsterdam, 1964, 109.

FW-BF   [Foster Wheeler-Bergbau-Forschung]   A dry *flue-gas desulfurization process that combines the sulfur removal system of Bergbau-Forschung, which uses a carbon adsorbent, with the Foster-Wheeler process for oxidizing adsorbed sulfur to sulfuric acid.

Habib, Y. and Bischoff, W. F., *Oil Gas J.,* 1975, **75**(8), 53.

Bischoff, W. F. and Habib, Y., *Chem. Eng. Prog.,* 1975, **71**(5), 59.

Sulphur, 1975, (119), 24.

Env. Sci. Technol., 1975, **9,** 712.

G

Galoter   A process for extracting oil and gas from shale, using a vertical retort. Operated in Estonia since 1964.

> Smith, J. W., in *Handbook of Synfuels Technology,* Meyers, R. A., Ed., McGraw-Hill, New York, 1984, 4-177.

Gardinier   A process for making monoammonium phosphate from gaseous ammonia and phosphoric acid by mixing them in a helical reactor. *See also* Swift.

Gardner   A process for making magnesium metal by reducing magnesium sulfide, made from magnesium oxide, with calcium carbide in the presence of a flux of calcium and aluminum chlorides. Invented by D. Gardner in 1895 and operated by Murex, UK, in the 1930s.

> British Patent 465, 421.

Garrett   *See* ORC.

Garrigue   A process for recovering glycerol from the residual liquor from the *kettle soapmaking process. After separating the solid soap, the liquor is treated with aluminum sulfate, thereby precipitating the residual carboxylic acids as their insoluble aluminum salts. After removing these by filtration, the liquor is concentrated by vacuum evaporation and the glycerol distilled out under vacuum. *See also* Van Ruymbeke.

> Martin, G. and Cooke, E. I., in *Industrial and Manufacturing Chemistry,* Cooke, E. I., Ed., Technical Press, Kingston Hill, Surrey, 1952, 126.

Gas   [Gas anti-solvent]   A process for separating dissolved materials by selective precipitation with added supercritical carbon dioxide. First used for recrystallizing the explosive RDX; subsequently used for recrystallizing other explosives, pharmaceuticals, fine chemicals, and food products.

> McHugh, M. A. and Krukonis, V. J., in *Supercritical Fluid Extraction: Principles and Practice,* 2nd. ed., Butterworth-Heinemann, Boston, 1994, 342.
> Brennecke, J. F., *Chem. Ind. London.,* 1996, (21), 831.

GasCat   A process for converting natural gas to liquid fuels. Essentially an updated *Fischer-Tropsch process, it is being developed by Energy International, a division of Williams Companies, Tulsa, OK. It first produces *syngas by *reforming natural gas and oxygen, and then passes this gas into a slurry bubble-column reactor containing a cobalt oxide catalyst. Not commercialized as of 1997.

> Singleton, A. H., *Oil & Gas J.,* 1997, **95** (31), 68.

Gasmaco   [Gas Machinery Company]   A process for making a substitute natural gas from petroleum fractions and residues by thermal cracking. Developed from the *Hall (2) process in the 1940s by the American Gas Association. *See also* Petrogas.

> Crane, K. J., *J. Inst. Fuel,* 1957, **30**, 661.
> Claxton, G., *Benzoles, Production and Uses*, National Benzole & Allied Products Association, London, 1961, 97.
> *Gasmaking,* British Petroleum Co., London, 1965, 70.

GAS/SPEC FT   *See* Econamine FG.

GAS/SPEC CS-Plus   A process for removing carbon dioxide and hydrogen sulfide from natural gas by washing with a solution of a special amine. Developed and offered by Dow Chemical Company. Operated since 1988.

Oil & Gas J., 1996, **94**(8), 38.

GASTAK *See* Purasiv HR.

Gastechnik A process for removing hydrogen sulfide and organic sulfur compounds from coal gas by absorption on formulated iron oxide pellets which flow by gravity down large absorption towers. The pellets are regenerated by removing their sulfur, either by atmospheric oxidation or by extraction with tetrachloroethylene. Developed by Gastechnik, Germany, in the 1950s and widely operated there and in England.

British Patents 433, 823; 683, 432.
Moore, D. B., *Gas World*, 1956, **143**, 153.
Egan, P. C., *Gas World*, 1957, **145**, 136.
Claxton, G., *Benzoles, Production and Uses*, National Benzole & Allied Products Association, London, 1961, 209.
Kohl, A. L. and Riesenfeld, F. C., *Gas Purification*, 4th ed., Gulf Publishing, Houston, TX, 1985, 429.

Gasynthan A process for making synthetic natural gas from naphtha by a two-stage *steam reforming process. Developed by Lurgi and BASF in the 1960s. In 1975, over 30 units were operating.

Jockel, H. and Triebskorn, B. E., *Hydrocarbon Process.*, 1973, **52**(1), 93.
Hydrocarbon Process., 1975, **54**(4), 114.

GEESI [General Electric Environmental Services] A process for making ammonium sulfate from the sulfur dioxide present in flue-gas.

Eur. Chem. News, 1996, **65**(1720), 23.

GEGas [General Electricity gas] A coal gasification process intended for producing gas for combined-cycle power generation. Powdered coal, with steam and air, is fed to a mechanically stirred gasifier. Piloted by the General Electric Company, United States, in the 1970s.

Geigy A one-stage process for making ethylenediamine tetra-acetic acid (EDTA) from ethylenediamine, hydrogen cyanide, and formaldehyde.

Anderson, E. V. and Gaunt, J. A., *Ind. Eng. Chem.*, 1960, **52**, 191.

GEMINI A family of gas separation processes based on selective adsorption. GEMINI 5 is for upgrading raw landfill gas to methane. GEMINI 9 is for producing hydrogen and carbon dioxide from *steam reformer off-gas and is used in Butler, PA. GEMINI Hyco separates carbon monoxide and hydrogen from *syngas. Developed by S. Sircar at Air Products & Chemicals in 1978.

U.S. Patents 4,171, 206; 5,073,356.
Kratz, W. C., Rarig, D. L., and Pietrantonio, J. M., in *Adsorption and Ion Exchange: Fundamentals and Applications,* LeVan, M. D., Ed., American Institute of Chemical Engineers, New York, 1998, 36.
Kumar, R. and Van Sloun, J. K., *Chem. Eng. Prog.*, 1989, **85**(1), 34.
Hydrocarbon Process., 1996, **75**(4), 118.

Geminox A direct process for converting butane to 1,4-butanediol. The butane is first oxidized in the gas phase to maleic anhydride, using BP's fluidized bed technology. The maleic anhydride is scrubbed with water and then catalytically dehydrogenated to butanediol. Developed in 1994 by BP Chemicals and Lurgi. Modifications of the process can be used to make tetrahydrofuran and γ-butyrolactone. The first plant will probably be built on BP's site at Lima, OH, for completion in 2000.

Chem. Eng. (N.Y.), 1995, **102**(8), 17.
Chem. Eng. (Rugby, England), 1997, (638), 26.

Generon Not a process, but a trademark used by BOC and Dow to designate their processes for separating nitrogen from air by either the *PSA process or a membrane process. The PSA process is based on the *Bergbau-Forschung process. The membrane process uses the GENERON HP membrane developed by the Dow Chemical Company. *See also* Novox.

GEODE [General Electric organic destruction] A development of the *Ultrox process in which a combination of ozone and ultraviolet radiation is used to oxidize traces of organic compounds in water. Developed by the General Electric Company and demonstrated at the Commonwealth Edison nuclear power plant at Dresden, IL, in 1989. The requirement was to reduce the concentration of total organic carbon in the process and makeup waters to the low parts-per-billion range.

   Head, R. A., Alexander, J. E., and Lezon, R. J., *Nucl. Eng. Int.*, 1989, **4**, 40.

GFETC A coal gasification process based on a slagging gasifier, developed for the U.S. Department of Energy, at Grand Forks, ND.

Ghaem A *DR process developed by the Esfahan Steel Company, Iran, in 1996.

GI *See* carbonization.

Giammarco-Vetrocoke Also known as G-V. Two processes are known by this name, both using an aqueous solution of sodium or potassium carbonate and arsenite to absorb acid gases. In one process, the solution is used to extract carbon dioxide from natural gas or synthesis gas. In the other, hydrogen sulfide is extracted from coke-oven or synthesis gas, yielding elemental sulfur from a complex sequence of reactions. The process was invented by G. Giammarco and developed by Giammarco-Vetrocoke, Italy. In 1992, more than 200 plants were operating. *See also* Benfield, Carsol, CATACARB, Hi-pure.

   U.S. Patents 2,840,450; 3,086,838.
   Giammarco, G., in *Ammonia*, Part 2, Slack, A. V., and James, G. R., Eds., Marcel Dekker, New York, 1974, 171.
   Hydrocarbon Process., 1975, **54**(4), 90.
   Kohl, A. L. and Riesenfeld, F. C., *Gas Purification,* 4th ed., Gulf Publishing, Houston, TX, 1985, 238, 505.
   Hydrocarbon Process., 1992, **71**(4), 103.

Gibbs A process for oxidizing naphthalene to phthalic anhydride, using air as the oxidant and catalyzed by vanadium pentoxide. Invented in 1917 by H. D. Gibbs and C. Conover.

   U.S. Patent 1,285,117.

Girbotol [Girdler Bottoms] Also spelled **Girbitol**. A gas scrubbing process using an alkanolamine as the absorbent. Used for removing hydrogen sulfide from refinery and natural gases, and carbon dioxide from hydrogen and combustion products. The gases are subsequently removed by steam stripping. Monoethanolamine, diethanolamine, and triethanolamine have all been used. The process depends on the reversible formation of an amine carbonate:

$$2HO \cdot C_2H_4 \cdot NH_2 + H_2O + CO_2 \rightleftharpoons (HO \cdot C_2H_4 \cdot NH_3)_2CO_3$$

Invented in 1930 by R. G. Bottoms at the Girdler Corporation, Louisville, KY. In 1950 it was the most commonly used process for removing hydrogen sulfide from refinery and natural gases.

   U.S. Patent 1,783,901.
   Reed, R. H. and Updegraff, N. C., *Ind. Eng. Chem.,* 1950, **42**, 2269.
   Unzelman, G. H. and Wolf, C. J., in *Petroleum Processing Handbook*, Bland, W. F. and Davidson, R. L., Eds., McGraw-Hill, New York, 1967, 3-132.

Girdler (1)   A process for removing acid gases from gas streams by scrubbing with aqueous triethanolamine. Developed by the Girdler Corporation, Louisville, KY, in the 1920s. *See also* Girbotol.

German Patent 549,556.

Girdler (2)   An improved *Claus process developed by the Girdler Corporation in 1948.

Sands, A. E. and Schmidt, L. D., *Ind. Eng. Chem.*, 1950, **42**, 2277.

GKT   A development of the *Koppers-Totzek coal gasification process.

Glanor   A Chlor-Alkali process using a bipolar diaphragm cell. Developed by PPG Industries and Oronzio de Nora Impianti Elettrochimic in the early 1970s.

Gluud   A process for removing hydrogen sulfide from gas streams by scrubbing with an aqueous solution containing ammonia and a nickel salt. Invented by W. Gluud in 1921.

U.S. Patent 1,597,964.
Gluud, W. and Schonfelder, R., *Chem. Metall. Eng.*, 1927, **34**(12), 742.
Kohl, A. L. and Riesenfeld, F. C., *Gas Purification*, 4th ed., Gulf Publishing, Houston, TX, 1985, 493.

GMD   [Gas to middle distillate]   A process for converting natural gas to diesel oil or synthetic crude oil. The catalyst is cobalt and rhenium on alumina, used in a slurry reactor. Developed by Statoil in the 1980s.

Jens, K.-J., *Appl. Catal. A: Gen.*, 1997, **152**(1), 161.

GM-IX   [Gas membrane-ion exchange]   A process for treating metal cyanide waste solutions, with recovery of both the metal and the cyanide. The solution is first passed through an anion exchange column, removing metal cyanide complexes and cyanide ion. Treatment of the column with 10 percent sulfuric acid releases the metal ions and hydrogen cyanide into solution. Passage of this solution through a gas membrane module, comprising thousands of hollow fibers of microporous polypropylene, strips the hydrogen cyanide gas from the solution. It not only removes the free hydrogen cyanide, it also forces the hydrolysis of metal cyanide complexes until all the cyanide has been removed from the system. The hydrogen cyanide is scrubbed by sodium hydroxide solution, with the resulting sodium cyanide solution being re-used. The metal is recovered from solution by electro-winning. Developed in 1988 at the Department of Civil and Mineral Engineering in the University of Minnesota, where a pilot plant was under construction in May 1989.

Semmens, M. J. and Chang, Y., in *Membrane Separation Processes*, Green, A., Ed., BHRA (Information Services), Cranfield, Bedford, UK, 1989, 167.

GO   [Glycol oxalate]   Also called UBE/UCC. A process for making ethylene glycol from carbon monoxide in three stages involving methyl nitrite and dimethyl oxalate:

$$4CH_3OH + 4NO + O_2 = 4CH_3ONO + 2H_2O$$

$$2CH_3ONO + 2CO = CH_3OOC \cdot COOCH_3 + 2NO$$

$$CH_3OOC \cdot COOCH_3 + 4H_2 = HOCH_2 \cdot CH_2OH + 2CH_3OH$$

All the methanol and nitric oxide are recycled. The process was developed jointly by UBE and Union Carbide Corporation in 1983 and piloted by the latter company.

Saunby, J. B., in *Oxygen and the Conversion of Future Feedstocks*, Royal Society of Chemistry, London, 1984, 235.

Goethite   A process for removing iron from leach liquors from hydrometallurgical leaching operations. Used in recovering zinc from the residues of the electrolytic zinc process. *See also* Jarosite, Haematite.

Morgan, S. W. K., *Zinc and Its Alloys and Compounds*, Ellis Horwood, Chichester, England, 1985, 121.

GO-fining [Gas-oil refining]   A *hydrodesulfurization process adapted for gas oil. The proprietary catalyst is regenerable. Developed by Esso Research & Engineering Company and the Union Oil Company of California and jointly licensed by them. First commercialized at Wakayama, Japan, in 1968; by 1972, nine units had been built.

Hydrocarbon Process., 1970, **49**(9), 210.
Hydrocarbon Process., 1994, **73**(11), 135.

GOLDOX [Gold oxidation]   A process for improving the extraction of gold from its ores by injecting oxygen into the cyanide solution by the *VITOX process. Developed by Afrox, a subsidiary of BOC, and now used widely in South Africa, Zimbabwe, Australia, Canada, and the United States.

Downie, N. A., *Industrial Gases*, Blackie Academic, London, 1997, 308.

Goldschmidt   The German company Th. Goldschmidt A.G., founded in 1847, has developed many processes, including the *Thermite process, and many for extracting and recovering tin. The process with which its name is usually associated is one for purifying tin metal by adding aluminum in order to precipitate out arsenic and antimony. Air is then blown into the molten metal, forming a surface scum containing these impurities, which is removed after cooling. A process for removing tin from tinplate by chlorination has also been known by this name.

German Patent 411,477.
Mantell, C. L., *Tin: Its Mining, Production, Technology, and Application*, Reinhold Publishing, New York, 1949, 147. Published in facsimile by Hafner Publishing, New York, 1970.

Gorham   *See* Paralene.

GPB [Geon process butadiene]   A process for extracting butadiene from C_4 petroleum cracking fractions, developed by Nippon Zeon. It was in use by more than 30 plants in 1985.

Weissermel, K. and Arpe, H.-J. *Industrial Organic Chemistry*, 3rd ed., VCH Publishers, Weinheim, Germany, 1997, 109.

GPP [Gas-phase polypropylene]   A gas-phase process for making polypropylene. Developed by Sumitomo and first commercialized in 1990.

Grainer   Originally called the Michigan process because it was widely used in Michigan for using the waste heat generated at sawmills. It is used for evaporating brine to yield salt. The brine is indirectly heated to below its boiling point and the crystals grow at the surface of the liquid.

Richards, R. B., in *Sodium Chloride,* Kaufmann, D. W., Ed., Reinhold Publishing, New York, 1960, Chap. 12.

Granodizing   *See* metal surface treatment.

Grätzel   An early process for making aluminum by electrolyzing a solution of alumina in molten calcium chloride. The process was also used for making beryllium, using a mixed fluoride melt.

Gravimelt   A coal cleaning process in which coal is heated with molten sodium hydroxide and then washed with acid. The process removes 90 percent of the sulfur and 95 percent of the ash. It was planned to be piloted in 1987.

Chem. Eng. News, 1986, 8 Sept, 32.

Gray   An early petroleum desulfurization process which used a fixed bed of an absorbent catalyst such as fuller's earth. A related process, **Gray Clay Treating,** removed di-olefins and other gum-forming constituents of thermally cracked gasolines by passing their vapors over hot fuller's earth. These processes, developed by T. T. Gray at the Gray Processing Corporation, were first used in the 1920s and abandoned in the 1950s with the advent of *catalytic cracking.

> British Patent 222,481.
> Unzelman, G. H. and Wolf, C. J., in *Petroleum Processing Handbook*, Bland, W. F. and Davidson, R. L., Eds., McGraw-Hill, New York, 1967, 3-61.

Green liquor   A papermaking process using a mixture of sodium hydroxide and sodium carbonate.

GreenOx   A pulp-bleaching process developed by Kemira in Finland.

GRH   A process for increasing the calorific value of a lean gas to make it suitable for use as a town gas, by thermal hydrogasification. Developed by British Gas.

Griesheim (1)   An early process for producing chlorine by electrolysis, developed by Chemische Fabrik Griesheim-Elektron, in Germany, and commercialized in 1890. The electrolyte was saturated potassium chloride solution, heated to 80 to 90°C. The byproduct potassium hydroxide was recovered. The process was superseded in the United States by several similar electrolytic processes before being ousted by the mercury cell, invented by H. Y. Castner and K. Kellner in 1892. *See* Castner-Kellner.

> Kircher, M. S., *Chlorine, Its Manufacture, Properties and Uses*, Sconce, J. S., Ed., Reinhold Publishing, New York, 1962, 85.
> Hocking, M. B., *Modern Chemical Technology and Emission Control*, Springer-Verlag, Berlin, 1985.

Griesheim (2)   A process for making potassium by reducing potassium fluoride with calcium carbide:

$$2KF + CaC_2 = CaF_2 + 2C + 2K$$

Grillo-Schröder   *See* Schröder-Grillo.

Grosvenor-Miller   A two-stage process proposed for making chlorine from hydrogen chloride. In the first stage, the hydrogen chloride reacts with ferric oxide to give ferric chloride:

$$Fe_2O_3 + 6HCl = 2FeCl_3 + 3H_2O$$

In the second, the ferric chloride is oxidized to ferric oxide and chlorine:

$$4FeCl_3 + 3O_2 = 2Fe_2O_3 + 6Cl_2$$

The iron catalyst, on an inert support, is contained in two fixed beds, which alternate in function. Invented by W. M. Grosvenor, Jr., and I. Miller and extensively studied in the 1940s but not commercialized.

> U.S. Patent 2,206,399.
> Redniss, A., in *Chlorine, Its Manufacture, Properties and Uses*, Sconce, J. S., Ed., Reinhold Publishing, New York, 1962, 259.

GRS   A process for making a synthetic rubber by co-polymerizing styrene and butadiene.

Grünstein   A process for hydrating acetylene to acetaldehyde, invented by N. Grünstein in 1910 and developed by Griesheim-Elektron at Rheinfelden, Germany.

Miller, S. A., *Acetylene, Its Properties, Manufacture and Uses*, Vol. 1, Academic Press, New York, 1965, 135.

Morris, P. J. T., *Chem. Ind. (London)*, 1983, (18), 711.

Grzymek   Also known as the sintering/self-disintegration process. A process for making both alumina and cement from aluminous ores and wastes. The ore is mixed with limestone and calcined in a coal-fired rotary kiln. The product spontaneously breaks up into small lumps suitable for leaching by aqueous sodium carbonate. Alumina is precipitated from the leachate by carbon dioxide. The residues are mixed with more limestone and calcined to produce cement clinker. Developed by J. Grzymek and his colleagues in Poland from the 1930s to the 1980s. A plant at Groszowice has been operating since 1966; by 1985 it had produced 70,000 tons of alumina and 600,000 tons of cement.

Grzymek, J., *Process Eng. (London)*, 1974, Feb., 43.

Grzymek, J., *Light Met. Mat. Ind.*, 1976, **2**, 29.

O'Connor, D. J., *Alumina Extraction from Non-bauxitic Materials*, Aluminium-Verlag, Dusseldorf, 1988, 234.

G-S   [Girdler sulphide]   A process for separating hydrogen isotopes, using the equilibrium between water and hydrogen sulfide:

$$HDO + H_2S \rightleftharpoons H_2O + HDS.$$

Several versions of this reaction have been used for separating deuterium and tritium; the G-S version operates in the liquid phase, without a catalyst, at two temperatures. Used at the Savannah River plant of the U.S. Atomic Energy Commission and at several Canadian plants.

Rae, H. K., in *Separation of Hydrogen Isotopes,* Rae, H. K., Ed., American Chemical Society, Washington, D.C. 1978, Chap. 1.

Benedict, M., Pigford, T. H., and Levi, H. W., *Nuclear Chemical Engineering,* 2nd. ed., McGraw-Hill, New York, 1980, 767.

GTG   [gas to gasoline]   A process for converting natural gas to gasoline by oxidative coupling. Ethylene, formed initially, is oligomerized to a gasoline-like distillate over a zeolite catalyst. Under development by the Arco Chemical Company in 1988.

Parkyns, N. D., *Chem. Br.,* 1990, **26**(9), 841

GTL   [Gas to liquids]   A general term for processes which convert natural gas to liquid fuels. The first such plant was that of *Sasol in South Africa. *See also* GTSC, Syntroleum.

Oil & Gas J., 1997, **95**(25), 16.

Oil & Gas J., 1997, **95**(30), 35.

GTSC   [Gas to syncrude]   A process for converting natural gas to a synthetic crude oil which may be mixed with natural crude oil and used in conventional oil refineries. Based on *F-T technology, but using a proprietary slurry bubble column reactor with a promoted cobalt catalyst. Developed by Syncrude Technology, Pittsburgh, PA, in the 1990s.

GUD   [Gas und Dampf—German, meaning gas and steam]   *See* ICG-GUD.

Guggenheim   A process for extracting sodium nitrate from caliche, a native sodium nitrate found in Chile. The ore is leached at 40°C with water containing controlled concentrations of magnesium and calcium sulfates. Operated on a large scale in Chile. *See also* Shanks.

Guillini   A process for making gypsum from the waste product from the *Wet Process for making phosphoric acid. The waste is heated with water in an autoclave; this removes impurities and converts the calcium sulfate dihydrate to the hemi-hydrate.

Guimet   A process for making ultramarine, a blue pigment. A mixture of clay, sodium carbonate, and sulfur is heated in the absence of air, and then slowly cooled while air is admitted over a long period. Invented in 1826, independently, in France by J. B. Guimet, and in Germany by C. B. Gmelin. Guimet won a prize for this invention. It changed the world's pigment industry and was soon operating in 22 European factories. Guimet's son Émile, who inherited the business, collected Asian art and founded the Musée Guimet, now in Paris.

Guimet, J. B., *Ann. Chim. Phys.,* 1831, **46** 431.

Mellor, J. W., *Comprehensive Treatise on Inorganic and Theoretical Chemistry,* Vol. 6, Longmans Green, London, 1925, 588.

Gulf   A process for making ethanol by the simultaneous hydrolysis and fermentation of cellulose.

Gulf HDS   A process for *hydrorefining and *hydrocracking petroleum residues in order to make fuels and feeds for *catalytic cracking. Developed by the Gulf Research & Development Company. *See also* hydrodesulfurization.

Unzelman, G. H. and Wolf, C. J., in *Petroleum Processing Handbook,* Bland, W. F. and Davidson, R. L., Eds., McGraw-Hill, New York, 1967, 3-23.

Gulfining   [**Gulf refining**]   A *hydrodesulfurization process adapted for heavy gas oils. Developed by the Gulf Research & Development Company in the early 1950s.

Hydrocarbon Process., 1970, **49**(9), 211.

Unzelman, G. H. and Wolf, C. J., in *Petroleum Processing Handbook,* Bland, W. F. and Davidson, R. L., Eds., McGraw-Hill, New York, 1967, 3-45.

Gulfinishing   A process for *hydrofining lubricating oils. Developed by Gulf Oil Corporation.

Hydrocarbon Process., 1964, **43**(9), 190.

Gulf Resid   A process for desulfurizing petroleum residues, developed by Gulf Oil Corporation.

Speight, J. G., *The Desulfurization of Heavy Oils and Residua,* Marcel Dekker, New York, 1981, 175.

Gulf SRC   *See* SRC.

Gutehoffnungshütte   [Named after the company of that name in Oberhausen, Germany, now a subsidiary of MAN A.G., Munich]   A process for oxidizing methane to formaldehyde, using nitrogen oxides as the oxidant and sodium borate as the catalyst. Operated at atmospheric pressure and 620°C. Developed in Germany during World War II and improved by P. Nashan in 1951.

U.S. Patent 2,757,201.

Sittig, M., *Combining Oxygen and Hydrocarbons for Profit,* Gulf Publishing, Houston, TX, 1962, 130.

Asinger, F., *Paraffins, Chemistry and Technology,* translated by B. J. Hazzard, Pergamon Press, Oxford, 1968, 582.

Guyot   A continuous sulfonation process for making phenol. Sodium sulfite is a byproduct and must be sold if the process is to be economic.

Molyneux, F., *Chem. Trade J.,* 1960, **147**(497), 518.

G-V   *See* Giammarco-Vetrocoke.

Gyro   Also called Gyro-cracking. An early vapor-phase thermal cracking process for refining petroleum.

The Petroleum Handbook, 3rd ed., Shell Petroleum, London, 1948, 170.

Asinger, F., *Mono-olefins: Chemistry and Technology,* translated by B. J., Hazzard, Pergamon Press, Oxford, 1968, 359.

Gyttorp [Named after a small Swedish town] A continuous process for nitrating polyols to form nitrate esters, used as explosives. Similar to the *Biazzi process.

H

H+ *See* Pechiney H+.

Haber Also called **Haber-Bosch,** and **Haber-Bosch-Mittasch.** A process for synthesizing ammonia from the elements, using high temperatures and pressures and an iron-containing catalyst. Invented by F. Haber at BASF in 1908. In 1909, C. Bosch of BASF built a pilot plant using an osmium-based catalyst; in 1913, a larger plant was built at Oppau, Germany. The process has been continuously improved and is still of major importance worldwide.

German Patents 235,421; 293,787.

Haber, F. and van Oordt, G., *Z. Anorg. Allg. Chem.,* 1905, **43,** 111.

Harding, A. J., *Ammonia, Manufacture and Uses,* Oxford University Press, London, 1959.

Haber, L. F., *The Chemical Industry 1900–1930,* Clarendon Press, Oxford, 1971, 198.

Vancini, C. A., *Synthesis of Ammonia,* translated by L. Pirt, Macmillan Press, Basingstoke, England, 1971, 234.

Jennings, J. R. and Ward, S. A., in *Catalyst Handbook,* 2nd. ed., Twigg, M. V., Ed., Wolfe Publishing, London, 1989, 384.

Travis, T. *Chem. Ind. (London),* 1993, (15), 581.

Haematite A process for removing iron from leach liquors from hydrometallurgical leaching operations. Used in recovering zinc from the residues of the electrolytic zinc process. *See also* Goethite, Jarosite.

Morgan, S. W. K., *Zinc and Its Alloys and Compounds,* Ellis Horwood, Chichester, England, 1985, 122.

Hänisch-Schröder A process for scrubbing sulfur dioxide from smelter gases using water. Developed in 1884 and still in use in 1950, although it was probably obsolete by 1990. It required 100 to 200 tons of water for each ton of sulfur recovered.

Hänisch, E. and Schröder, M., *J. Soc. Chem. Ind.,* 1884, **3,** 570.

Katz, Z. and Cole, R. J., *Ind. Eng. Chem.,* 1950, **42,** 2263.

Haifa Also called IMI. One of the two *Wet Processes for producing phosphoric acid by the acidulation of phosphate rock; the other is the *Dorr process. The Haifa process uses hydrochloric acid for the acidulation and solvent extraction for the purification. It is economic only where by-product hydrochloric acid is available. The overall reaction is:

$$Ca_5(PO_4)_3F + 10HCl = 3H_3PO_4 + 5CaCl_2 + HF$$

Various organic extractants may be used, including: butyl and amyl alcohols, di-isopropyl ether, tri-*n*-butyl phosphate, and tri-2-ethylhexyl phosphate. The fluoride remains with the calcium chloride solution, from which it may be recovered for sale or disposal. The process was invented in 1957 by A. Baniel and R. Blumberg at Israel Mining Industries, Haifa.

U.S. Patent 2,880,063.
British Patents 805,517; 1,051,521.
Baniel, A., Blumberg, R., Alon, A., El-Roy, M., and Goniadski, C., *Chem. Eng. Prog.,* 1962, **58**(11), 100.
Baniel, A. and Blumberg, R., in *Phosphoric Acid,* Vol. 1, Part 2, Slack, A. V., Ed., Marcel Dekker, New York, 1968, 889.
Blumberg, R., Gonen, D., and Meyer, D., in *Recent Advances in Liquid-Liquid Extraction,* Hanson, C., Ed., Pergamon Press, Oxford, 1971, 93.

Haines   A process for recovering sulfur from natural gas, using a zeolite adsorbent. The hydrogen sulfide in the gas is adsorbed on the zeolite; when the bed is saturated, hot sulfur dioxide is passed through it. The zeolite catalyzes the reaction between hydrogen sulfide and sulfur dioxide to form elemental sulfur, which sublimes out and is condensed. The process was invented by H. W. Haines in 1960; it was developed by Krell Associates and piloted in Canada from 1961 to 1962, but not commercialized because of problems caused by fouling of the zeolite with heavy hydrocarbons.

Pet. Refin., 1960, **39**(12), 208.
Haines, H. W., Van Wielingen, G. A., and Palmer, G. H., *Pet Refin.,* 1961, **40**(4), 123.
Hydrocarbon Process., 1961, **40**(11), 291.

HAL   [Hot acid leaching]   A process for purifying silica sand or zircon by leaching out surface iron compounds with hot sulfuric acid. Derived from an earlier process, invented in 1955 by British Industrial Sand, in which silica sand was treated with hot, gaseous hydrogen chloride. The process for cleaning zircon sand was developed jointly by Hepworth Minerals & Chemicals, UK, and Metallurgical Services Pty, Australia, in 1991.

British Patents 845,745; 1,223,177.
Ind. Miner. (London), 1992, (296), 101.

Halcon (1)   Halcon International (later The Halcon SD Group) designed many organic chemical processes, but is perhaps best known for its process for making phenol from cyclohexane. Cyclohexane is first oxidized to cyclohexanol, using air as the oxidant and boric acid as the catalyst, and this is then dehydrogenated to phenol. Invented in 1961 by S. N. Fox and J. W. Colton, it was operated by Monsanto in Australia for several years.

U.S. Patents 3,109,864; 3,239,552; 3,256,348; 3,932,513.

Halcon (2)   A process for oxidizing ethylene to ethylene oxide, using atmospheric oxygen, and catalyzed by silver. Developed by Halcon International in the late 1940s and early 1950s and first commercialized at Lavera, France. *See* Halcon (1).

Landau, R., *Pet. Refin.,* 1953, **32** (9), 146.
Porcelli, J. V., *Catal. Rev. Sci. Eng.,* 1981, **23,** 151.

Halex   [Halogen exchange]   A process for making fluoro-aromatic compounds by reacting the corresponding chloro- or bromo-aromatic comounds with an inorganic fluoride, usually potassium fluoride. Widely used for the manufacture of fluoro-intermediates.

Siegemund, G., in *Ullmann's Encyclopedia of Industrial Chemistry,* 5th ed., Vol. A11, VCH Publishers, Weinheim, Germany, 1989, 379.
Dolby-Glover, L., *Chem. Ind. (London),* 1986, 518.

Hall (1)   Also called the dry puddling process. An early ironmaking process, invented by J. Hall in 1830.

Hall (2)   A process for making fuel gas from petroleum fractions and residues by thermal cracking. Developed by W. A. Hall in 1913, later forming the basis of the *Gasmaco process. *See also* Petrogas.

U.S. Patents 105,772; 1,175,909; 1,175,910.

Hall (3)   An early process for making aromatic hydrocarbons by thermally cracking petroleum naphtha. *See also* Rittman.

Ellis, C., *The Chemistry of Petroleum Derivatives*, Chemical Catalog Co., New York, 1934, 164.
Asinger, F., *Mono-olefins: Chemistry and Technology*, translated by B. J., Hazzard, Pergamon Press, Oxford, 1968, 137.

Hall (4)   A process for making alumina by reducing bauxite with coke in an electric furnace. The co-product is an alloy of iron-silicon-aluminum-titanium. Invented by C. M. Hall in 1901 and later developed and commercialized by ALCOA.

O'Connor, D. J., *Alumina Extraction from Non-bauxitic Materials*, Aluminium-Verlag, Düsseldorf, 1988, 304.

Hall-Héroult   An electrolytic process for making aluminum metal from alumina, invented in 1886 independently by C. M. Hall in the United States, and P. L. Héroult in France. The alumina, made by the *Bayer process, is dissolved in fused cryolite, Na_3AlF_6, and electrolyzed at approximately 1,000°C. Because of the large requirement for electricity, the process is operated only where hydroelectric power is available. The cryolite was originally obtained from a deposit in Greenland but is now made synthetically from alumina, hydrofluoric acid, and sodium hydroxide:

$$Al_2O_3 + 6NaOH + 12HF = 2Na_3AlF_6 + 9H_2O$$

Hall's process was first operated by the Pittsburgh Reduction Company, a predecessor of ALCOA, in 1889. Héroult's process was first operated by the Société Metallurgique Suisse at Neulaissen, Switzerland, in 1887, using electric power generated at the Rhine Falls.

Grjotheim, K., Krohn, C., Malinovsky, M., Matiasovsky, K., and Thonstad, J., *Aluminium Electrolysis—Fundamentals of the Hall-Héroult Process*. Aluminium-Verlag GmbH, Düsseldorf, 1982.
Palmear, I. J., *The Chemistry of Aluminium, Gallium, Indium, and Thallium*, Downs, A. J., Ed., Blackie, London, 1993, 85.
Chem. Eng. News, 1997, **75** (39), 42.

Halomet   [**Halo**gen **met**al]   A process for reducing halides to metals by reaction with metallic aluminum or magnesium in a closed vessel. Invented in 1968 by R. Nowak and W. Schuster at Halomet, Basle.

U.S. Patents 3,244,509; 3,466,169.
Canadian Patent 899,631.

HALT   [**H**ydrate **a**ddition at **l**ow **t**emperature]   A *flue-gas desulfurization process in which lime slurry is injected into the combustion gases after they have been cooled in the heat exchanger. Developed in Canada in the mid-1970s.

Ashley, M. J. and Greaves, R. A., *Chem. Ind., (London)*, 1989, 6 Feb, 62.

Hansgirg   Also called Radenthein. A process for making magnesium metal by reducing magnesium oxide with carbon in an electric arc furnace at > 2,000°C and shock-chilling the vapor with hydrogen. The product was a fine dust that had to be converted to ingot. Operated on a pilot scale at Radenthein, Austria, by Österr.-Amerik Magnesit in the 1930s. Made obsolete by the invention of the electrolytic process for making magnesium. *See* Elektron.

Hanson-Van Winkle-Munning   *See* metal surface treatment.

Hargreaves-Bird   An early process for electrolyzing brine. It used a vertical diaphragm, unlike earlier designs. Developed by J. Hargreaves and T. Bird, around 1890.

> Hardie, D. W. F., *A History of the Chemical Industry in Widnes,* Imperial Chemical Industries, Widnes, England, 1950, 193.

Hargreaves-Robinson   A process for making sodium sulfate and hydrochloric acid by passing a mixture of wet sulfur dioxide and air through a series of vertical chambers containing briquettes of salt lying on a perforated floor, the temperature being maintained at approximately 500°C.

$$4NaCl + 2SO_2 + O_2 + 2H_2O = 2Na_2SO_4 + 4HCl$$

The addition of a small amount of nitric acid vapor or nitric oxide accelerates the process. Invented by J. Hargreaves and T. Robinson in Widnes in 1870, in order to provide sodium sulfate for the *Leblanc process, circumventing the need for the sulfuric acid used in the salt-cake section of that process. Reportedly still in use in 1984.

> British Patents 46 (1870); 3,045 (1870); 1,733 (1871); 3,052 (1872).
> Hardie, D. W. F., *A History of the Chemical Industry in Widnes,* Imperial Chemical Industries, Widnes, England, 1950, 80,130.

Harloff   A process for purifying sugar syrup by the addition of lime and sulfur dioxide. The precipitated calcium sulfite carries down many of the impurities in the syrup.

> Spencer, G. L. and Meade, G. P., *Cane Sugar Handbook,* 8th ed., John Wiley & Sons., New York, 1945, 110.

HARP   [**H**ybrid **ar**gon **r**ecovery **p**rocess]   A process for extracting argon from the hydrogen recycle stream in ammonia synthesis. Both *PSA and a cryogenic process are used.

> Krishnamurthy, R., Lerner, S. L., and MacLean, D. M., *Gas Sep. Purif.,* 1987, **1,** 16.

Harris   A softening process for removing antimony, arsenic, and tin from lead. The mixed metals are heated with a molten mixture of sodium hydroxide and sodium nitrate. Invented by H. Harris at H. J. Enthoven & Sons.

HARVEST   [**H**ighly **A**ctive **R**esidue **V**itrification **E**xperimental **S**tudies]   A process for immobilizing nuclear waste by incorporation in a borosilicate glass. Developed from *FINGAL. Piloted by the UK Atomic Energy Authority at Sellafield, in the late 1970s, but abandoned in 1981 in favor of *AVM, the French vitrification process.

> Lutze, W., in *Radioactive Waste Forms for the Future,* Lutze, W. and Ewing, R. C., Eds., North-Holland, Amsterdam, 1988, 15.

Hasenclever   An improvement to the *Deacon process for oxidizing hydrogen chloride to chlorine, in which the hydrogen chloride is first dried with concentrated sulfuric acid.

> Hardie, D. W. F., *A History of the Chemical Industry in Widnes,* Imperial Chemical Industries, Widnes, England, 1950, 69.

Hass (1)   A process for making aliphatic nitro-compounds from aliphatic hydrocarbons. The hydrocarbon vapor, mixed with nitric acid vapor, is passed through a narrow tube at 420°C. Invented by H. B. Hass in 1933.

> U.S. Patents 1,967,667; 2,071,122.
> Hass, H. B. and Riley, E. F., *Chem. Rev.,* 1943, **32,** 373.

Hass (2)   A process for making cyclopropane by reacting 1, 3-dichloropropane with zinc.

> Hass, H. B., McBee, E. T., Hinds, G. E., and Gluesenkamp, E. W., *Ind. Eng. Chem.,* 1936, **28,** 1178.
> Gustavson, G., *J. Prakt. Chem.,* 1887, **36,** 300.

Hass-McBee A thermal, vapor-phase process for chlorinating aliphatic hydrocarbons. The chlorine and the hydrocarbon vapor are separately heated to >250°C and then mixed. Propane is thus converted to 1, 3-dichloropropane. Invented in 1934 by H. B. Hass and E. T. McBee at Purdue University.

U.S. Patent 2,004,073.
Hass, H. B., McBee, E. T., and Weber, P., *Ind. Eng. Chem.,* 1935, **27,** 1192.
McBee, E. T., Hass, H. B., Burt, W. E., and Neher, C. M., *Ind. Eng. Chem.,* 1949, **41,** 799.
Asinger, F., *Paraffins, Chemistry and Technology,* translated by B. J. Hazzard, Pergamon Press, Oxford, 1968, 219.

HAT [Homogeneous alkylation technology] An *alkylation process using an alkyl aluminium catalyst. Developed by Kerr-McGee in 1994.

Rhodes, A. K., *Oil & Gas J.,* 1994, **92**(34), 52.

HBN *See* carbonization.

HBNPC *See* carbonization.

HB Unibon [Hydrogenation of benzene] The revised name for *Hydrar.

HCM [High carbon monoxide] Also called HICOM. A process for making methane from coal, based on the British Gas/Lurgi slagging gasifier.

Tart, K. R. and Rampling, T. W. A, *Hydrocarbon Process.,* 1981, **60**(4), 114.

H-Coal A coal gasification process. Crushed coal is mixed with process-derived oil and catalytically hydrogenated in an ebullated bed under pressure at 455°C. The catalyst is a mixture of cobalt and molybdenum oxides on alumina. Developed by Hydrocarbon Research from the 1960s and piloted in Catlettsburg, KY, from 1980 to 1982. *See also* CSF, H-Oil, CSF, Synthoil.

U.S. Patent 3,321,393.
Alpert, S. B., Johanson, E. S., and Schuman, S. C., *Hydrocarbon Process.,* 1964, **43**(11), 193.
Eccles, R. M., DeVaux, G. R., and Dutkiewicz, B., in *The Emerging Synthetic Fuel Industry,* Thumann, A., Ed., Fairmont Press, Atlanta, GA, 1981, Chap. 4.
Papso, J. E., in *Handbook of Synfuels Technology,* Meyers, R. A., Ed., McGraw-Hill, New York, 1984.

HC Platforming [Hydrocracking] A version of the *Platforming process which uses different catalyst systems before the reforming catalyst in order to partially hydrocrack the feed before converting it to aromatic hydrocarbons.

HCR (1) [High capacity reactor] *See* Otto Aqua-Tech HCR.

HCR (2) [High Claus ratio] A variation of the *RAR process developed by KTI.

HC Unibon [Hydrocracking] A version of the *hydrocracking process for simultaneously hydrogenating and cracking various liquid petroleum fractions to form branched-chain hydrocarbon mixtures of lower molecular weight. The catalyst is dual-functional, typically silica and alumina with a base metal, in a fixed bed. Developed by UOP. By 1988, 46 licenses had been granted. Currently offered under the name Unicracking.

Tajbl, D. G., in *Handbook of Petroleum Refining Processes,* Meyers, R. A., Ed., McGraw-Hill, New York, 1986.
Hydrocarbon Process, 1988, **67**(9), 73.

H-D (1) *See* DR.

H-D (2) *See* Huron-Dow.

HDA [Hydrodealkylation] A proprietary *dealkylation process for making benzene from toluene, xylenes, pyrolysis naphtha, and other petroleum refinery intermediates. The catalyst, typically chromium oxide or molybdenum oxide, together with hydrogen gas, removes the methyl groups from the aromatic hydrocarbons, converting them to methane. The process also converts cresols to phenol. Developed by Hydrocarbon Research with Atlantic Richfield Corporation and widely licensed worldwide.

HDH (1)   [Hemi dihydrate]   A *Wet Process for making phosphoric acid. The calcium sulfate is first produced as the hemihydrate and then in another stage this is converted to the dihydrate. Developed by Fisons, UK, and operated in Yugoslavia and the United Kingdom.

> Blumrich, W. E., Koening, H. J., and Schwer, E. W., *Chem. Eng. Prog.,* 1978, **74,** 58.

HDH (2)   [Hydrocracking-distillation-hydrotreatment]   A process for *hydrotreating and hydroconverting petroleum residues.

> Marcos, F. and Rosa-Brussin, D., *Catal. Rev., Sci. Eng.,* 1995, **37**(1), 3.

HDM   *See* hydrodemetallation.

HDN   *See* hydrodenitrogenation.

HDS   *See* hydrodesulfurization. Not to be confused with *TETRA HDS.

Heat-Fast   *See* DR.

Heinrich Koppers   A fractional distillation process for separating the constituents of coal tar. In 1981 it was in use in two plants in the United Kingdom.

> McNeil, D., *Chemistry of Coal Utilization,* Vol. 2., Elliott, M. A., Ed., John Wiley & Sons, New York, 1981, 1070.

Heiskenskjold   A process for making an animal feed by growing the yeast *Saccharomyces cerevisae* on the waste liquor from paper manufacture by sulfite pulping. Developed in Finland in 1936.

> Prescott, S. C. and Dunn, G. G., *Industrial Microbiology,* 3rd ed., McGraw Hill, New York, 1959, Chap. 3.
> Litchfield, J. H., *CHEMTECH,* 1978, **8,** 218.

Henderson   A wet process for extracting copper from its ores. Operated in Widnes in 1860.

> Hardie, D. W. F., *A History of the Chemical Industry in Widnes,* Imperial Chemical Industries, Widnes, England, 1950, 78.

Henkel   Also named Raecke, after the inventor. A process for making terephthalic acid from potassium benzoate by disproportionation:

$$2C_6H_5(COOK) = C_6H_4(COOK)_2 + C_6H_6$$

The reaction is conducted in the melt, in the presence of carbon dioxide under pressure, catalyzed by zinc or cadmium compounds. Invented by B. Raecke at Henkel, Germany. Improved variations of this process, known as Henkel I and Henkel II, were operated by several other companies, but by 1975 had been abandoned in favor of various other methods of oxidizing *p*-xylene.

> German Patents 936,036; 958,920.
> U.S. Patents 2,794,830; 2,823,229; 2,891,992; 2,905,709.
> Raecke, B., *Angew. Chem.,* 1958, **70,** 1.
> Weissermel, K. and Arpe, H.-J. *Industrial Organic Chemistry* 3rd ed., VCH Publishers, Weinheim, Germany, 1997, 397.

Heraeus   A process for making clear, fused quartz by passing powdered quartz crystals through an oxy-hydrogen flame and collecting the product on a rotating tube of fused quartz. Invented by H. R. Heraeus in 1952 and developed by Heraeus Quartzschmelze, Germany.

> U.S. Patent 2,904,713.

Hercosett *See* Chlorine/Hercosett.

Hercules (1)   An organic nitration process, similar to the *Bofors process but using a tee-joint for mixing. Developed by the Hercules Powder Company, Wilmington, DE, from 1957.

U.S. Patents 2,951,746; 2,951,877.

Hercules (2)   A modification of the *Witten process for making dimethyl terephthalate. Further improvements have been made by Glitsch Technology.

Hercules-BP   A process for making phenol from cumene, based on processes first developed by Hercules and BP and engineered by Kellogg International Corporation, and first installed at Montreal, PQ, in 1953. By 1993, more than half of the world's production of phenol was made by this process.

Weissermel, K. and Arpe, H.-J. *Industrial Organic Chemistry,* 3rd ed., VCH Publishers, Weinheim, Germany, 1997, 354.

Herz   A process for making *o*-amino thiophenols by heating the hydrochlorides of aromatic amines with sulfur monochloride at 50 to 75°C. The products are used as intermediates in the manufacture of thio-indigo dyestuffs. Invented by R. Hertz in Germany in 1914.

German Patent 360,690.
U.S. Patents 1,637,023; 1,699,432.
Warburton, W. K., *Chem. Rev.,* 1957, **57**, 1011.

HETACAT   An alkylation process using a solid acid catalyst. Not commercialized as of 1997.

Lerner, B.A., *Chem. Ind. (London),* 1996, (1), 16.

Hexall   A process for making hexane by dimerizing propylene. The reaction takes place in the liquid phase in a fixed bed of catalyst. Developed by UOP, but not commercialized as of 1992.

Ward, D. J., Friedlander, R. H., Frame, R., and Imai, T., *Hydrocarbon Process.,* 1985, **64**(5), 81.
Hydrocarbon Process., 1987, **66**(11), 77.

Hexone *See* Redox.

Heyden-Wacker   A process for making phthalic anhydride by the catalytic oxidation of naphthalene or *o*-xylene. Offered by Lurgi.

HF   A predecessor of the *DHD petroleum reforming process, operated in Germany in 1939. The catalyst was a mixed $MoO_3/ZnO/Al_2O_3$ system.

Weisser, O. and Landa, S., *Sulphide Catalysts, Their Properties and Applications,* Pergamon Press, Oxford, 1973, 379.

HFC   [Hydrocracking with fine catalyst]   A Japanese process for *hydrotreating and hydroconverting petroleum residues.

Marcos, F. and Rosa-Brussin, D., *Catal. Rev., Sci. Eng.,* 1995, **37**(1), 3.

HIB   An ironmaking process in which the powdered ore is reduced in a fluidized bed; the reducing gas is made by reforming natural gas. *See* DR.

Hibernia   A process for making formaldehyde by the partial oxidation of methane by ozonized oxygen. The catalyst is barium peroxide activated with silver oxide. Developed in Germany during World War II but not commercialized.

Hann, V., *Chem. Ind. (London),* 1950, 67, 386.
Asinger, F., *Paraffins, Chemistry and Technology,* translated by B. J. Hazzard, Pergamon Press, Oxford, 1968, 582.

Hichlor   A family of processes for making aluminum trichloride and other volatile metal chlorides by chlorinating fly ash and other coal wastes in fixed and fluidized beds. Developed by Ames Laboratory and Iowa State University from the mid-1970s to the mid-1980s. Not commercialized.

O'Connor, D. J., *Alumina Extraction from Non-bauxitic Materials,* Aluminium-Verlag, Düsseldorf, 1988, 297.

Hi-Chloroff   A thermal (non-catalytic) process for removing chlorine from chlorinated hydrocarbon wastes containing either low or high concentrations of chlorine. Developed by Kinetics Technology International. *See also* Chloroff.

HICOM   [High Carbon monoxide]   *See* HCM.

High Productivity   Also called HP. An improved method of operating gas-phase plants for making polyethylene. It involves removing gas and liquid from the fluidized bed, separating them, and returning the liquid to the bed via a patented nozzle system. Developed by BP in 1994 and operated in its Grangemouth refinery since 1995. *See also* Supercondensed Mode.

Chem. Br., 1995, **31**(4), 278.
Chem. Br., 1997, **33**(5), 31.

Hilgenstock   *See* Hoerde.

Hiperion   A process for removing hydrogen sulfide and organic sulfur compounds from hydrocarbons. Similar to the *Takahax process but using a solution of chelated iron and naphthaquinone. The elemental sulfur produced is removed by filtration. Licensed by Ultra-systems, CA.

Chem. Eng. (N.Y.), 1987, **94**(2), 159.
Dalrymple, D. A., Trofe, T. W., and Evans, J. M., *Chem. Eng. Prog.,* 1989, **85**(3), 43.

HiPure   A variation on the *Benfield process, using two stages of scrubbing by hot potassium carbonate solution in order to reduce the carbon dioxide contents of gases to very low levels. *See also* Carsol, CATACARB, Giammarco-Vetrocoke.

Benson, H. E. and Parrish, R. W., *Hydrocarbon Process.,* 1974, **53**(4), 81.
Kohl, A. L. and Riesenfeld, F. C., *Gas Purification,* 4th ed., Gulf Publishing, Houston, TX, 1985, 214.

Hirohax   A process for removing hydrogen cyanide from gas streams. It is first converted to thiocyanate, and this is oxidized at 350°C with oxygen under pressure. Developed in 1972 by Nippon Steel, Tokyo.

U.S. Patent 3,855,390.

H-Iron   [Hydrogen iron]   A process for making iron by reducing powdered iron oxides from ores or scrap, using hydrogen. A variation on the process will convert iron/titanium ores to a titanium concentrate and metallic iron. Developed by Hydrocarbon Research and United States Steel Corporation, and used in Pennsylvania and California. *See also* DR.

HIsmelt   A direct iron smelting process in which noncoking coal, fine iron ore, and fluxes and gases, are injected into a molten iron bath; the carbon monoxide produced is used to pre-reduce the ore in a fluidized bed. Under development by CRA, Australia, since the early 1980s, joined by Midrex Corporation in 1988. Their joint venture company, Hismelt Corporation, commissioned a pilot plant at Kwinana, near Perth, Australia, in 1993.

Hitachi Wet   A *flue-gas desulfurization process using activated carbon.

Speight, J. G., *Gas Processing,* Butterworth Heinemann, Oxford, 1993, 293.

Hivalloy   A process for grafting styrenic polymers on to polyolefines, using a *Ziegler-Natta catalyst. The products combine the physical properties of both polymer types. Developed by Montell and commercialized in the United States in 1997. *See also* Catalloy.

Oxley, D. F., *Chem. Ind. (London)*, 1998, (8), 307.

Hock   Also known as the Hock Lang process, and the cumene peroxidation process. A process for converting isopropyl benzene (cumene) to a mixture of phenol and acetone; *m*-di-isopropyl benzene likewise yields resorcinol, and *p*-di-isopropyl benzene yields hydroquinone. The basis of the process is the liquid-phase air oxidation of cumene to cumene hydroperoxide:

$$C_6H_5CH(CH_3)_2 + O_2 \rightarrow C_6H_5C(CH_3)_2OOH$$

followed by the cleavage of the hydroperoxide by hot acid:

$$C_6H_5C(CH_3)_2OOH \rightarrow C_6H_5OH + (CH_3)_2CO$$

The second stage can be carried out in the liquid phase by sulfuric acid, or in the gas phase by phosphoric acid adsorbed on a carrier.

Invented in Germany during World War II by H. Hock and S. Lang in the course of developing cumene hydroperoxide for initiating the polymerization of butadiene-styrene mixtures. After the war the process was developed by the Distillers Company in England and Allied Chemical Corporation in the United States. Since 1954 this has been the main commercial process for the production of phenol and acetone. By 1987, 97 percent of the phenol made in the United States was produced via this route. In 1990, both resorcinol and hydroquinone were produced commercially by this route as well. *See also* Cumox.

Hock, H. and Lang, S., *Ber. Dtsch. Chem. Ges.*, 1944, **77**, 257.
Armstrong, G. P., Hall, R. H., and Quin, D. C., *J. Chem. Soc.*, 1950, 666.
Weissermel, K. and Arpe, H.-J., *Industrial Organic Chemistry*, 3rd ed., VCH Publishers, Weinheim, Germany, 1997, 353.

Hoechst coker   A continuous process for pyrolyzing various oils to yield olefins. The heat is carried by coke spheres. First operated in 1956, later supplanted by *Hoechst HTP.

Asinger, F., *Mono-olefins: Chemistry and Technology*, translated by B. J., Hazzard, Pergamon Press, Oxford, 1968, 128.
Stokes, R. G., *Opting for Oil*, Cambridge University Press, Cambridge, 1994, 187.

Hoechst HTP   [High temperature pyrolysis]   Also called **Hoechst-Uhde.** A two-stage process for making a mixture of acetylene and ethylene by cracking higher hydrocarbons. In the first stage, a fuel is burnt with the stoichiometric quantity of oxygen. In the second stage, the hot combustion products from the first stage are contacted with the vapor of the hydrocarbon to be cracked. The process was developed and operated by Farbwerke Hoechst in Germany in the 1950s until 1975. In Czechoslovakia it was operated until the 1990s. *See also* Wulff.

Kamptner, H. K., Krause, W. R., and Schilken, H. P., *Hydrocarbon Process.*, 1966, **45**(4), 187.
Asinger, F., *Mono-olefins: Chemistry and Technology*, translated by B. J., Hazzard, Pergamon Press, Oxford, 1968, 178.
Stokes, R. G., *Opting for Oil*, Cambridge University Press, Cambridge, 1994, 190.
Weissermel, K. and Arpe, H.-J., *Industrial Organic Chemistry*, 3rd ed., VCH Publishers, Weinheim, Germany, 1997, 97.

Hoechst-Shawinigan   A catalytic process for oxidizing acetaldehyde directly to acetic anhydride, using oxygen. *See* Shawinigan.

Hoechst-Uhde (1)   An electrolytic process for generating hydrogen and chlorine from hydrochloric acid. Widely used.

Hoechst-Uhde (2)   A variation of the *Wacker process, which makes vinyl acetate from ethylene and acetic acid. The catalyst is an aqueous solution of palladium and copper chlorides.

> Dumas, T. and Bulani, W., *Oxidation of Petrochemicals: Chemistry and Technology,* Applied Science Publishers, London, 1974, 20.

Hoechst-Wacker   *See* Wacker (1).

Hoechst-WLP   A process for making acetylene by injecting a liquid hydrocarbon into a hydrogen plasma. It is an improvement on the *Hüls process in its ability to use heavy hydrocarbons without forming carbon.

> Miller, S. A., *Acetylene, Its Properties, Manufacture and Uses,* Vol. 1, Academic Press, New York, 1965, 407.

Hoerde   Also called Massener, and Hilgenstock. A process for removing sulfur from molten iron or steel by adding manganese. This produces manganese sulfide, which forms an upper phase that can be skimmed off. Invented in 1891.

Hoesch   *See* Bertrand Thiel.

Hoffman   A process for making hydrazine by reacting urea and sodium hypochlorite in water:

$$OC(NH_2)_2 + NaOCl + 2NaOH = N_2H_4 + NaCl + Na_2CO_3 + H_2O$$

> *Riegel's Handbook of Industrial Chemistry,* Kent, J.A. Ed., 9th ed., Van Nostrand Reinhold, New York, 1992, 1123.

H-Oil   A process for demetallizing, desulfurizing, and hydrocracking heavy fuel oil with the production of lighter distillates and gas. It is intended for heavy hydrocarbons including residues, tar, and shale oil. The oil is hydrogenated in an "ebullated" fluidized bed containing catalyst granules. Invented by Hydrocarbon Research in the early 1950s and developed jointly with Cities Service Research & Development Company. First demonstrated at Lake Charles, LA, in 1963; commercialized in Kuwait in 1968, and later installed in Mexico and the United States. Six units were built between 1963 and 1996. *See also* H-Coal.

> Unzelman, G. H. and Wolf, C. J., in *Petroleum Processing Handbook,* Bland, W. F. and Davidson, R. L., Eds., McGraw-Hill, New York, 1967, 3-19.
> Mounce, W. and Rubin, R. S., *Chem. Eng. Prog.,* 1971, **67**(8), 81.
> Papso, J. E., in *Handbook of Synfuels Technology,* Meyers, R. A., Ed., McGraw-Hill, New York, 1984, 6-19.
> *Hydrocarbon Process.,* 1996, **75**(11), 126.

Hojanas   Also called Siurin. An iron extraction process. Magnetite, mixed with carbon-coke breeze and limestone, is heated in a ceramic retort by passage through a tunnel kiln at 1,200°C. Used commercially in Sweden since 1911. *See also* DR.

> Dennis, W. H., *Metallurgy of the Ferrous Metals,* Sir Isaac Pitman & Sons, London, 1963, 102.

HOKO   A process for making nitric acid simultaneously at two concentrations.

> Hellmer, L., *Chem. Eng. Prog.,* 1972, **68**(4), 67.

Holmes-Manley   An early mixed-phase petroleum cracking process.

Holmes-Maxted   A process for removing organic sulfur compounds from coal gas. The gas, mixed with hydrogen, is passed over a metal thiomolybdate catalyst at 300 to 380°C, which converts the sulfur compounds to hydrogen sulfide which is then absorbed by iron oxide. Developed by E. B. Maxted at W. C. Holmes & Company, UK, based on an invention made in 1937. More than 50 units were in operation by 1985.

British Patent 490,775.

Wedgewood, W., *Inst. Gas Engrs.,* 1958, 525; *Chem. Abstr.,* **52,** 14138.

Kohl, A. L. and Riesenfeld, F. C., *Gas Purification,* 4th ed., Gulf Publishing, Houston, TX, 1985, 726.

Holmes-Stretford A version of the *Stretford process, developed by W.C. Holmes & Company, Huddersfield, England.

Moyes, A. J. and Vasan, S., *Oil & Gas J.,* 1974, **72**(35), 56.

Speight, J. G., *Gas Processing,* Butterworth Heinemann, Oxford, 1993, 229.

Holst A batch process for making chlorine dioxide by reducing sodium chlorate with sulfur dioxide in the presence of sulfuric acid. The sodium hydrogen sulfate byproduct can be used in the *Kraft papermaking process. Developed by Moch Domsjo in Sweden, in 1946.

Partridge, H. de V., in *Chlorine, Its Manufacture, Properties, and Uses,* Sconce, J. S., Ed., Reinhold Publishing, New York, 1962, 275.

Holter A *flue-gas desulfurization process in which the sulfur dioxide is absorbed in an aqueous suspension of calcium hydroxide and calcium chloride, yielding gypsum. Operated in an experimental plant at the Weiherr III power station in Quierscheid, Germany, in 1988.

Hooker-Raschig *See* Raschig (2).

Hoopes An electrolytic process for refining aluminum metal. The electrolyte is a mixture of fluoride salts. The cell is constructed of graphite. The electrolyte in contact with the side-walls of the cell is frozen, thus preventing short-circuiting of electricity through the walls. Developed by W. Hoopes and others at Aluminum Company of America in the 1920s.

U.S. Patents 1,534,321; 1,534,322.

Frary, F. C., *Trans. Am. Electrochem. Soc.,* 1925, **47,** 275.

Hornsey *See* steelmaking.

Hot Acid An improved version of the *Cold Acid process, which increases the yield of octanes. It operates at 60 to 90°C.

The Petroleum Handbook, 3rd ed., Shell Petroleum, London, 1948, 229.

Hot-Water A process for separating bitumen from tar sands. Developed by SUNCOR for treating the sands from Athabasca, Alberta, based on an invention made by K. A. Clarke in 1932. Two large plants were operating in Canada in 1984.

Clark, K. A. and Pasternack, D. S., *Ind. Eng. Chem.,* 1932, **24,** 1410.

Camp, F. W., *The Tar Sands of Alberta,* 2nd. ed., Cameron Engineers, Denver, 1974.

Rider, D. K., *Energy: Hydrocarbon Fuels and Chemical Resources,* John Wiley & Sons, New York, 1981.

Erskine, H. L., in *Handbook of Synfuels Technology,* Meyers, R. A., Ed., McGraw-Hill, New York, 1984, 5-13.

Houdresid A catalytic petroleum cracking process, similar to *Houdriflow, adapted to processing residues.

Unzelman, G. H. and Wolf, C. J., in *Petroleum Processing Handbook,* Bland, W. F. and Davidson, R. L., Eds., McGraw-Hill, New York, 1967, 3-10.

Houdriflow A catalytic petroleum cracking process in which the beads of catalyst move continuously through the reactor and the catalyst regenerator.

Enos, J. L., *Petroleum Progress and Profits,* MIT Press, Cambridge, MA, 1962, 177.

Unzelman, G. H. and Wolf, C. J., in *Petroleum Processing Handbook,* Bland, W. F. and Davidson, R. L., Eds., McGraw-Hill, New York, 1967, 3-9.

Houdriforming   A continuous catalytic reforming process for producing aromatic concentrates and high-octane gasoline. It used a fixed bed of a platinum catalyst. Developed in the 1950s by the Houdry Process Corporation.

Unzelman, G. H. and Wolf, C. J., in *Petroleum Processing Handbook,* Bland, W. F. and Davidson, R. L., Eds., McGraw-Hill, New York, 1967, 3-28.
Asinger, F., *Mono-olefins: Chemistry and Technology,* translated by B. J., Hazzard, Pergamon Press, Oxford, 1968, 389.
Oil & Gas J., 1971, **69**(51), 45.

Houdry   The first catalytic petroleum cracking process, based on an invention by E. J. Houdry in 1927, which was developed and commercialized by the Houdry Process Corporation. The process was piloted by the Vacuum Oil Company, Paulsboro, NJ, in the early 1930s. The catalyst was contained in a fixed bed. The first successful catalyst was an aluminosilicate mineral. Subsequently, other related catalysts were developed by Houdry in the United States, by I. G. Farbenindustrie in Germany, and by Imperial Chemical Industries in England. After World War II, the clay-based catalysts were replaced by a variety of synthetic catalysts, many based on alumino-silicates. Later, these too were replaced by zeolites.

U.S. Patents 1,837,963; 1,957,648; 1,957,649.
Houdry, E. J., Burt, W. F., Pew, A. E., and Peters, W. A., *Oil & Gas J.,* 1938, **37**(28), 40.
Haensel, V. and Sterba, M. J., *Ind. Eng. Chem.,* 1948, **40**, 1662
Enos, J. L., *Petroleum Progress and Profits,* MIT Press, Cambridge, MA, 1962, Chap. 4.
The Petroleum Handbook, 6th ed., Elsevier, Amsterdam, 1963, 284.

Houdry-Litol   *See* Litol.

Howard   A process for separating lignosulfonates from sulfite liquor from papermaking. The lignosulfonates are precipitated as their calcium salts. The product is used in the manufacture of vanillin.

Sarkanen, K. V. and Ludwig, C. H., *Lignins,* Wiley-Interscience, New York, 1971, 800.

Howden   An early *flue-gas desulfurization process using a lime or chalk slurry in wooden grid-packed towers. The calcium sulfate/sulfite waste product was intended for use in cement manufacture, but this was never commercialized. The key to the process was the use of a large excess of calcium sulfate in suspension in the scrubbing circuit, which minimized the deposition of scale on the equipment. The process was developed by Imperial Chemical Industries and James Howden & Company in the 1930s and operated for several years at power stations at Fulham, London, and Tir John, South Wales, being finally abandoned during World War II.

British Patents 420,539; 433,039.
Pearson, J. L., Nonhebel, G., and Ulander, P. H. N., *J. Inst. Fuel,* 1935, **8**, 119.
Katz, M. and Cole, R. J., *Ind. Eng. Chem.,* 1950, **42**, 2266.
Achillades, B., *Chem. Ind., (London),* 1975, 19 April, 337.

Howe-Baker   A process for removing sulfur compounds from petroleum fractions by treatment with sulfuric acid.

HP   *See* High Productivity.

HPC   [Hot potassium carbonate]   A generic name for a process for absorbing acid gases by the use of hot aqueous potassium carbonate. Developed by H. E. Benson and J. H. Field at the U.S. Bureau of Mines in the 1950s. Further developed by other organizations, it became the basis for the *Benfield, *CATACARB, and other gas purification processes.

U.S. Patent 2,886,405.

H-Plus   A process for extracting aluminum from clays by successive treatment with sulfuric and hydrochloric acids. The product is aluminum trichloride hexahydrate.

HPM [Hydrochloric acid hydrogen peroxide mix] *See* RCA.

HPO [Hydroxylamine phosphate oxime] A process for making caprolactam, an intermediate in the manufacture of polyamides. It differs from related processes, such as *HSO, in producing less of the ammonium sulfate by-product. Developed by DSM Research, The Netherlands, operated by DSM Polymers and Hydrocarbons, and offered for license by Stamicarbon.

HR [Heat recycle] *See* UTI.

HS A family of gas purification processes developed by Union Carbide Corporation, based on the use of proprietary solvents known as UCARSOLs. UCARSOL HS-101, is based on methyl diethanolamine and is used for removing hydrogen sulfide and carbon dioxide from other gases. Ucarsol LH-101 is used in its *Cansolv system for *flue-gas desulfurization.

> Kohl, A. L. and Riesenfeld, F. C., *Gas Purification,* 4th ed., Gulf Publishing, Houston, TX, 1985, 119.

HSC [High-conversion Soaker Cracking] A *visbreaking process, developed and offered by Toyo Engineering Corporation, Japan. Demonstrated from 1988 to 1989 in the Schwedt oil refinery, Germany.

> Washimi, K. and Limmer, H., *Hydrocarbon Process.,* 1989, **68**(9), 69.

HSO [Hydroxylamine sulphate oxime] A process for making caprolactam, the intermediate for making polyamide 6. Developed by DSM Research, The Netherlands, and used by DSM Chemicals at Geleen, The Netherlands, and in Augusta, GA. Offered for license by Stamicarbon. *See also* HPO.

HSR *See* Biobor HSR.

HTC (1) [High temperature chlorination] A general term for the process for making 1,2-dichloroethane from ethylene and chlorine by processes operated above the boiling point of the product (83°C). *See also* CER.

HTC (2) [Hydrogen transfer catalysis] A catalytic process for reducing aromatic nitro-compounds. Developed by Rohner in 1993.

HTP *See* Hoechst HTP. Also used as an abbreviation for High Test Peroxide (*i.e.,* concentrated hydrogen peroxide).

HTR A catalytic process for making gas from oil, designed for meeting peak loads. Developed by North Western Gas, UK, in the 1960s.

H.T.S. Carter *See* Carter.

HTW [High temperature Winkler] A process for gasifying coal, using oxygen and steam in a fluidized bed. An improved version of the *Winkler process, differing from it by being operated under pressure, at a higher temperature, and with dust recycle. Developed at Aachen Technical University, and piloted near Cologne from 1978 to 1984. A large-scale demonstration plant was built at Ville-Berrenrath in the mid-1980s. Uhde built a plant at Wesseling, Germany for the Rheinische Braunkohlenwerke (completed in 1989), and one in Finland.

> Schrader, L., Teggers, H., and Theis, K.-A., *Chem. Ing. Tech.,* 1980, **52**, 794.
> Theis, K.-A. and Nitschke, E., *Hydrocarbon Process.,* 1982, **61**(9), 233.
> Thomas, G. and Nitschke, E., in *Handbook of Petroleum Refining Processes,* Meyers, R. A., Ed., McGraw-Hill, New York, 1986, 11-3.
> Cornils, B., in *Chemicals from Coal: New Processes,* Payne, K. R., Ed., John Wiley & Sons, Chichester, England, 1987, 17.

Hüls   Also called the Hüls Flaming Arc process. A process for making acetylene by passing hydrocarbon gases through a d.c. electric arc. Electrode erosion is minimized by rotating the arc roots by swirling the gas. Operated by Chemische Werke Hüls in Germany from 1939 to 1993, and in Romania.

Gladisch, H., *Hydrocarbon Process.,* 1962, **41**(6), 159.

Miller, S. A., *Acetylene: Its Properties, Manufacture and Uses,* Vol. 1, Ernest Benn, London, 1965, 394.

Weissermel, K. and Arpe, H.-J. *Industrial Organic Chemistry,* 3rd ed., VCH Publishers, Weinheim, Germany, 1997, 95.

Humphrey   A catalytic process for hydrogenating rosin acids.

Hunter   A process for making titanium metal by reducing titanium tetrachloride with sodium:

$$TiCl_4 + 4Na = Ti + 4NaCl$$

The process is operated in heated, batch reactors under an inert atmosphere. Two companies (Deeside Titanium, North Wales, and New Metals Industries, Nihongi, Japan) operate a one-stage process. Reactive Metals Industries Company, Ashtabula, OH, operates a two-stage process: in the first stage, at 230°C, the trichloride and dichloride are formed. In the second, more sodium is added and the temperature is raised to 1,000°C. The sponge product is mixed with sodium chloride, which is leached out with dilute hydrochloric acid. Based on the work by M. A. Hunter at Rensselaer Polytechnic, New York in 1910. *See also* Kroll.

Hunter, M. A., *J. Am. Chem. Soc.,* 1910, **32**, 330.

Minkler, W. W., *The Production of Titanium, Zirconium and Hafnium,* Metal. Treatises, 1981, 171 (*Chem. Abstr.,* **96**, 184821).

Eveson, G. F., in *Speciality Inorganic Chemicals,* Thompson, R., Royal Society of Chemistry, London, 1981, 231.

Huntingdon-Heberlein   A lead smelting process, succeeded by the *Dwight-Lloyd process.

Cocks, E. J. and Walters, B., *A History of the Zinc Smelting Industry in Britain,* George G. Harrap, London, 1968, 59.

Huntsman   Also called the Crucible process. A method for purifying blister steel, made by *cementation, by melting it in a closed clay or graphite crucible and pouring the melt into a cast iron mold. Developed by B. Huntsman, a clockmaker, around 1740, initially in Doncaster, later in Sheffield, UK. The process was not patented but was operated in secrecy for some years. It was subsequently widely operated in the UK, Europe (where the combination of this with the cementation process was known as "Les Procédés Anglais"), and in the United States (from 1830), until the introduction of the *Bessemer process. Thereafter its usage declined, but it continued in use in some countries until the 1920s. The growth of the town of Sheffield in the 19th century was largely due to this development.

Barraclough, K. C., *Steelmaking Before Bessemer, Vol. 2. Crucible Steel,* The Metals Society, London, 1984.

Huron-Dow   Also called H-D. A process for making hydrogen peroxide by electrolyzing alkaline water. Developed in the 1980s, initially by Huron and Dow Chemical, subsequently as H-D Technologies, and commercialized in 1991. Intended for on-site production of dilute alkaline hydrogen for direct use in the pulp and paper industries. The electrolytic cell has a porous cathode through which oxygen is passed, a Pt-coated titanium anode, and a separator containing an ion-exchange membrane.

Hess, W. T., in *Kirk-Othmer's Encyclopedia of Chemical Technology,* 4th ed., Vol. 13, John Wiley & Sons, New York, 1991–1998, 978.

Hurter   *See* Deacon.

HYAN   [Hybrid anaerobic]   A process for treating the supernatant liquor from sewage sludge treatment plants by anaerobic digestion. The methane produced is burnt to provide heat for the treatment plant and to destroy odors. Developed by a Canadian consortium and operated at the Lakeview Water Pollution Control Plant, Mississauga, Ontario, from 1991.

Hybinette   A process for extracting nickel from sulfide ores. The nickel ore that occurs in Canada is a mixture of the sulfides of nickel, copper, and iron. Several methods have been used to separate these metals. In the Hybinette process, the ore is first smelted in a blast furnace, yielding a nickel-copper matte (*i.e.,* a mixture of their lower sulfides). This is roasted to remove sulfur and leached with dilute sulfuric acid to remove copper. The resulting crude nickel oxide is used as the anode of an electrochemical cell. The nickel deposits on the cathode, which is contained in a cloth bag. Precious metals collect in the anode slime. The process was invented by N. V. Hybinette in 1904 and operated at the Kristiansand refinery, Norway, from 1910.

> U.S. Patents 805,555; 805,969.
> Archibald, F. R., *J. Met.,* 1962, **14,** 648.
> Dennis, W. H., *A Hundred Years of Metallurgy,* Gerald Duckworth, London, 1963, 204.
> Burkin, A. R., *Extractive Metallurgy of Nickel,* John Wiley & Sons, Chichester, England, 1987, 121.

Hycar (1)   A *reforming process for making *syngas from light hydrocarbons, differing from the standard process in using two reactors. The second reactor (a convective reformer), operated in parallel with the primary reformer, preheats the feedstock. Developed by Uhde.

> *Chem. Eng., (N.Y.),* 1992, **99**(5), 33.

Hycar (2)   A hydrovisbreaking process for oil sands. *See* Visbreaking.

> Marcos, F. and Rosa-Brussin, D., *Catal. Rev., Sci. Eng.,* 1995, **37**(1), 3.

Hy-C Cracking   A *hydrocracking process. The catalyst is nickel/tungsten on alumina. Developed by Cities Service Research and Development Company and Hydrocarbon Research.

> Johnson, A. R. and Rapp, L. M., *Hydrocarbon Process.,* 1964, **43**(5), 165.
> Unzelman, G. H. and Wolf, C. J., in *Petroleum Processing Handbook,* Bland, W. F. and Davidson, R. L., Eds., McGraw-Hill, New York, 1967, 3-20.

HyChlor   Formerly called DHC. A catalytic *hydrodechlorination process which converts organic chlorides to hydrogen chloride and saturated hydrocarbons. The UOP HyChlor process recovers and recycles both the organic and inorganic reaction products in order to minimize waste disposal requirements and maximize yield. Used for treating wastes from the production of chlorinated petrochemicals such as vinyl chloride. Developed by UOP but not commercialized as of 1992.

> Johnson, R. W., Youtsey, K. J., Hifman, L., and Kalnes, T., in *Management of Hazardous and Toxic Wastes in the Process Industries,* Kolaczkowsti, S. T. and Cuttenden, B. D., Eds., Elsevier Applied Science, Amsterdam, 1987.
> Kalnes, T. N. and James, R. B., *Environmental Prog.,* 1988, **7**(3), 185.
> Kalnes, T. N. and James, R. B., *Effluent Treatment and Waste Disposal,* Institution of Chemical Engineers, Rugby, 1990, 179.

Hyco   *See* GEMINI Hyco.

Hycon (1)   [Hydrogen conversion]   A hydrogenation process for converting high-sulfur petroleum residues to clean transport fuels. Sulfur is removed as hydrogen sulfide. Two catalytic stages are used. The first achieves demetallization with a regenerable catalyst. The second achieves *HDS, *HDN, and partial *hydrocracking. Developed by Shell and used in a semi-industrial plant at Pernis from 1990.

Chem. Eng. (Rugby, England), 1989, Nov, 11.
Chauvel, A., Delmon, B., and Hölderich, W. F., Appl. Catal. A: Gen., 1994, 115, 178.

Hycon (2)   A process for making nitric acid which can provide a range of concentrations. Developed in 1968 by the Chemical Construction Corporation.

U.S. Patent 3,542,510.
Newman, D. J. and Klein, L. A., Chem. Eng. Prog., 1972, 68(4), 62.

Hydeal [Hydrodealkylation]   A process for making benzene by de-alkylating other aromatic hydrocarbons. Generally similar to the *Litol process. Developed in the 1950s by UOP and Ashland Oil Company, but abandoned in favor of UOP's *THDA process. See dealkylation.

Hydrocarbon Process., 1963, 42(11), 180.

HYDECAT [Hypochlorite destruction catalyst]   A continuous process for destroying un- wanted hypochlorite streams. A heterogeneous catalyst containing nickel converts the hypochlorite ion to chloride ion and oxygen gas:

$$2NaOCl = 2NaCl + O_2$$

Developed by ICI Katalco; first demonstrated at Billingham in 1989 and first operated on a large scale at Yarwun, Queensland, in 1991. Three versions were subsequently developed: HYDECAT TD (total destruction), HYDECAT PD (partial destruction), and HYDECAT ID (integrated destruction). See also ODORGARD.

European Patent 397,342.
WPCT Patent WO92/18235.
Chem. Eng. (Rugby, England), 1991, (493), 22.
Chem. Br., 1991, 27(5), 402.
Eur. Chem. News, 1991, 18 Mar, 24.
King, F. and Hancock, F. E., Catal. Today, 1996, 27(1-2), 203.

Hydrane   A coal gasification process in which coal is hydrogenated directly to methane, using hydrogen gas under pressure, preheated to a high temperature. Developed by the U.S. Bureau of Mines; as of 1980 it had not been fully piloted.

Hydrocarbon Process., 1975, 54(4), 122.

Hydrar   A catalytic process for hydrogenating benzene to cyclohexane. It is conducted in the vapor phase with a fixed-bed reactor. The catalysts are based on platinum and modified by lithium; an alternative nickel-based catalyst is also used. Developed by UOP, subsequently renamed HB Unibon.

Hydrocarbon Process., 1963, 42(11), 181.
Franck, H.-G. and Stadelhofer, J. W., Industrial Aromatic Chemistry, Springer-Verlag, Berlin, 1988, 192.

Hydrisom   A selective hydrogenation process offered by Phillips Petroleum. It is used in Argentina to upgrade C_4 olefins before *alkylation.

Oil & Gas J., 1998, 96(11), 66.

Hydrobon   A catalytic petroleum hydrotreating process, developed by UOP and formerly licensed jointly with the Union Oil Company of California under the name *Unifining.

Davidson, R. L., Pet. Process., 1956, 11(11), 115.
Unzelman, G. H. and Wolf, C. J., in Petroleum Processing Handbook, Bland, W. F. and Davidson, R. L., Eds., McGraw-Hill, New York, 1967, 3-44.

Hydrocol   A process for making gasoline from natural gas. Partial combustion with oxygen yields *syngas, which is catalytically converted to gasoline in a fluidized bed containing an iron catalyst. Developed by Hydrocarbon Research, and commercialized by Carthage Hydrocol in Brownsville, TX, in 1950.

Keith, P. C., *Am. Gas J.,* 1946, **164**(6), 11 (*Chem. Abstr,* **40,** 4873).

Asinger, F., *Paraffins, Chemistry and Technology,* translated by B. J. Hazzard, Pergamon Press, Oxford, 1968, 161.

hydrocracking A general term for any catalytic hydrogenation process for upgrading heavy petroleum fractions to produce fractions of lower molecular weight which can be used as fuels. The vaporized feed is mixed with hydrogen at high temperature and pressure and allowed to react in a fixed- or fluidized-catalyst bed. The forerunner of today's hydrocracking processes was the *Bergius process. Those hydrocracking processes with special names that are described in this dictionary are: ABC, Chiyoda ABC, Canmet Hydrocracking, HC-Unibon, HDDV, H-G, H-Oil, Hy-C Cracking, Isocracking, Isomax, LC-Fining, Lomax, Unicracking, Veba.

The Petroleum Handbook, 6th ed., Elsevier, Amsterdam, 1983, 294.

Froment, G. F. and Marin, G. B., *Hydrocracking: Science and Technology,* Elsevier, Amsterdam, 1987.

Coonradt, H. L. and Garwood, W. E., *Ind. Eng. Chem., Process Des. Dev.,* 1964, **3**(1), 38.

hydrocyanation A complex process for making adiponitrile by adding hydrogen cyanide to butadiene. The homogeneous catalyst used is $Ni[P(OC_6H_5)_3]_4$. Developed by Du Pont in 1971.

Chem. Eng. News, 1971, **49**(17), 30.

hydrodealkylation The use of hydrogen to convert an alkyl benzene (typically toluene) to benzene. The reaction takes place at high temperatures and pressures and may or not be catalyzed. Named processes described in this dictionary are: DETOL, HDA, Hydeal, Litol, MBE, THD, THDA.

Weissermel, K. and Arpe, H.-J. *Industrial Organic Chemistry,* 3rd ed., VCH Publishers, Weinheim, Germany, 1997, 329.

hydrodechlorination A general term for processes which convert organic chlorine compounds to hydrogen chloride and saturated hydrocarbons. *See* DCH (2), HyChlor.

hydrodemetallation Often abbreviated to HDM. A general term for processes for removing metal compounds from petroleum fractions by catalytic reduction with hydrogen.

Reynolds, J. G., *Chem. Ind. (London),* 1991, 570.

hydrodenitrogenation Often abbreviated to HDN; also called hydrodenitrification. An essential stage in the *hydrocracking of petroleum fractions in which organic nitrogen compounds are removed by reaction with hydrogen to produce ammonia. The compounds are removed in order to prevent them from poisoning the catalyst used in the following stage.

Reynolds, J. G., *Chem. Ind. (London),* 1991, 570.

hydrodesulfurization Often abbreviated to HDS. A general term for processes which convert sulfur compounds in petroleum fractions to hydrogen sulfide, and simultaneously convert high molecular weight hydrocarbons to more volatile ones. The process operates in the liquid phase under hydrogen pressure, using a heterogeneous catalyst. The catalyst is typically a mixture of cobalt and molybdenum oxides on alumina. Such processes with special names that are described in this dictionary are: Alkacid, Alkazid, Autofining, Cycloversion, Diesulforming, GO-FINING, Gulfining, Hycon, Hyperforming, RDS Isomax, Residfining, Trickel, Ultrafining, VGO Isomax, VRDS Isomax.

Reynolds, J. G., *Chem. Ind. (London),* 1991, 570.

Startsev, A. N., *Catal. Rev. Sci. Eng.,* 1995, **37**(3), 353.

Hydrofining   A process for desulfurizing and hydrotreating a wide range of petroleum fractions. Licensed by the Esso Research and Engineering Company.

> Unzelman, G. H. and Wolf, C. J., in *Petroleum Processing Handbook,* Bland, W. F. and Davidson, R. L., Eds., McGraw-Hill, New York, 1967, 3-40.
>
> *Hydrocarbon Process.,* 1994, **73**(11), 135.

Hydrofinishing   The final stage in a number of petroleum hydrorefining processes, commonly used in the manufacture of lubricating oil.

Hydroforming   [**Hydrog**en re**forming**]   A *catalytic reforming process, operated in an atmosphere of hydrogen. The catalyst was molybdena/alumina in a fixed bed. Developed jointly by Standard Oil of New Jersey, Standard Oil of Indiana, and MW Kellogg, and first operated in 1940 in Texas City. An improved version was developed by Standard Oil of California in 1943. The name has also been used as a general term for the catalytic dehydrogenation of aliphatic hydrocarbons to aromatic hydrocarbons.

> Ciapetta, F. G., Dobres, R. M., and Baker, R. W., *Catalysis,* in Vol. 6, Emmett, P. H., Ed., Reinhold, New York, 1958, 495.
>
> Asinger, F., *Mono-olefins: Chemistry and Technology,* translated by B. J., Hazzard, Pergamon Press, Oxford, 1968, 388.
>
> Spitz, P. H., *Petrochemicals, the Rise of an Industry,* John Wiley & Sons, New York, 1988, 159.

hydroformylation   A general name for processes which extend the chain lengths of aliphatic compounds by the catalytic addition of carbon monoxide and hydrogen. An olefin is thus converted to an aldehyde, whose molecule contains one more carbon atom than the original olefin. The *OXO process was the first such process to be developed. Other related processes described in this dictionary are: Aldox, Kuraray, RCH/RP.

> Weissermel, K. and Arpe, H.-J., *Industrial Organic Chemistry,* 3rd ed., VCH Publishers, Weinheim, Germany, 1997, 125.

hydrogenation, catalytic   *See* catalytic hydrogenation.

Hydrogen Polybed PSA   A version of the *Polybed process, for purifying hydrogen from various industrial processes. Developed by the Union Carbide Corporation in 1975 and now licensed by UOP. More than 400 units were operating worldwide in 1992.

> *Hydrocarbon Process.,* 1996, **75**(4), 120.

hydroisomerization   A general name for processes which isomerize aromatic hydrocarbons, operated in a hydrogen atmosphere. Exemplified by *Hysomer, *Isarom, *Isoforming, *Isomar, *Isopol, and *Octafining. The name is used also for processes which convert less-branched hydrocarbons to more-branched ones, to increase their octane numbers.

> Ravishankar, R. and Sivasenker, S., *Appl. Catal.,* 1996, **142**, 47.

Hydromag   A process for removing sulfur dioxide from industrial gas streams by absorption in magnesia. Developed in Japan by Nissan Chemical Industries.

Hydropol   A process for co-hydrogenating *n*-butenes with olefinic gasoline fractions. Developed by the Institut Français du Pétrole as part of its "polymer gasoline" process.

> *Hydrocarbon Process.,* 1980, **59**(9), 219.

Hydroprocessing   A general term for a family of petrochemical processes including *hydrotreating, *hydrorefining, and *hydrocracking. They are all heterogeneous catalytic processes for removing sulfur- and nitrogen-containing compounds from crude oil fractions and they may or may not induce *cracking. Oil and Gas Journal classifies these processes according to the amount of cracking induced. Hydrotreating induces no cracking; hydrorefining converts approximately 10 percent of the feed to lower molecular weight products; and hydrocracking converts approximately 50 percent to lower molecular weight products.

Oil & Gas J., 1998, **96**(2), 13.

Hydropyrolysis A catalytic process for converting coal into a mixture of liquid and gaseous products. It is operated at high temperatures and pressures, with a residence time in the pyrolysis zone of only a few seconds.

Hydrorefining A general name for petroleum refining processes using hydrogen gas. *See also* Hydroprocessing.

Hydrosulfreen A process for removing sulfur compounds from the tail gas from the *Claus process. It combines the *Sulfreen process with an upstream hydrolysis/oxidation stage, which improves efficiency and optimizes the emission control. Developed jointly by Lurgi and Société National Elf Aquitaine, and installed in 1990 in the Mazovian Refining and Petrochemical Works, near Warsaw, Poland. *See also* Oxysulfreen.

Ind. Miner. (London), 1990, (274), 14.

hydrotreating A general name for a family of catalytic petroleum refining processes which use hydrogen. They may remove organic sulfur-, nitrogen-, metal-, and oxygen-compounds, or hydrogenate olefins. *See also* Hydroprocessing.

The Petroleum Handbook, 6th ed., Elsevier, Amsterdam, 1983, 306.
Hydrotreating Catalysts, Occelli, M. L. and Anthony, R. G., Eds., Elsevier, Amsterdam, 1989.
Topsoe, H., Clausen, B. S., Massoth, F. E., *Hydrotreating Catalysis: Science and Technology,* Springer-Verlag, Berlin, 1996.

HYDROZONE [**Hydr**ogen peroxide **ozone**] A process for destroying organic wastes in aqueous streams by oxidation with a combination of hydrogen peroxide, ozone, and UV illumination. Developed by EA Technology, Chester, UK, in the 1980s. *See also* Ultrox.

HYFLEX A process for making fuel gases and liquids by hydrogenating coal. Powdered coal is reacted with hydrogen at high temperature and pressure, in a reactor with a reaction time of only two seconds. The products are mostly methane, ethane, and aromatic hydrocarbons. An advantage over other coal gasification processes is that no heavy tars are produced. The process will also accept other plant-derived raw materials such as eucalyptus wood, sugar cane, and pineapple wastes. Developed and piloted since 1976 by the Institute of Gas Technology, Chicago.

HYGAS [**Hy**dro**gas**ification] Also known as **IGT HYGAS,** after the Institute of Gas Technology where it was developed. A coal gasification process, intended to maximize the production of methane for use as SNG (substitute natural gas). Crushed coal is slurried in light oil and dried in a fluidized bed. It is then hydrogenated by one of three processes: steam treatment over an iron catalyst (S-I), steam and oxygen treatment, or electrothermally. Laboratory work began in 1946 under sponsorship from AGA. The U.S. Office of Coal Research took over sponsorship in 1964 and continued until 1976. Recent funding has come from the Department of Energy and the American Gas Association. Several pilot units were operated in the 1960s and 70s. The electrothermal method was abandoned in 1974 as it was uneconomic; the steam-oxygen process was still being tested in 1976; and sections of the steam-iron process were being piloted in 1976. One version was piloted in Chicago in 1981 and a commercial plant was designed.

Blair, W. G., Leppin, D., and Lee, A. L., *Methanation of Synthesis Gas,* American Chemical Society, Washington, D.C., 1975, 123.
Dainton, A. D., in *Coal and Modern Coal Processing,* Pitt, G. J., and Millward, G. R., Eds., Academic Press, London, 1979, 150.

Hebden, D. and Stroud, H. J. F., in *Chemistry of Coal Utilization*, 2nd. Suppl. Vol., Elliott, M. A., Ed., John Wiley & Sons, New York, 1981, 1679.

HyL [Hojalata y Lamina]   A direct reduction ironmaking process in which pellets or lumps of ore are reduced in a batch reactor using a mixture of hydrogen and carbon monoxide. Used in countries which have natural gas and cannot afford to invest in blast furnaces. Developed in the 1950s in Mexico by the Hojalata y Lamina Steel Company (now Hylsa) and the MW Kellogg Company, and now operated in nine other countries too. *See* DR.

Hyperforming   A *hydrodesulfurization process in which the catalyst moves by gravity down the reactor and is returned to the top by a solids conveying technique known as "hyperflow." Developed by the Union Oil Company of California in 1952 and first operated commercially in 1955.

Berg, C., *Pet. Refin.*, 1952, **31**(12), 131 (*Chem. Abstr.*, **49**, 15223).

Unzelman, G. H. and Wolf, C. J., in *Petroleum Processing Handbook*, Bland, W. F. and Davidson, R. L., Eds., McGraw-Hill, New York, 1967, 3-33.

Little, D. M., *Catalytic Reforming*, PennWell Publishing, Tulsa, OK, 1985, xv.

Hypersorption   A continuous chromatographic separation process using a moving bed. Invented in 1919 by F. D. Soddy (famed for his work on isotopes) at Oxford and developed commercially for petroleum refinery separations by the Union Oil Company of California in 1946. Six plants were built in the late 1940s, using activated carbon as the adsorbent. The process was abandoned because attrition of the bed particles proved uneconomic.

U.S. Patents 1,422,007; 1,422,008.

Berg, C., *Trans. Inst. Chem. Eng.*, 1946, **42**, 665.

Berg, C., *Chem. Eng. Prog.*, 1951, **47**(11), 585.

Yang, R. T., *Gas Separation by Adsorption Processes*, Butterworths, Guildford, England, 1987, 217.

Hypol   A process for making polypropylene, generally similar to *Spheripol. Developed by Mitsui Petrochemical Company, Japan.

Hypotreating   A process for desulfurizing and hydrogenating petroleum fractions. Developed by the Houdry Process and Chemical Company.

Unzelman, G. H. and Wolf, C. J., in *Petroleum Processing Handbook*, Bland, W. F. and Davidson, R. L., Eds., McGraw-Hill, New York, 1967, 3-42.

Hypro   A process for making hydrogen by catalytically decomposing hydrocarbons to carbon and hydrogen. The carbon is burnt to provide the heat for the reaction. Developed by UOP.

Pohlenz, J. B. and Stine, L. O., *Hydrocarbon Process.*, 1962, **41**(5), 191.

Hydrocarbon Process., 1964, **43**(9), 232.

Unzelman, G. H. and Wolf, C. J., in *Petroleum Processing Handbook*, Bland, W. F. and Davidson, R. L., Eds., McGraw-Hill, New York, 1967, 3-147.

Hy-Pro   A process for recovering propane and heavier components from natural gas or refinery off-gases. Licensed by ABB Randall Corporation.

Hydrocarbon Process., 1996, **75**(4), 124.

HYSEC   A process for purifying hydrogen from coke-oven gas by *PSA, developed by Mitsubishi Kakoki Kaisha and the Kansai Coke & Chemicals Company.

Suzuki, M., in *Adsorption and Ion Exchange: Fundamentals and Applications*, LeVan, M. D., Ed., American Institute of Chemical Engineers, New York, 1998, 122.

Hysomer [**H**ydro**isomer**ization]   A process for converting *n*-pentane and *n*-hexane into branched-chain hydrocarbons. Operated in the vapor phase, in the presence of hydrogen, in a fixed bed of a mordenite catalyst loaded with a platinum. Developed by Shell Oil Company and licensed worldwide through UOP. Used in the *Total Isomerization process.

> Kouwenhoven, H. W., Langhout, W. C., Van Zijll, *Chem. Eng. Prog.* 1971, **67**(4), 65.
> Symoniac, M. F., *Hydrocarbon Process.,* 1980, **59**(5), 110.
> Cusher, N. A., in *Handbook of Petroleum Refining Processes,* Meyers, R. A., Ed., McGraw-Hill, New York, 1986, 5-15.

Hysulf   A process for converting hydrogen sulfide to hydrogen and sulfur. The two steps are conducted in a polar organic solvent. In the first, the H_2S reacts with a quinone to form sulfur and the corresponding hydroquinone. In the second, the hydroquinone solution is catalytically processed to yield hydrogen and the original quinone. Developed by Marathon Oil, with funding from the Institute of Gas Technology, Chicago.

> *Chem. Eng. (N.Y.),* 1994, **101**(11), 19.
> Quinlan, M. P., Echterhof, L. W., Leppin, D., and Meyer, H. S., *Oil & Gas J.,* 1997, **95**(29), 54.

Hytanol   A *peak shaving process for making fuel gas from methanol. Offered by Lurgi.

HyTex   [**Hydrogen Tex**aco]   A process for making pure hydrogen from waste gases in oil refineries. There are three stages. In the first, partial combustion with oxygen in a noncatalytic reactor yields a mixture of carbon monoxide and hydrogen (*syngas). In the second, the shift reaction with steam converts the carbon monoxide to carbon dioxide and more hydrogen. The third stage uses *PSA to separate the hydrogen from the carbon dioxide and various impurities. Developed by Texaco Development Corporation and announced in 1991; the first unit, in Anacortes, WA, was scheduled for completion in 1993.

> *Chem. Eng. News,* 1991, **69**(19), 29.
> *Hydrocarbon Process.,* 1992, **71**(4), 112.

Hytoray   [**Hydrogenation Toray**]   A process for hydrogenating benzene to cyclohexane. Developed by Toray.

> Weissermel, K. and Arpe, H.-J., *Industrial Organic Chemistry,* 3rd ed., VCH Publishers, Weinheim, Germany, 1997, 346.

HYTORT   A process for making gaseous and liquid fuels from oil shale. Developed by the Institute of Gas Technology, Chicago, in 1959. It uses high-pressure hydrogenation, which recovers more of the carbon from shale than does pyrolysis. In 1981 a joint venture of IGT with the Phillips Petroleum Company was formed in order to make a feasibility study.

> *Hydrocarbon Process.,* 1986, **65**(4), 99.

HYVAHL   A *hydrotreating process for upgrading petroleum residues. Developed by the Institut Français du Pétrole. Three units were operating in 1996. *See also* TERVAHL.

> *Hydrocarbon Process.,* 1988, **67**(9), 78.
> *Hydrocarbon Process.,* 1996, **75**(11), 130.

HYVAHL F   An improved version of *HYVAHL

> *Hydrocarbon Process.,* 1994, **73**(11), 130.

HYVAHL S   An improved version of *HYVAHL which uses two reactors containing two different catalysts, operated in a swing mode.

> *Appl. Catal. A: Gen.,* 1993, **102**(2), N19.
> *Hydrocarbon Process.,* 1994, **73**(11), 130.

I

Ibuk   A process for removing free sulfur and sulfur dioxide from "benzole" (a mixture of aromatic hydrocarbons used as a motor fuel) by passing the vapor through sulfuric acid and then hot aqueous sodium hydroxide. Invented by T. Troniseck and used in Germany in the 1930s.

> British Patent 364,778.
> Claxton, G., *Benzoles, Production and Uses,* National Benzole & Allied Products Assoc., London, 1961, 417.

ICAR   [Intermediate Catalytic Accumulation of Ionizing Radiation Energy]   A process for converting nuclear energy to chemical energy. Energy from a nuclear reactor is used to promote the catalytic reforming of methane to *syngas. Proposed by Yu. A. Aristov in 1993. *See also* EVA-ADAM.

> Parmon, V. N., *Catal. Today,* 1997, **35** (1–2), 153.

ICEM   *See* DR.

ICG-GUD   [Integrated coal gasification—Gas und Dampf]   An integrated coal gasification and power generation system developed by Siemens, Germany. The coal is gasified by exposure to steam and oxygen. The gases are then burnt in a gas turbine, producing electric power, and the hot combustion products then generate more power in a steam turbine. Removal of sulfur dioxide, nitrogen oxides, and dusts is easier than in conventional coal-fired power stations. A demonstration plant was scheduled to be built in The Netherlands in 1990.

IChPW   *See* carbonization.

ICI Low Pressure Methanol   A process for making methanol from methane and steam. The methanol is first converted to *syngas by *steam reforming at a relatively low pressure. The syngas is then converted to methanol over a copper-based catalyst:

$$CO + 2H_2 = CH_3OH$$
$$CO_2 + 3H_2 = CH_3OH + H_2O$$
$$CO + H_2O = CO_2 + H_2O$$

Conversion per pass is limited by reaction equilibrium; after cooling to condense the product methanol, the unreacted gas is recycled to the reactor. Developed by Imperial Chemical Industries in the late 1960s, since when it has been the leading process. As of 1991, 41 plants had been commissioned and a further 7 were under contract or construction.

> *Hydrocarbon Process.,* 1991, **70**(3), 164.
> Weissermel, K. and Arpe, H.-J. *Industrial Organic Chemistry,* 3rd ed., VCH Publishers, Weinheim, Germany, 1997, 29.

ICI Steam Naphtha Reforming   A *steam reforming process, adapted for processing high-boiling naphtha. A highly selective nickel-based catalyst is used. Developed by Imperial Chemical Industries in the 1960s, based on its early experience of the steam reforming of methane. Following the development, over 400 reformers were built in over 30 countries. When North Sea gas and other natural gas reserves around the world were discovered, the use of naphtha as a feedstock declined; nevertheless, the process is still used in many countries.

> Weissermel, K. and Arpe, H.-J. *Industrial Organic Chemistry,* 3rd ed., VCH Publishers, Weinheim, Germany, 1997, 18.

ICON [Integrated chlorination and oxidation]   An improved version of the *Chloride Process for making titanium dioxide pigment. It operates at above atmospheric pressure and is claimed to be cheaper to build. Chlorine from the oxidation section, under pressure, is introduced directly to the chlorinator. Developed by Tioxide Group, and first operated at its plant at Greatham, UK, in 1990.

Chem. Eng. (Rugby, England), 1991, (497), 13.
Ind. Miner. (London), 1991, (286), 17.

IDAS   A process for making isoprene (for the manufacture of a synthetic rubber) by the oxidative dehydrogenation of isopentene. Iodine is the initial reactant; the hydrogen iodide produced in the reaction is reconverted to elemental iodine via nickel iodide.

Dumas, T. and Bulani, W., *Oxidation of Petrochemicals: Chemistry and Technology*, Applied Science Publishers, London, 1974, 139.

Idemitsu   A process for making C_6–C_{18} α-olefins from ethylene, catalyzed by a modified homogeneous *Ziegler-Natta catalyst containing a zirconium chloride. Developed by the Idemitsu Petroleum Company in 1988.

European Patent 328,728.
Al-Sa'doun, A. W., *Appl. Catal. A: Gen.*, 1993, **105**(1), 3.

IDR (1)   [Isobaric double recycle]   A process for making urea from ammonia and carbon dioxide, via ammonium carbamate:

$$NH_3 + CO_2 \rightarrow NH_2COONH_4 \rightarrow NH_2CONH_2$$

Developed by Fertimont (a subsidiary of Montedison) and operated in Italy since 1988 and in China since 1989.

Eur. Chem. News, 1982, **39**(1044), 15.
Zardi, U., *Nitrogen*, 1982, (135), 33.

IDR (2)   [Integrated dry route]   A process for making ceramic-grade uranium dioxide from uranium hexafluoride. Uranium hexafluoride vapor is first reacted with superheated steam to form uranyl fluoride; this is then reduced with hydrogen to form uranium dioxide. Developed at Springfields, UK, by British Nuclear Fuels in the 1970s. By 1989, more than 7,500 tons of uranium had been processed there in this way. Licenses have been granted to Westinghouse Electric Corporation, United States, and FBFC, France.

ifawol   A process for removing phenol from aqueous wastes by solvent extraction into a high-boiling organic solvent mixture made by the *OXO process.

Martinez, D., *Chemical Waste Handling and Treatment*, Muller, K. R., Ed., Springer-Verlag, Berlin, 1986, 182.

I-Forming   [isobutane reforming]   A process for selectively converting heavy naphtha to isobutane. Developed by UOP.

IFP   The Institut Français du Pétrole has developed many processes, but the one most associated with its name is that for removing residual sulfur dioxide and hydrogen sulfide from the tail gases from the *Claus process. This was invented by P. Renault in 1966. The gases are passed through a solvent such as tributyl phosphate at a temperature higher than that of the melting point of sulfur. A soluble catalyst promotes a Clause-type reaction and the resulting liquid sulfur is separated from the base of the vessel. In 1976, 27 plants were either in operation or in design or construction.

US Patent 3,441,379.
Barthel, Y., Bistri, Y., Deschamps, A., Renault, P., Simadoux, J. C., and Dutriau, R., *Hydrocarbon Process.*, 1971, **50**(5), 89.

Kohl, A. L. and Riesenfeld, F. C., *Gas Purification,* 4th ed., Gulf Publishing, Houston, TX, 1985, 542.

Ifpexol   A process for removing water and acid gases from hydrocarbon gas streams by washing with a proprietary solvent at a very low temperature. Developed by the Institut Français du Pétrole.

Hydrocarbon Process., 1996, **75**(4), 124.
Chem. Technol. Europe,, 1996, **3**(3), 10.

IFP Oxypyrolysis   Also called *NGOP. A process for converting natural gas to gasoline, based on the oxidative coupling of methane to ethane in a fixed-bed reactor. Developed in 1991 by the Institut Français du Pétrole.

Mimoun, H., Robine, A., Bonnaudet, S., and Cameron, C. J., *Appl. Catal. A: Gen.,* 1990, **58**, 269.
Hutchings, G. H. and Joyner, R. W., *Chem. Ind. (London),* 1991, (16), 575.
Raimbault, C. and Cameron, C. J., in *Natural Gas Conversion,* Holmen, A., Jens, K.-J., and Kolboe, S., Eds., Elsevier, Amsterdam, 1991, 479.

IFP-SABIC   *See* Alphabutol.

IFP Stackpol   *See* Stackpol.

IG   IG-Farbenindustrie in Germany developed many processes before World War II, but the one most associated with its name is probably the *Aldol process for making butadiene for synthetic rubber. The name has been used also for the *Bergius-Pier process.

IGCC   [Integrated gasification combined cycle]   A family of coal gasification processes in which the hot gases from the combustion are used to generate electrical power, in either a gas turbine or a high-temperature fuel cell. The need to remove hydrogen sulfide from the gases without cooling them has led to the development of special high-temperature adsorbents such as zinc ferrite and zinc titanate.

Chem. Eng. (Rugby, England), 1989, Nov, 35.
Woods, M. C., Gangwal, S. K., Harrison, D. P., and Jothimurugesan, K., *Ind. Eng. Chem. Prod. Res. Dev.,* 1991, **30**, 100.
Bissett, L. A. and Strickland, L. D., *Ind. Eng. Chem. Prod. Res. Dev.,* 1991, **30**, 170.
Chem. Eng. (N.Y.), 1992, **99**(1), 39.
Mills, G. A. and Rostrup-Nielsen, J., *Cat. Today,* 1994, **22**(2), 337.
Konttinen, J. T., Zevenhoven, C. A. P., and Hupa, M. M., *Ind. Eng. Chem. Res.,* 1997, **36**(6), 2332.
Jothimurugesan, K. and Gangwal, S. K., *Ind. Eng. Chem. Res.,* 1998, **37**, 1929.

IG-Hydrogenation   An advanced version of the *Bergius process in which the initial product of coal gasification is refined by centrifugation and the slurry residue is carbonized.

IGI   *See* carbonization.

IG-NUE   A coal liquifaction process developed by Bergbau-Forschung in Germany during World War II. Catalytic metal salts were impregnated in, or precipitated on, the coal. A pilot plant was to have been built in Westphalia in 1977.

Alpert, S. B. and Wolk, R. H., *Chemistry of Coal Utilization,* 2nd Suppl. Vol., Elliott, M. A., Ed., John Wiley & Sons, New York, 1981, 1963.

IGT HYGAS   *See* HYGAS.

Imatra   Also called the Solid Lime Process. A method for desulfurizing steel made by an electric arc process. Additions are made of "burnt lime" (calcium oxide), fluorspar (mineral calcium fluoride), and ferro-silicon.

IMI   [Israel Mining Industries]   *See* Haifa.

Imperial Smelting   A process for simultaneously extracting zinc and lead from sulfide ores, developed and commercialized by the Imperial Smelting Corporation at Avonmouth, UK, after World War II, and now widely used. Based on an invention by L. J. Derham in which the vapors emerging from a reducing kiln are rapidly quenched in a shower of droplets of molten lead. The first trial was made in 1943 but most of the development work was done from 1945 to 1947. Eleven plants were operating in 1973.

> Cocks, E. J. and Walters, B., *A History of the Zinc Smelting Industry in Britain,* George G. Harrap, London, 1968, 166.

INCO   [International Nickel Company]   An electrolytic process for extracting nickel from nickel sulfide matte. The matte is melted and cast into anodes. Electrolysis with an aqueous electrolyte containing sulfate, chloride and boric acid dissolves the nickel and leaves the sulfur, together with precious metals, as an anode slime. Operated in Manitoba by International Nickel Company of Canada.

> Spence, W. W. and Cook, W. R., *Trans. Can. Inst. Min. Met.,* 1964, **67**, 257 (*Chem. Abstr.,* **62**, 4951).
> Gupta, C. K. and Mukherjee, T. K., *Hydrometallurgy in Extraction Processes,* Vol. 1, CRC Press, Boca Raton, FL, 1990, 24.
> Hill, J., in *Insights into Speciality Inorganic Chemicals,* Thompson, D., Ed., Royal Society of Chemistry, Cambridge, 1995, 18.

Indirect   An alternative name for the *French process for making zinc oxide.

INICHAR   *See* carbonization.

INIEX   *See* carbonization.

Innovene   A proprietary process for polymerizing lower olefins in the gas-phase, incorporating *condensed phase technology. Various types of catalysts may be used, but the *Insite family of catalysts is now preferred. Developed by BP in the 1990s and first used commercially in 1998.

Insite   Not a process, but a range of constrained-geometry metallocene catalysts for polymerizing olefins. Developed by Dow Chemical.

> *Eur. Chem. News,* 1997, **67**(1770), 28.

InTox   A process for destroying toxic wastes in aqueous solution by oxidation with oxygen at high temperatures and pressures in a pipe reactor. No catalyst is required. The reactions take place at approximately 300°C and 120 atm. Developed by InTox Corporation, UK, based on a process for extracting aluminum from bauxite developed by Lurgi in the 1960s. *See also* Zimpro.

Iodide   *See* van Arkel and de Boer.

Ionics   A *flue-gas desulfurization process using aqueous sodium hydroxide. The resulting sodium sulfate solution is electrolyzed to yield sodium hydroxide, sodium bisulfate, sulfuric acid, oxygen, and hydrogen.

> Speight, J. G., *Gas Processing,* Butterworth Heinemann, Oxford, 1993, 296.

Iotech   A process for preparing wood for enzymatic digestion. The wood is heated with aqueous alkali under pressure and then rapidly decompressed.

> Jurasek, L., *Dev. Ind. Microbiol.,* 1979, **78**, 177.

IPA   [Interpass absorption]   Also called Double absorption, and Double catalysis. An improved version of the *Contact process for making sulfuric acid, by which the efficiency of the conversion of sulfur to sulfuric acid is increased from 98 percent to over 99.5 percent.

Phillips, A., in *The Modern Inorganic Chemicals Industry,* Thompson, R., Ed., Royal Society of Chemistry, London, 1977, 184.

Iron Sponge   Also called Dry box. An obsolete process for removing hydrogen sulfide from gas streams by reaction with iron oxide monohydrate. The ferric sulfide that is formed is periodically re-oxidized to regenerate ferric oxide and elemental sulfur. When this process becomes inefficient because of pore-blockage, the sulfur is either oxidized to sulfur dioxide for conversion to sulfuric acid, or is extracted with carbon disulfide.

Kohl, A. L. and Riesenfeld, F. C., *Gas Purification,* 4th ed., Gulf Publishing, Houston, TX, 1985, 421.

Speight, J. G., *Gas Processing,* Butterworth Heinemann, Oxford, 1993, 296.

ISAL   A *hydrotreating process for removing sulfur and nitrogen compounds from petroleum fractions without reducing their octane values. Developed by Intevep SA, the research and technology arm of Venezuela's state petroleum company PDVSA. A proprietary zeolite catalyst first saturates the olefins and then isomerizes them to higher octane-value compounds.

Chem. Eng. (N.Y.), 1997, **104**(4), 19.

Isarom   A catalytic process for isomerizing the xylene isomers, developed by Institut Français du Pétrole. The catalyst is aluminum trifluoride.

Isasmelt   A two-stage lead smelting process. In the first stage, the lead concentrate is oxidized by injecting air down a "Sirosmelt" lance, using helical vanes to swirl the gas. In the second, the high-lead slag is reduced with coal. The process was developed by Mount Isa Mines and the CSIRO, Australia, and due for startup at Mount Isa in 1991.

Chem. Eng. (N.Y.), 1990, **97**(4), 55.

ISCOR   *See* carbonization.

ISEP   [Ion separation exchange process]   A continuous ion-exchange process for purifying large volumes of water. The granular ion-exchange material is contained in a rotating carousel. Developed by the Progress Water Technologies Corporation, St. Petersburg, FL, and demonstrated for the removal of ammonium ion from municipal water supplies at the Florida South Water Reclamation facility in 1989, using clinoptilolite as the ion-exchanger.

Chem. Eng. (N.Y.), 1989, **96**(10), 197.

Iso-CDW   [Isomerization and catalytic dewaxing]   A general term for dewaxing processes which include these two processes. Exemplified by Isodewaxing, MSDW.

Oil & Gas J., 1997, **95**(35), 64.

Isocracking   A hydrocracking process developed and licensed by Chevron Research Company. The catalyst is nickel or cobalt sulfide on an aluminosilicate. First commercialized in 1962; more than 45 units had been built by 1994. *See also* Isomax.

U.S. Patents 2,944,005; 2,944,006.

Unzelman, G. H. and Wolf, C. J., in *Petroleum Processing Handbook,* Bland, W. F. and Davidson, R. L., Eds., McGraw-Hill, New York, 1967, 3-18.

Weisser, O. and Landa, S., *Sulphide Catalysts, Their Properties and Applications,* Pergamon Press, Oxford, 1973, 308.

Hydrocarbon Process., 1994, **73**(11), 122.

Bridge, A. G., in *Handbook of Petroleum Refining Processes,* Meyers, R. A., Ed., McGraw-Hill, New York, 1997, 7.21.

Isocure   A process for making foundry molds developed by Ashland Chemical Company. In 1990 it was announced that a pilot plant was to be built in cooperation with the USSR and that the process had been licensed in China. *See also* Pep Set.

Isodewaxing A catalytic dewaxing process developed by Chevron Research & Technology. It incorporates catalysts that achieve both wax isomerization and shape-selective cracking.

Oil & Gas J., 1997, **95**(35), 64.

ISOFIN [**Iso** ole**fin**s] A cataylytic process for making iso-olefins from normal olefins by skeletal isomerization. The principle example converts *n*-butenes to isobutylene, needed as a feedstock for making methyl *t*-butyl ether. Developed by BP Oil Company, Mobil Corporation, and MW Kellogg from 1992.

Chem. Week, 1992, **151**(2), 8.
Chem. Week, 1994, **155**(6), 35.
Hydrocarbon Process., 1994, **73**(11), 138.
Hydrocarbon Process., 1996, **75**(11), 140.

Isoforming A process for increasing the octane rating of thermally cracked gasolines by catalytic isomerization over silica/alumina. Terminal alkenes are thus converted to nonterminal alkenes. Developed by Standard Oil Company of Indiana in the 1940s.

U.S. Patent 2,410,908.
Weissermel, K. and Arpe, H.-J., *Industrial Organic Chemistry,* 3rd ed., VCH Publishers, Weinheim, Germany, 1997, 331.

Iso-Kel [**Iso**merization-**Kel**logg] A fixed-bed, vapor-phase isomerization process for making high-octane gasoline from aliphatic petroleum fractions. The catalyst is platinum on alumina. Developed by MW Kellogg.

Unzelman, G. H. and Wolf, C. J., in *Petroleum Processing Handbook,* Bland, W. F. and Davidson, R. L., Eds., McGraw-Hill, New York, 1967, 3-49.

Isolene II A catalytic process for converting ethylbenzene to mixed xylenes. The catalyst is platinum on an acidic support. Developed by Toray Industries, Japan. *See also* Isomar.

Otani, S., *Chem. Eng. (N.Y.),* 1973, **80**(21), 106.

Isomar [**Isom**erization of **ar**omatics] A catalytic process for isomerizing xylene isomers and ethylbenzene into equilibrium isomer ratios. Usually combined with an isomer separation process such as *Parex (1). The catalyst is a zeolite-containing alumina catalyst with platinum. Developed by UOP and widely licensed by them. It was first commercialized in 1967; by 1992, 32 plants had been commissioned and 8 others were in design or construction. *See also* Isolene II.

Hydrocarbon Process., 1991, **70**(3), 192.
Jenneret, J. J., in *Handbook of Petroleum Refining Processes,* Meyers, R. A., Ed., McGraw-Hill, New York, 1997, 2.37.

Isomate A continuous, nonregenerative process for isomerizing $C_5 - C_8$ normal paraffins, catalyzed by aluminum trichloride and hydrogen chloride. Developed by Standard Oil of Indiana.

Unzelman, G. H. and Wolf, C. J., in *Petroleum Processing Handbook,* Bland, W. F. and Davidson, R. L., Eds., McGraw-Hill, New York, 1967, 3-49.
Asinger, F., *Paraffins, Chemistry and Technology,* translated by B. J. Hazzard, Pergamon Press, Oxford, 1968, 712.

Isomax Originally a general trade name for a family of petroleum processes developed jointly by UOP and Chevron Research Company, including RCD Isomax, RDS Isomax, VGO Isomax, and VRDS Isomax. The name was later used by UOP. It has been applied to the family of UOP hydrotreating and Unibon processes. *See also* Unicracking.

Isomerate   A continuous hydrocarbon isomerization process for converting pentanes and hexanes to highly branched isomers. Developed by the Pure Oil Company, a division of the Union Oil Company of California. The catalyst, unlike those used in most such processes, does not contain a noble metal.

Unzelman, G. H. and Wolf, C. J., in *Petroleum Processing Handbook,* Bland, W. F. and Davidson, R. L., Eds., McGraw-Hill, New York, 1967, 3-48.

Isomerization   The following named isomerization processes are described elsewhere in this dictionary. Many of them take place in the presence of hydrogen and may therefore also be called hydroisomerization processes; but it is not always clear whether the hydrogen is essential, and no distinction is made here between isomerization and hydroisomerization: Anglo-Jersey, Butamer, Butomerate, Catstill, Chevron (1), Cold acid, Cold Hydrogenation, Hysomer, Isarom, Isarom, Isoforming, Iso-Kel, Isolene II, Isomar, Isomate, Isomerate, Koch, MHTI, MLPI, MVPI, Octafining, Octol, Penex, Pentafining, T2BX, Tatoray, TDP, TIP, Xylofining.

Unzelman, G. H. and Wolf, C. J., in *Petroleum Processing Handbook,* Bland, W. F. and Davidson, R. L., Eds., McGraw-Hill, New York, 1967, 3-46.

ISOMPLUS   A process for isomerizing *n*-butenes to isobutene. Developed by CD Tech and Lyondell Petrochemical. One unit was operating in 1996.

Eur. Chem. News, 1993, **60**(1597), 28.
Hydrocarbon Process., 1996, **75**(11), 138.

IsoPlus   Not a process but a range of zeolite catalysts for making iso-olefins by *FCC. Developed by Englehard.

Benton, S., *Oil & Gas J.,* 1995, **93**(18), 98.

Iso-Plus Houdriforming   A complex petroleum *reforming process, based on the *Houdriforming process. Developed by the Houdry Process and Chemical Company.

Asinger, F., *Mono-olefins: Chemistry and Technology,* translated by B. J., Hazzard, Pergamon Press, Oxford, 1968, 391.
Unzelman, G. H. and Wolf, C. J., in *Petroleum Processing Handbook,* Bland, W. F. and Davidson, R. L., Eds., McGraw-Hill, New York, 1967, 3-36.

Isopol   A *hydroisomerization process for converting 1-butene to 2-butene. Developed by the Institut Français du Pétrole.

Hydrocarbon Process., 1980, **59**(9), 219.

IsoSiv   [Isomer separation by molecular sieves]   A process for separating linear hydrocarbons from naphtha and kerosene petroleum fractions. It operates in the vapor phase and uses a modified 5A zeolite molecular sieve, which selectively adsorbs linear hydrocarbons, excluding branched ones. Developed by Union Carbide Corporation and widely licensed, now by UOP. The first plant was operated in Texas in 1961. By 1990, more than 30 units had been licensed worldwide. *See also* Total Isomerization.

Avery, W. F. and Lee, M. N. Y., *Oil & Gas J.,* 1962, **60**(23), 121.
Symoniak, M. F., *Hydrocarbon Process.,* 1980, **59**(5), 110.
Yang, R. T., *Gas Separation by Adsorption Processes,* Butterworths, Guildford, England, 1987, 242.
Cusher, N. A., in *Handbook of Petroleum Refining Processes,* Meyers, R. A., Ed., McGraw-Hill, New York, 1997, 10.53.

Iso-Synthesis   A version of the *Fischer-Tropsch process developed in Germany during World War II.

Lane, J. C., *Pet. Refin.,* 1946, **25**(8), 87; (9), 423; (10), 493 (11), 587 (*Chem. Abstr.,* **42,** 9118).

Isotex A process for isomerizing olefines, catalyzed by a novel zeolite. It is intended for making methyl *t*-butyl ether from C_4 hydrocarbons.

Chem. Week Internat., 1995, **156**(8), 9.

Ispra Mark 13A A *flue-gas desulfurization process developed at the Joint Research Centre of the European Community at Ispra, Italy, from 1979. It uses a novel electrochemical method to regenerate the solution used for absorbing the sulfur dioxide. The products are concentrated sulfuric acid and hydrogen. The absorbent is a dilute aqueous solution of sulfuric and hydrobromic acids, containing a small amount of elemental bromine. Sulfur dioxide reacts with the bromine thus:

$$SO_2 + 2H_2O + Br_2 = H_2SO_4 + 2HBr$$

The bromine is regenerated by electrolysis:

$$2HBr = H_2 + Br_2$$

A large pilot plant was built at the SARAS oil refinery in Sardinia in 1989.

Langenkamp, H. and Van Velzen, D., *Flue-gas Desulphurization by the Ispra Mark 13 Process.* The European Commission, Ispra, Italy, 1988.

ITmk3 [**mark 3** indicates that this is a third generation ironmaking process, marks one and two being the blast furnace and direct reduction] A modification of the *Fastmet process, for making molten iron. Pelleted iron ore fines are reduced with a solid reductant. The iron in the reduced pellets separates as molten metal, uncontaminated by gangue. Developed in 1996 by Midrex Corporation and Kobe Steel. Commercialization is expected in 2003.

J

Jacobs-Dorr An alternative name for the *Dorr-Oliver process for making phosphoric acid, adopted since the technology was acquired by the Jacobs Company in 1974. Twenty seven plants were operating in 1989.

Gard, D. R., in *Encyclopedia of Chemical Processing and Design,* McKetta, J. J. and Cunningham, W. A., Eds., Marcel Dekker, New York, 1990, **35**, 455.

James An early process for making mixed oxygenated organic compounds by the catalytic oxidation of petroleum fractions. The products were aldehydes, alcohols, and carboxylic acids. Developed by J. H. James at the Carnegie Institute of Technology, Pittsburgh.

James, J. H., *Chem. Metall. Eng.,* 1922, **26**(5), 209.

Asinger, F., *Paraffins, Chemistry and Technology,* translated by B. J. Hazzard, Pergamon Press, Oxford, 1968, 632.

Jarosite [Named after the mineral, first recognized at Jarosa, Spain] A process for removing iron from the leach liquors from hydrometallurgical operations. First used in 1964 in processing zinc sulfate liquors at Asturiana de Zinc, Spain. Also used for recovering zinc from the residues from the electrolytic zinc process. *See also* Goethite, Haematite.

Steinveit, G., in *Advances in Extractive Metallurgy and Refining,* Jones, M. J., Ed., Institution of Mining & Metallurgy, London, 1972, 521.

Morgan, S. W. K., *Zinc and Its Alloys and Compounds,* Ellis Horwood, Chichester, England, 1985, 117.

Jenkins   An early liquid-phase thermal cracking process. *See also* Dubbs.

Jet Smelting   *See* DR.

JGCC   [Japan Gas-Chemical Company]   *See* MGCC.

Jindal   A direct reduction ironmaking process, using coal as the reductant. Two plants were operating in India in 1997 and two more were under construction. *See* DR.

Jones   A regenerative process for making carbon black by pyrolyzing petroleum fractions. The gaseous co-product can be added to town gas.

British Petroleum Co., *Gas Making and Natural Gas,* British Petroleum Co., London, 1972, 66.

JPL Chlorinolysis   [Jet Propulsion Laboratory]   A process for desulfurizing coal by oxidation with chlorine. The sulfur becomes converted to sulfur monochloride, S_2Cl_2. Developed by the Jet Propulsion Laboratory of the California Institute of Technology from 1976 to 1981.

IEA Coal Research, *The Problems of Sulphur,* Butterworths, London, 1989, 30.

Joosten   Also known as the two-shot system. A chemical grouting system for solidifying permeable sandy masses and masonry composed of sandy materials. Successive injections of sodium silicate and calcium chloride solutions are made through a pipe, which terminates in the ground to be hardened. Calcium silicate precipitates and binds the soil particles together. Invented by H. Joosten in 1928. *See also* Siroc.

U.S. Patent 1,827,238.
British Patent 322,182.

Juratka   A process for forming a protective oxide film on aluminum. *See also* Alzac, metal surface treatment.

K

KA   A process for making the explosive RDX by nitrating hemamethylene tetramine in acetic anhydride. Developed in Germany during World War II.

KAAP   [Kellogg advanced ammonia process]   The first high-pressure process developed for synthesizing ammonia from its elements which does not use an iron-containing catalyst. The reformer gas for this process is provided by the *KRES process. The catalyst was developed by BP; it contains ruthenium supported on carbon. Developed by MW Kellogg Company in 1990 and first installed by the Ocelot Ammonia Company (now Pacific Ammonia) at Kitimat, British Columbia, from 1991 to 1992. Another plant was installed at Ampro Fertilizers in Donaldsonville, LA, in 1996.

Chem. Week, 1991, 3 Apr, 13.
Eur. Chem. News, 1992, **58**(1524), 42.
Eur. Chem. News, 1993, **60**(1592), 27.
Oil & Gas J., 1996, **94**(47), 37.

Kaldnes   A variation of the *Activated Sludge process for sewage treatment, in which the biological matter is immobilized within short lengths of plastic pipe. It is very effective for removing nitrogenous compounds. Developed in 1987 by Trondheim Technical University and commercialized by Kaldnes, a Norwegian engineering company. Tested in a full-scale plant in Oslo in 1990.

Pollution Prevention, 1994, **4**(5), 50.

Kaldo   [**Kal**ling **Do**mnarvets]   An oxygen steelmaking process, first operated in Domnarvets, Sweden, in 1956. The furnace rotates at approximately 30 rpm around an axis tilted 17°C to the vertical. Variation in the rate of oxygen supply and speed of rotation permit close control of the steel composition. Invented by B. Kalling. The name is now used as a general name for both ferrous and nonferrous metallurgical processes using rotating furnaces, developed by the Boliden group of companies in Sweden.

Kalling   *See* DR.

Kalthydrierung   [German, meaning cold hydrogenation]   A process for selectively hydrogenating "pyrolysis gasoline," a petroleum refining byproduct, at temperatures below 100°C. A palladium catalyst is used.

Krönig, W., *Erdoel Kohle,* 1965, **18**, 432.

Kalunite   [from **K**, potassium, and **alunite**, the ore]   A process for extracting aluminum from alunite, a naturally occurring basic sulfate of aluminum and potassium having the idealized formula $KAl_3(SO_4)_2(OH)_6$. Based on an invention made by G.S. Tilley in 1924. The ore is first dehydrated at up to 600°C. It is then leached with a solution of sulfuric acid and potassium sulfate. After clarification of the leachate, potassium alum is crystallized out. Hydrothermal treatment of potassium alum precipitates a basic potassium alum, $K_2SO_4 \cdot 3Al_2O_3 \cdot 4SO_3 \cdot 9H_2O$, simultaneously regenerating potassium sulfate and sulfuric acid. Calcination of this potassium alum yields a mixture of alumina and potassium sulfate, which is leached out. Piloted by Kalunite in Salt Lake City, UT, in 1943 but later abandoned. *See also* Alumet.

U.S. Patent 1,591,798.

O'Connor, D. J., *Alumina Extraction from Non-bauxitic Materials,* Aluminium-Verlag, Düsseldorf, 1988, 196.

Kalvar   A reprographic process in which microscopic gas bubbles provide white pigmentation over a dark background. The photosensitive layer is a solid solution of a blowing agent in a resin. Exposure to ultraviolet radiation liberates nitrogen from the blowing agent, and heat softens the resin, permitting the nitrogen to collect in the form of sub-micron sized bubbles. The optical resolution of such a process can approach the ideal value since the bubbles can be as small as the wavelength of light. Commercialized by the Kalvar Corporation, New Orleans, in the 1950s, but later largely abandoned except for special applications.

U.S. Patents 2,911,299; 3,032,414.

Kanigen   An "electroless" process for plating metals with nickel, i.e., a process not using electrolysis. A nickel solution in contact with the metal is reduced with sodium hypophosphite. Developed by the General American Transportation Company.

Kastone   A process for destroying cyanide ion in solution by oxidizing it with a mixture of hydrogen peroxide and formaldehyde. Invented by Du Pont in 1970 and licensed to Degussa.

U.S. Patent 3,617,582.

Lawes, B. C., Fournier, C. B., and Mathre, D. B., *Plating,* 1973, **60**, 902,909.

KATAPAK   Not a process but a range of catalysts and catalyst supports using the principle of static mixing. Developed by Sulzer Chemtech since 1991. Used in *catalytic distillation processes.

Katasorbon   A process for removing carbonyl sulfide and other organic sulfur compounds from *syngas by combined catalysis and adsorption. Offered by Lurgi.

Katasulf   A process for removing hydrogen sulfide and ammonia from coke-oven gas, developed by IG Farbenindustrie in Germany in the 1920s. The basic reaction involved is the catalyzed oxidation of hydrogen sulfide to sulfur dioxide and water at approximately 420°C:

$$H_2S + \tfrac{3}{2}O_2 = H_2O + SO_2$$

Various catalysts have been used, including activated carbon, bauxite, and bimetallic oxides. The sulfur dioxide is then absorbed in a solution of ammonium sulfite and bisulfite; acidulation of this yields ammonium sulfate and elemental sulfur.

British Patent 310,063.
U.S. Patents 1,678,630; 1,889,942; 2,152,454.
Bühr, H., *Chem. Fabrik.*, 1938, **11**(1/2), 10.

KATOX   A wet-oxidation process for destroying organic residues in aqueous effluents. Oxidation takes place at the surface of active carbon granules.

Wysocki, G. and Hoeke, B., *Wasser Luft Betrieb.*, 1974, **18**, 311.
Martinez, D., in *Chemical Waste Handling and Treatment*, Muller, K. R., Ed., Springer-Verlag, Berlin, 1986, 229.

Katsobashvili   A low-pressure, catalytic process for the destructive hydrogenation of petroleum residues. Piloted in the USSR in the 1950s.

Weisser, O. and Landa, S., *Sulphide Catalysts, Their Properties and Applications*, Pergamon Press, Oxford, 1973, 297.

Katzschmann   *See* Witten.

Kawasaki   *See* DR.

Kawasaki Kasei   A process for oxidizing naphthalene to naphthaquinone. It is operated in the gas phase at 400°C, using air as the oxidant, and uses vanadium pentoxide on silica as the catalyst. Phthalic acid is a co-product.

Kaysam   A process for making rubber articles from rubber latex by flocculating it with a mixed electrolyte and then casting in a rotating, porous mold.

KBW   [Koppers Babcock & Wilcox]   A coal gasification process developed jointly by the Koppers Company and Babcock & Wilcox, intended to supply the synthetic fuels industry. The product is a mixture of carbon monoxide and hydrogen. Dry, powdered coal, oxygen, and steam are injected into the reactor. The reaction temperature is sufficiently high that the ash is molten; it runs down the reactor walls, is tapped out as a molten slag, and is quenched in water before disposal. In 1984, seven commercial synfuels projects planned to use this process but it is not known whether any was commercialized.

Dokuzoguz, H. Z., Kamody, J. F., Michaels, H. J., James, D. E., and Probert, P. B., in *Handbook of Synfuels Technology*, Meyers, R. A., Ed., McGraw-Hill, New York, 1984, 3-87.

Kel-Chlor   [Kellogg Chlorine]   A non-catalytic version of the *Deacon process for making chlorine by oxidizing hydrochloric acid, in which nitrosyl sulfuric acid and nitrosyl chloride are intermediates and concentrated sulfuric acid is used as a dehydrating agent:

$$HCl + NO \cdot HSO_4 = NOCl + H_2SO_4$$

$$HCl + NOCl + O_2 = Cl_2 + H_2O + NO_2$$

Developed by MW Kellogg and by Du Pont and operated by the latter company in Texas from 1974. *See also* Deacon, MT-Chlor.

Chem. Eng. News, 1969, **47**(19), 14.

van Dijk, C. P. and Schreiner, W. C., *Chem. Eng. Prog.,* 1973, **69**(4), 57.

Bostwick, L. E., *Chem. Eng. (N.Y.),* 1976, **83**(21), 86.

Tozuka, Y., in *Science and Technology in Catalysis,* Izumi, Y., Aral, H., and Iwamoto, M., Eds., Elsevier, Amsterdam, 1994, 44.

Van Dijk, C. P. and Schreiner, W. C., in *Inorganic Chemicals Handbook,* Vol. 2., McKetta, J. J., Ed., Marcel Dekker, New York, 1993, 759.

Kellogg-Hydrotreating   A two-stage hydrogenation process for converting olefins in petroleum fractions to benzene. Developed by the MW Kellogg Company.

Griffiths, D. J., James, J. L., and Luntz, D. M., *Erdoel Kohle Erdgas Petrochem.,* 1968, **21**(2), 83.

Kenox   A wet-air oxidation system for destroying municipal and industrial organic wastes. It uses a complex mixing system to accelerate the reactions. The process conditions are: pH 4; 250°C; 50 bar. Invented by R. P. McCorquodale in 1984, developed by Kenox Corporation, Mississauga, Ontario, and demonstrated at a drum recycling plant in Toronto. First commercialized by Leigh Environmental near Birmingham, England, in 1992. *See also* Zimpro.

Canadian Patent 1,224,891.

U.S. Patents 4,604,215; 4,793,919.

Chem. Eng. (Rugby, England), 1991, (508), 16.

Kepro   [Kemira process]   A process for recovering valuable products from municipal sewage sludge. It makes four products: crude iron phosphate, a biofuel, water treatment chemicals, and a carbon source for denitrification in the sewage plant. Developed by Kemira Chemicals in the 1990s and first installed in Helsingborg, Sweden.

Eur. Chem. News (Finland Suppl.), 1997, **68**, 8.

Kerpely   An early coal gasification process.

Kesting   A process for making chlorine dioxide by reducing sodium chlorate with hydrochloric acid. Chlorine is also produced and the usual equation given is:

$$2NaClO_3 + 4HCl = 2ClO_2 + Cl_2 + 2NaCl + 2H_2O$$

However, the process is more complex than this and the usual molar ratio of chlorined dioxide to chlorine produced is about 1:1. The process is integrated with an electrolytic process for making the sodium chlorate, such that the liquor from the reduction step is recirculated to the electrolytic step. The product gas, a mixture of chlorine with chlorine dioxide, is washed with water, which preferentially dissolves the chlorine dioxide. The resulting solution is used for pulp bleaching.

Invented by E. E. Kesting in the 1940s at Elektrochemische Werke München and first operated commercially by the Brown Company at Berlin, New Hampshire. A similar process was patented at about the same time by G. A. Day and E. F. Fenn in the United States, so the process has also been called the Day-Kesting process. Later, development of the process by H. Fröhler, who used titanium metal for the construction and closed the recycle loops, led to what is now known as the *Munich process.

Kesting, E. E., Day, G. A. and Fenn, E. F., German Patents 831,542; 841,565; 924,689; 971,285.

U.S. Patents 2,484,402; 2,736,636.

Kesting, E. E., *Das Papier,* 1952, **6**, 155 (*Chem. Abstr.,* **46**, 10557).

Ketazine   A process for making hydrazine by oxidizing ammonia with chlorine in the presence of an aliphatic ketone. A ketazine is an intermediate.

Audrieth, L. F. and Ogg, B. A., *The Chemistry of Hydrazine,* John Wiley & Sons, New York, 1951, 115.

Kettle   A simple batch process for making soap by boiling fat with an aqueous alkali solution in a "kettle."

Keyes   A process for separating water from ethanol, using azeotropic distillation with benzene. Invented in 1922 by D. B. Keyes.

U.S. Patents 1,676,735; 1,830,469.
Keyes, D. B., *Ind. Eng. Chem.,* 1929, **21**, 998.

KHD-Contop   [from the German company, **KHD** Humboldt Wedag, and **Con**tinuous **top** blowing]   A steelmaking process.

Kiener-Goldshöfe   A process for pyrolyzing solid wastes, used in Germany.

Martinez, D., in *Chemical Waste Handling and Treatment,* Muller, K. R., Ed., Springer-Verlag, Berlin, 1986, 148.

Kiflu   A process for making sodium hydroxide from sodium chloride, using sodium fluorosilicate.

KILnGAS   A coal gasification process, conducted in a rotary kiln. Developed by Allis-Chalmers Coal Gas Corporation in 1971 and piloted on a large scale at the Wood River power station, IL, since 1980.

Kinglor-Metor   A *DR process. Lump iron ore, mixed with coke or coal, and lime, is passed through a heated rectangular chamber and there reduced to sponge iron. First operated in Italy in 1973 and now operating in Burma. *See* DR.

KIP   *See* steelmaking.

Kiss   An obsolete process for extracting silver from its ores. The ores were roasted with sodium chloride, producing silver chloride, and this was leached out with a solution of calcium thiosulfate. The process was replaced by the *cyanide process.

Kivcet   [A Russian acronym meaning "vortex oxygen electric smelting"]   A flash smelting process for sulfide ores, using oxygen. In 1990, three plants had been built in Russia, one in Italy, and one in Bolivia.

Warner, N. A., in *Oxygen in the Metal and Gaseous Fuel Industries,* Royal Society of Chemistry, London, 1977, 227.
Morgan, S. W. K., *Zinc and Its Alloys and Compounds,* Ellis Horwood, Chichester, England, 1985, 96.
Chem. Eng. (N.Y.), 1990, **97**(4), 57.

Kiviter   A process for extracting oil and gas from shale, using a vertical kiln. The first unit was under construction in Estonia in the 1980s.

Smith, J. W., in *Handbook of Synfuels Technology,* Meyers, R. A., Ed., McGraw-Hill, New York, 1984, 4-175.

Kjellgren   A process for making beryllium by reducing beryllium fluoride with magnesium. Invented in 1941 by B. R. F. Kjellgren at the Brush Beryllium Company, OH, and now the principle commercial method for making beryllium metal.

U.S. Patent 2,381,291.
Kjellgren, B. R. F., *Trans. Electrochem. Soc.,* 1948, **93**(4), 122.

Kjellgren-Sawyer   *See* Fuse-quench.

KK   [Kunugi and Kunii]   A process for cracking crude petroleum or heavy oil in a fluidized bed, using coke as the heat carrier. Developed originally by Kunigi and Kunii, subsequently improved by the Japanese Agency of Industrial Science with five Japanese companies. Piloted between 1979 and 1982.

Hu, Y.C., in *Chemical Processing Handbook,* Marcel Dekker, New York, 1993, 776.

KLP [Dow K Catalyst liquid phase] A selective hydrogenation process for removing acetylenes from crude C_4 hydrocarbons from ethylene cracking, with no loss of butadiene. The catalyst is based on either copper metal or alumina. Developed by Dow Chemical Company and first commercialized at its plant in Terneuzen, The Netherlands. The KLP licensing business was sold to UOP in 1991.

U.S. Patent 4,440,956.
Eur. Chem. News, 1991, 22 Apr, 37.

Knapsack A process for making acrylonitrile from lactonitrile, itself made from acetaldehyde and hydrogen cyanide. Operated in Germany from 1958. Not industrially significant today for the production of acrylonitrile, although part of the process is still used for making lactic acid in Japan.

Dumas, T. and Bulani, W., *Oxidation of Petrochemicals: Chemistry and Technology,* Applied Science Publishers, London, 1974, 141.
Weissermel, K. and Arpe, H.-J. *Industrial Organic Chemistry,* 3rd ed., VCH Publishers, Weinheim, Germany, 1997, 303.

Knauf A process for making gypsum, suitable for use as plaster, from the waste from the *Wet Process for making phosphoric acid. Developed by Research-Cottrell.

Knietsch An early version of the *Contact process for making sulfuric acid. Developed by R. Knietsch at BASF, Ludwigshaven.

Knietsch, R., *Ber. Dtsch. Chem. Ges.,* 1901, **34,** 4069.

Knox An early vapor-phase thermal cracking process for refining petroleum.

Koch A family of processes for making polymethyl benzenes by isomerization, alkylation, and disproportionation in the presence of a Friedel Crafts catalyst. Invented in 1968 by the Sun Oil Corporation and developed and commercialized by the Koch Corporation.

U.S. Patent 3,542,890.

Koho *See* DR.

Kolbel-Rheinpreussen A process for converting *syngas to gasoline. The gas was passed through a suspension of an iron catalyst in an oil. Developed by H. Kolbel at Rheinpreussen, Germany, from 1936 until the 1950s when it was supplanted by the *Fischer-Tropsch process.

Asinger, F., *Paraffins, Chemistry and Technology,* translated by B. J. Hazzard, Pergamon Press, Oxford, 1968, 153.

Kombi [**Kombi**nations-Verfahren] A liquid-phase petroleum hydrogenation process which combined *hydrogenation with *hydrorefining. The catalyst contained molybdenum and tungsten on an aluminosilicate. Developed by BASF.

Urban, W., *Erdoel Kohle,* 1955, **8,** 780.

KOMBISORBON A process for removing toxic vapors from waste gases by adsorption. The adsorbant is a composite of activated carbon with an inert material derived from a volcanic rock. Developed by Lurgi Bamag in the 1990s and used first in a sewage sludge incineration plant.

Konox A process for removing hydrogen sulfide from industrial gases by absorption in aqueous sodium ferrate (Na_2FeO_4) solution. The ferrate is reduced to ferrite ($NaFeO_2$) and the sulfide is oxidized to elemental sulfur. The main reactions are:

$$4Na_2FeO_4 + 6H_2S = 4NaFeO_2 + 4NaOH + 6S + 4H_2O$$

$$4NaFeO_2 + 4NaOH + 3O_2 = 4Na_2FeO_4 + 2H_2O$$

Developed in Japan and licensed by Sankyo Process Services, Kawasaki.

Kasai, T., *Hydrocarbon Process.*, 1975, **54**(2), 93.

Kohl, A. L. and Riesenfeld, F. C., *Gas Purification,* 4th ed., Gulf Publishing, Houston, TX, 1985, 519.

Kontisorbon   A process for removing and recovering soluble solvents from water. Developed and offered by Lurgi.

Koppers   *See* Heinrich Koppers.

Koppers Hasche   A cyclic process for converting methane to *syngas by partial oxidation over an alumina catalyst:

$$2CH_4 + O_2 = 2CO + 4H_2$$

Operated in a pair of horizontal catalyst chambers which alternate their functions at one-minute intervals.

Koppers Kontalyt   A gas-making process.

British Petroleum Co., *Gas Making and Natural Gas,* British Petroleum Co., London, 1972, 113.

Koppers phenolate   A process for removing hydrogen sulfide from coal gas by absorption in aqueous sodium phenolate. Invented in 1931 by J. A. Shaw at the Koppers Company, Pittsburgh, but possibly never used.

U.S. Patent 2,028,124.

Koppers-Totzek   A coal gasification process using an entrained bed. The coal is finely ground and injected in a jet of steam and oxygen into a circular vessel maintained at 1,500°C. Reaction is complete within one second. The ash is removed as a molten slag. The process was invented by F. Totzek at Heinrich Koppers, Essen, and further developed by Koppers Company in Louisiana, MO, under contract with the U.S. Bureau of Mines. The first commercial operation was at Oulu, Finland, in 1952; by 1979, 53 units had been built. Most of the plants are operated to produce a hydrogen-rich gas for use in ammonia synthesis. Developed by Lurgi. *See also* PRENFLO.

Totzek, F., *Chem. Eng. Prog.*, 1954, **50**(4), 182.

Dainton, A. D., in *Coal and Modern Coal Processing,* Pitt, G. J. and Millward, G. R., Eds., Academic Press, London, 1979, 137.

Firnhaber, B. and Wetzel, R., in *Coal Chem 2000,* Institution of Chemical Engineers, Rugby, England, 1980, K1.

Cornils, B., in *Chemicals from Coal: New Processes,* Payne, K. R., Ed., John Wiley & Sons, Chichester, England, 1987, 13.

Korte   A process for retting flax or hemp by treatment with hydrochloric acid, then with hypochlorous acid, and then neutralizing.

Kossuth   An electrochemical process for extracting bromine from brines. The cell had bipolar electrodes and no diaphragm. It was developed in Germany in 1897 but abandoned in favor of the *Kubierschky process. *See also* Wunsche.

German Patent 103,644.

Yaron, F., *Bromine and Its Compounds,* Jolles, Z. E., Ed., Ernest Benn, London, 1966, 16.

KPA   A process for making phosphoric acid from low-grade ores, developed by the Occidental Chemical Company.

Eur. Chem. News, 1983, **41**(1110), 19.

KPEG   [K (potassium) polyethylene glycol]   A process for destroying polychlorinated biphenyls in contaminated soil by heating to 150°C, under pressure, with potassium hydroxide, a polyethylene glycol, and a sulfoxide. Developed by Galson Research Corporation, New York, and first demonstrated in 1988. *See also* CDP.

U.S. Patents 4,447,541; 4,574,013.

K-Process [Kalocsai or Kaljas]   A process for extracting gold from ores, concentrates, tailings, and scrap by means of a proprietary solution containing a bromide and an oxidizing agent. Invented in 1983 by G. I. Z. Kalocsai and developed by Kaljas Pty, Australia. A pilot plant was under construction in 1987.

PCT Patent WO 85/00384.
German Patent 3,424,460.

KR   [Kellogg-Rust]   A fluidized-bed process for gasifying coal, initially developed by the Westinghouse Corporation.

Kraft   [From the German, meaning strength]   An alkaline papermaking process, also known as sulfate pulping. Wood chips are digested in an aqueous solution of sodium sulfate, which becomes reduced to sodium sulfide by the organic matter, and sodium hydroxide. The overall chemical process is the attack of the bisulfide ion on the lignin molecule, depolymerizing it, with the formation of lignosulfonates. Originally developed by C. F. Dahl in Danzig in the 1870s.

Sawyer, F. G., Beals, C. T., and Neubauer, A. W., in *Modern Chemical Processes,* Vol. 2, Reinhold Publishing, New York, 1952, 267.
Grant, J., *Cellulose Pulp and Allied Products,* Leonard Hill, London, 1958, Chap. 9.
Chem. Eng. News, 1980, **58**(21), 26.

Kramfors   A two-stage variation on the *Sulfite papermaking process. The first stage uses slightly alkaline sodium sulfite; the second uses acid calcium bisulfite. See also Stora.

Higham, R. R. A., *A Handbook of Papermaking,* Business Books, London, 1963, 261.

Kranz MWS   A *flue-gas desulfurization system based on activated carbon. One carbon bed removes most of the sulfur dioxide. Ammonia is then injected for the *SCR process to occur in the second bed, which also removes the residual sulfur dioxide. The carbon is regenerated off-site. Developed by Krantz & Company, Germany. In 1986, three plants were operating in Germany.

KRES   [Kellogg reforming exchanger system]   A *reforming process for providing *syngas to the *KAAP process.

Eur. Chem. News, 1993, **60**(1592), 27.
Oil & Gas J., 1996, **94**(47), 37.

Kroll (1)   A process for making a metal by reducing its halide with another metal. Thus titanium is prepared by reducing titanium tetrachloride with magnesium:

$$TiCl_4 + 2Mg = Ti + 2MgCl_2$$

Also used commercially for making tantalum, niobium, and zirconium. The reduction takes place in a batch reactor under an inert gas atmosphere. Invented by W. J. Kroll in Luxembourg in 1937, first commercialized by Du Pont in 1948, and now widely used. *See also* Hunter.

U.S. Patent 2,205,854.
Kroll, W. J., *Trans. Electrochem. Soc.,* 1940, **78**, 35.
McQuillan, A. D. and McQuillan, M. K., *Titanium,* Butterworths, Guildford, England, 1956, 57.

Kroll (2)   *See* Abgas-Turbo-Wascher von Kroll.

Kroy   A variation on the *Hercosett process for making wool fire-resistant. The chlorination step is effected by hypochlorous acid. Invented in Toronto in 1975 by F. Mains of Kroy Unshrinkable Wools.

British Patent 1,524,392.

Krupp-Kohlechemie   A process for making hard paraffin wax from *water gas by a variant of the *Fischer-Tropsch process. The products were called "Ruhrwachse." Developed by Ruhr Chemie and Lurgi Ges. fur Warmetechnie.

Ziesecke, K. H., *Fette, Seifen, Astrichm.,* 1957, **59**(6), 409.
Asinger, F., *Paraffins, Chemistry and Technology,* translated by B. J. Hazzard, Pergamon Press, Oxford, 1968, 175.

Krupp-Koppers (1)   A process for separating *p*-xylene from its isomers by crystallization. In 1979, eight plants were operating.

Hydrocarbon Process., 1979, **58**(11), 253.

Krupp-Koppers (2)   A process for separating butane and butene isomers from their mixtures by extractive distillation. The added solvent (Butenex) is a morpholine derivative, possibly N-formyl morpholine.

Krupp-Lurgi   *See* carbonization.

Krupp-Renn   *See* DR.

Krupp sponge iron   *See* DR.

Krutzsch   A vapor-phase process for making hydrogen peroxide from a mixture of hydrogen and oxygen, saturated with water vapor, in a silent electric discharge. Invented and developed by J. Krutzsch at the Elektrochemische Werke München from 1931 to 1944. The electrodes were of silica coated with aluminum. The electric discharge was at 12,000 volts, 9,500 Hz. The product was an aqueous solution containing 10 percent hydrogen peroxide. The process was not commercialized; the *Pietzsch and Adolph process continued in use at Munich until the introduction of the *AO process.

British Patent 453,458.
French Patent 790,916.
Wood, W. S., *Hydrogen Peroxide,* Royal Institute of Chemistry Lectures, London, 1954, 11.
Schumb, W. C., Satterfield, C. N., and Wentworth, R. L., *Hydrogen Peroxide,* Reinhold Publishing, New York, 1955, 52.

Kryoclean   A process for removing volatile organic compounds from effluent gas streams by low-temperature condensation. The refrigerant is liquid nitrogen, used subsequently in various ways. Developed by BOC in the 1990s. A simplified version of the process was announced in 1997.

Chem. Eng. (Rugby, England), 1996, (614), 23.

Kryosol   An adsorptive process for purifying methane from landfill gas. Operated at high pressure. The overall methane recovery is 90 to 95 percent.

Kumar, R. and Van Sloun, J. K., *Chem. Eng. Prog.,* 1989, **85**(1), 36.

K-T   A coal gasification process, used as a source of synthesis gas for making ammonia.

Chemistry of Coal Utilization, Elliott, M. A., Ed., John Wiley & Sons, New York, 1981, 1759.

Kubierschky   A process for extracting bromine from brines. Chlorine gas is passed in, and the liberated bromine is removed by steaming out.

Bromine and Its Compounds, Yarron, F., Joller, L. E., Ernest Benn, London, 1966, 17.

Kubota   A process for treating municipal wastes, incorporating a membrane through which the liquor is recycled to a bioreactor. Eight plants were operating in 1996. Developed in Japan.

Brindle, K. and Stephenson, T., *Water Waste Treat.,* 1996, **12**(39), 18.

KURASEP [**Kura**ray **Sep**aration]   A process for separating nitrogen from air by a variant of the *PSA process, using carbon molecular sieve as the adsorbent. Developed by Kuraray Chemical Company.

> Suzuki, M., in *Adsorption and Ion Exchange: Fundamentals and Applications,* LeVan, M. D., Ed., American Institute of Chemical Engineers, New York, 1998, 120.

Kureha   A process for making di-isopropyl naphthalene mixtures from naphthalene and propylene by transalkylation. It operates at 200°C, using a silica/alumina catalyst. Operated in 1988 at the Rutgerswerke plant in Duisberg-Meiderich, Germany. The name has also been used for a process for making acetylene from petroleum.

> Franck, H.-G. and Stadelhofer, J. W., *Industrial Aromatic Chemistry,* Springer-Verlag, Berlin, 1988, 330.

Kureha/Union Carbide   A process for cracking crude oil to olefins and aromatic hydrocarbons, using steam superheated to 2,000°C. Reaction time is only 15 to 20 milliseconds.

Kurtz   A process for making acrylonitrile by reacting hydrogen cyanide with acetylene in the presence of aqueous cuprous chloride. Invented by P. Kurtz at I. G. Farbenindustrie in the 1940s. The process was widely used, but by 1970 had been abandoned in the United States in favor of the *ammoxidation processes.

> German Patent 728,767.
> Kurtz, P., *Pet. Refin.,* 1953, **32**(11), 142.

Kvaerner   Also called Union Carbide/Kvaerner. A improved version of the *OXO process using an enhanced catalyst. To be used at Petrochemical Corporation's plant in Shandong Province, China, from 1999.

> *Chem. Mark. Rep.,* 1996, **250**(22), 9.

L

Lacell   A zinc extraction process in which zinc sulfide is converted to zinc chloride and molten sulfur, and the molten zinc chloride is electrolyzed.

Lachmann   An early process for refining gasoline by treatment with aqueous zinc chloride.

Lacy-Keller   A process for removing hydrogen sulfide and mercaptans from natural gas by absorption in a proprietary solution. Elemental sulfur precipitates as a colloid and is separated from the solution by means of an electrolytic flotation cell. The process does not remove carbon dioxide.

> Kohl, A. L. and Riesenfeld, F. C., *Gas Purification,* 4th ed., Gulf Publishing, Houston, TX, 1985, 546.

La-Mar   [Named after the inventors, R. J. **La**gow and J. L. **Mar**grave]   A process for fluorinating organic compounds, using fluorine gas at low partial pressures. Commercialized by the 3M company.

> Bedford, C. T., Blair, D., and Stevenson, D. E., *Nature (London),* 1977, **267,** 35.
> Lagow, R. J. and Margrave, J. L., *Prog. Inorg. Chem.,* 1979, **26,** 161.
> Lagow, R. J., *J. Fluorine Chem.,* 1986, **33,** 321.

Lampblack   The original process for making carbon black by the incomplete combustion of oils. Superseded by the *Acetylene Black, *Channel Black, *Furnace Black, and *Thermal processes.

> Kühner, G. and Voll, M., in *Carbon Black Science and Technology,* Donnet, J.-B., Bansai, R. C., and Wang, M.-J., Eds., Marcel Dekker, New York, 1993, 54.

Lane   A process for making hydrogen by passing steam over sponge iron at approximately 650°C. The iron becomes converted to magnetite.

LANFILGAS   An integrated process for dealing with municipal waste which stabilizes the solid residue and generates methane by bacteriological innoculation. Developed by the Institute of Gas Technology, Chicago.

Lanxide   A process for making composites of metals with oxides. A molten metal reacts with an adjacent oxidant and is progressively drawn through its own oxidation product so as to yield a ceramic/metal composite. Fibres or other reinforcing materials can be placed in the path of the oxidation reaction and so incorporated in the final product. The Lanxide Corporation was founded in 1983 in Newark, DE, to exploit this invention. In 1990 it formed a joint venture with Du Pont to make electronic components by this process. Variations are: Dimox (directed metal oxidation), for making ceramic metal composites, and Primex (pressureless infiltration by metal), for making metal matrix composites.

> Newkirk, M. S., Urquart, A. W., and Zwicker, H. R., *J. Mater. Res.,* 1986, **1**(1), 81.
> Chiang, Y.-M., Haggerty, J. S., Messner, R. P., and Demetry, C., *Am. Ceram. Soc. Bull.,* 1989, **68**(2), 423.

LAR   [Low air ratio]   A process for oxidizing *o*-xylene or naphthalene to phthalic anhydride, using a titania/vanadia catalyst containing molybdenum. Developed by Alusuisse Italia in the 1980s. A plant was operated at Valdarno, Italy, in 1984.

> Verde, L. and Neri, A. *Hydrocarbon Process.,* 1984, **11**, 83.

LARAN   [Linde anaerobic methane]   An anaerobic process for treating industrial waste waters, generating methane for use as fuel. The process uses a fixed-bed loop reactor. Developed by Linde in the early 1980s, first commercialized in 1987.

> European Patent 161,469.

Larkin   An early direct process for reducing iron ore to iron metal by heating with carbon. *See also* DR.

Laxal   *See* metal surface treatment.

Laux   Also called the Aniline process. A process for making red iron oxide pigment in the course of making aniline by reducing nitrobenzene with scrap iron:

$$4C_6H_5NO_2 + 9Fe + 4H_2O = 4C_6H_5NH_2 + 3Fe_3O_4$$

Invented in 1926 by J. Laux at Bayer and used commercially thereafter. It was an improvement on the *Bechamp process.

> Buxbaum, G. and Printzen, H., in *Industrial Inorganic Pigments,* Buxbaum, G., Ed., VCH Publishers, Weinheim, Germany, 1993, 91.
> *Kirk-Othmer's Encyclopedia of Chemical Technology,* 4th ed., Vol. 19, John Wiley & Sons, New York, 1991–1998, 24.

LBE   [Lance-bubbling-equilibrium]   A steelmaking process in which nitrogen or argon is injected at the base of the furnace and oxygen is introduced at the top. Introduced in the 1970s. *See* steelmaking.

LC   [Lummus-Crest]   A coal gasification process developed by ABB Lummus Crest.

LCA [Leading Concept Ammonia]   A process for making ammonia from air and natural gas. Essentially a simplified form of the standard ammonia synthesis process, more suitable for smaller plants. Thermal economies are achieved in the steam reforming section. Developed by ICI in the mid-1980s. Two units began operating at the ICI plant in Severnside in 1988. The first non-ICI installation was designed by KTI for Mississippi Chemicals, Yazoo City, MS.

> *Chem. Eng. (N.Y.),* 1989, **96**(7), 43.
> *Chem. Eng. (Rugby, England),* 1990, (471), 21.
> *Hydrocarbon Process.,* 1991, **70**(3), 134.

LC-Fining [Lummus Cities re**fining**]   A *hydrocracking process using an ebullated catalyst bed. Developed by Lummus Crest and Cities Service Research and Development Company since the 1960s, initially for upgrading bitumen from tar sands. Three units were operating in 1996.

> Van Driessen, R. P., Caspers, J., Campbell, A. R., and Lunin, G., *Hydrocarbon Process.,* 1979, **58**(5), 107.
> Chillingworth, R. S., Potts, J. D., Hastings, K. E., and Scott, C. E. in *Handbook of Synfuels Technology,* Meyers, R. A., Ed., McGraw-Hill, New York, 1984, 6-47.

LCM [Leading Concept for Methanol]   A process for making methanol, combining the *ICI Low Pressure Methanol process with the *steam reforming section of the *LCA ammonia process. Developed by ICI in 1990 and piloted in Melbourne, Australia, from 1994. Envisaged for floating factories in off-shore gas fields.

> *Chem. Br.,* 1991, **27**(12), 1100.

L-D [Linz, Austria; and either **D**usenverfahren (nozzle process), or **D**onawitz, the other Austrian town where it was developed]   A basic steelmaking process in which oxygen is used instead of air to remove most of the carbon from the molten pig iron. Developed in Austria by the Vereinigte Österreichisch Eisen und Stahlwerke of Linz, and Österreichisch Alpine of Donawitz, in the 1930s and 40s; commercialized in 1952, and now widely adopted. The furnace is essentially a Bessemer converter, modified with a water-cooled oxygen injector. *See also* Bessemer.

> Finniston, M., *Chem. Ind. (London),* 1976, 19 June, 501.
> Dennis, W. H., *A Hundred Years of Metallurgy,* Gerald Duckworth, London, 1963, 121.

LD/AC [Named after *L-D, ARBED (a company in Luxembourg), and CNRM (a Belgian metallurgical research laboratory)]   Also called the *OCP process. A version of the *L-D steelmaking process in which powdered lime is introduced with the oxygen in order to remove phosphorus from the steel. *See also* OLP.

> Dennis, W. H., *A Hundred Years of Metallurgy,* Gerald Duckworth, London, 1963, 121.
> Jackson, A., *Oxygen Steelmaking for Steelmakers,* Newnes-Butterworths, London, 1969, 165.
> Boltz, C. L., *Materials & Technology,* Vol. 3, Longman, London, and J. H. de Bussy, Amsterdam, 1970, Chap. 3.

LDF   *See* DF.

LEAD   An integrated ammonia synthesis process, developed by Humphreys & Glasgow.

> Saviano, F., Lagana, V., and Bisi, P., *Hydrocarbon Process.,* 1981, **60**(7), 99.

Lebedev   A one-step process for converting ethanol, derived from carbohydrates, to butadiene, using a mixed alumina/zinc oxide catalyst at approximately 400°C:

$$2CH_3CH_2OH \rightarrow CH_2{=}CH{-}CH{=}CH_2$$

Invented by S. V. Lebedev in Leningrad in 1929 and used in Germany during World War II. In 1997 it was still in use in the CIS, Poland, and Brazil.

154

Leblanc

British Patent 331,482.

Lebedev, S. V., *J. Gen. Chem. USSR*, 1933, **3**, 698.

Egloff, G. and Hulla, G., *Chem. Rev.*, 1945, **36**, 67.

Corson, B. B., Stahly, E. E., Jones, H. E., and Bishop, H. D., *Ind. Eng. Chem.*, 1949, **41**, 1012.

Weissermel, K. and Arpe, H.-J., *Industrial Organic Chemistry*, 3rd ed., VCH Publishers, Weinheim, Germany, 1997, 106.

Leblanc   (Also written LeBlanc and Le Blanc). An obsolete, two-stage process for making sodium carbonate from sodium chloride. In the first stage, the "salt cake" process, salt was heated with sulfuric acid, yielding sodium sulfate (salt cake) and gaseous hydrogen chloride:

$$2NaCl + H_2SO_4 = Na_2SO_4 + 2HCl$$

In the second stage, the "black ash process," the sodium sulfate was reduced to sodium sulfide and then converted to sodium carbonate by calcining with limestone and coal in a rotating kiln known as a black ash furnace or revolver:

$$Na_2SO_4 + 4C = Na_2S + 4CO$$

$$Na_2S + CaCO_3 = Na_2CO_3 + CaS$$

The black product was extracted with water and the sodium carbonate in it was recovered by concentration and crystallization. The residue, chiefly calcium sulfide, known as "galigu," was dumped on land and created an environmental nuisance for many years because it never hardened. The process was invented by N. Leblanc in France in 1789, in response to a competition organized by the French Academy of Sciences. Operation of the first factory was delayed for several years because of the French Revolution. The process was operated widely until it was progressively superseded by the *Ammonia-soda process in 1872. But it was still in use in Bolton, UK, until 1938, and the last plant in Europe closed in 1992. *See also* Black ash.

Taylor, F. S., *A History of Industrial Chemistry*, Heinemann, London, 1957, 183.

Hardie, D. W. F. and Pratt, J. D., *A History of the Modern British Chemical Industry*, Pergamon Press, Oxford, 1966, 21.

Smith, J. G., *The Origins and Early Development of the Heavy Chemical Industry in France*, Clarendon Press, Oxford, 1979, 209.

Campbell, W. A., in *Recent Developments in the History of Chemistry*, Russell, C. A., Ed., Royal Society of Chemistry, London, 1985, 244.

Lord Todd, *Chem. Ind. (London)*, 1989, 519.

Brown, A. H., *Chem,. Br.*, 1993, **29**(10), 866.

Leckie   *See* steelmaking.

LEDA   [Low energy de-asphalting]   A process for removing the asphalt fraction from petroleum residues by liquid–liquid extraction in a special rotating disc contactor. The extractant is a C_3–C_6 aliphatic hydrocarbon or a mixture of such hydrocarbons. Developed in 1955 by Foster Wheeler USA Corporation and still widely used; 42 units were operating in 1996.

Hydrocarbon Process., 1996, **75**(11), 106.

Ledgemont   A process for removing sulfur from coal by an oxidative leach with lime and ammonia. Developed by Hydrocarbon Research.

IEA Coal Research, *The Problems of Sulphur*, Butterworths, London, 1989, 24.

Lefort   A process for making ethylene oxide by oxidizing ethylene in the presence of a silver catalyst. Invented and developed in the 1930s by T. E. Lefort at the Société Française de Catalyse. For many years, refinements of this basic process were operated in competition with the ethylene chlorohydrin process, but by 1980 it was the sole process in use.

French Patent 794,751.

US Patent 1,998,878.

Françon, J., *Chim. et Ind.*, 1933, **29**, 869.

Leidie   A process for extracting the platinum metals from their ores by fusion with sodium peroxide, followed by a complex separation process. Developed by A. Quennessen, a leading French manufacturer of platinum in the 19th century, and E. Leidie. The process is still used for extracting precious metals, and in chemical analysis.

McDonald, D., *A History of Platinum*, Johnson Matthey, London, 1960.

Leming   A process for removing sulfur compounds from coal gas by reaction with iron oxide. Invented in 1847.

Lenze   A process for removing naphthalene from coal gas by washing with cold, aqueous ammonia. Operated in Germany in the 1930s.

Lenze, F. and Rettenmaier, A., *Gas Wasserfach.*, 1926, **69**, 689 (*Chem. Abstr.*, **20**, 33556).

Leonard   A process for making mixed methylamines by reacting ammonia with methanol over a silica-alumina catalyst at elevated temperature and pressure. Developed and licensed by the Leonard Process Company. In 1993, the installed worldwide capacity of this process was 270,000 tonnes/y.

Weissermel, K. and Arpe, H.-J. *Industrial Organic Chemistry*, 3rd ed., VCH Publishers, Weinheim, Germany, 1997, 49.

Le Seur   An early process for electrolyzing brine. Developed in 1891 by E. A. Le Seur in Ottawa, Canada, and commercialized in Rumford, ME, in 1893.

British Patent 5,983 (1891).

LETS   A process for making triple superphosphate (a calcium hydrogen phosphate). Developed by the J. R. Simplot Company in 1976.

Bierman, L. W., in *Sulphuric/Phosphoric Acid Plant Operations*, American Institute of Chemical Engineers, New York, 1982, 81.

Levinstein   A process for making mustard gas, $(ClCH_2CH_2)_2S$, by reacting sulfur monochloride with ethylene:

$$2CH_2 = CH_2 + S_2Cl_2 \rightarrow ClCH_2CH_2SCH_2CH_2Cl + S$$

LF   *See* steelmaking.

LFC   [Liquids From Coal]   A general term for such processes.

Lidov   A process for chlorinating cyclopentadiene to octachloropentadiene, which is then thermally dechlorinated to hexachlorocyclopentadiene (HCCP), used as an intermediate in the manufacture of insecticides and flame retardants. The initial chlorination is catalyzed by phosphorus pentachloride or arsenious oxide. Invented by R. E. Lidov in The Netherlands and commercialized by the Shell Chemical Company.

British Patent 703,202.

LIFAC   [Limestone in-furnace and Added Calcium]   A dry *flue-gas desulfurization process in which limestone is injected into the furnace and calcium hydroxide is injected after it. Developed by Tampella in 1984 and used in a power station in Finland. A demonstration plant was built for Saskatchewan Power, Canada, in 1990.

Kenakkala, T. and Valimaki, E., *Desulphurisation in Coal Combustion Systems*, Institution of Chemical Engineers, Rugby, England, 1989, 113.

Lightox   A photochemical process for destroying organic materials in aqueous solution by oxidation with chlorine, activated by ultraviolet radiation. Developed by the Taft Water Research Center, United States in the 1960s.

Martinez, D., in *Chemical Waste Handling and Treatment*, Muller, K. R., Ed., Springer-Verlag, Berlin, 1986, 253.

Lignite Ash   A *flue-gas desulfurization process which uses ash from lignite combustion as the adsorbent.

Speight, J. G., *Gas Processing,* Butterworth Heinemann, Oxford, 1993, 300.

Lignol   [**Lign**in phe**nol**]   A catalytic process for hydrogenating lignin to a mixture of phenol, benzene, and fuel gas. Developed by Hydrocarbon Research. *See also* Noguchi.

Lignox   [**Lign**in **ox**idation]   A pulp-bleaching process using hydrogen peroxide as the oxidant, and a chelating agent. Developed in Sweden in 1990 by Eka Nobel. *See also* Acetox.

LIMB   [**L**ime/l**i**mestone injection into a **m**ulti-stage **b**urner]   A *flue-gas desulfurization process used in Germany and Finland. Dry, ground limestone is injected directly into the combustion chamber. This reacts with the sulfur dioxide, and the dry particulate product is collected downstream together with the ash. The process is suitable only for those systems which limit the maximum combustion temperature by staging, in order to minimize the production of oxides of nitrogen.

Lime-soda   *See* causticization.

Linde   Also called Hampson-Linde. A process for separating oxygen and nitrogen from air by liquifation followed by fractional distillation. Developed by K. P. G. von Linde in Germany and W. Hampson in England at the start of the 20th century.

Linde-Frank-Caro   A process for extracting hydrogen from *water-gas by liquifaction.

Linde/Yukong   A two-stage, catalytic process for making 1,4-butanediol from acetylene. The first stage, catalyzed by palladium on alumina, makes 1,4-butynediol. The second stage, catalyzed by nickel on a silicate, hydrogenates this to 1,4-butanediol. Developed by Linde, Germany, and Yukong, Korea, in 1993 and now offered for license.

Heidegger, E. and Schödel, N. *Reports on Science and Technology,* Linde AG, Wiesbaden, Germany, 1997, **59,** 17.

LINDOX   [**Lind**e **ox**idation]   A variation of the *Activated Sludge sewage treatment system, using industrial oxygen (90 to 98 percent) instead of air. The liquor passes through several closed tanks in series and the oxygen is absorbed through the surface of the liquor. It is particularly suitable for treating effluents from the food processing industry. Developed by Linde, Munich, in the 1970s and first operated at a meat rendering plant in Oberding in 1974. It was superseded by the *Unox process in 1980.

Linear-1   A process for making linear $C_6 - C_{10}$ alpha-olefins from ethylene. Developed by UOP in 1996 but not commercialized as of 1997.

Eur. Chem. News, 1996, **67**(1755), 16.
Eur. Chem. News, 1997, **68**(1778), 2.

LINPOR   [**Lin**de **por**ous medium]   A biological waste water treatment process, using an open-pore plastic foam for retaining the biomass. Its use enables the capacity of an activated sludge plant to be increased without adding extra tanks. Invented at the Technische Universität, Munich, and further developed by Linde, Munich. *See also* CAPTOR.

European Patent 92,159.
Cooper, P. F., in *Topics in Wastewater Treatment,* Sidgewick, J. M., Ed., Blackwells, Oxford, 1985, 49.

Linz-Donan   *See* L-D.

Linz Donawitz   *See* L-D.

LIPP-SHAC   [**L**iquid **p**olymerization of **p**ropylene with **s**uper **h**igh **a**ctivity **c**atalyst]   A process for making polypropylene. Developed by the Shell Chemical Company and used at Pernis, The Netherlands; Carrington, England; and Geelong, Australia.

Chem. Eng. (Rugby, England), 1990, Nov, 7.

liquation A metallurgical process for separating metals by partial melting. Used for purifying zinc, tin, and in conjunction with the *Parkes process for desilvering lead.

Liquicel A liquid–liquid extraction process in which the two liquids are separated by a permeable membrane in the form of hollow plastic fibers. Developed by Hoechst Celanese Corporation.

Chem. Eng. (Rugby, England), 1992, (513), 10.

Liritan A leather tanning process, introduced in 1960 and now widely used worldwide. The leather is first pickled in a solution of sodium hexametaphosphate and then soaked in baths of various vegetable products.

Shuttleworth, S. G., *J. Soc., Leather Trades' Chem.*, 1963, **47**, 143.

Litol Also called **Houdry-Litol.** A process for making benzene by dealkylating other aromatic hydrocarbons. It is a complex process which achieves desulfurization, removal of paraffins and naphthenes, and saturation of unsaturated compounds, in addition to dealkylation. The catalyst contains cobalt and molybdenum. Developed by the Houdry Process and Chemical Company and Bethlehem Steel Corporation. First installed by the Bethlehem Steel Corporation in 1964. Subsequently used at British Steel's benzole refinery, Teesside, England.

Tarhan, M. O. and Windsor, L. H., *Chem. Eng. Prog.*, 1966, **62**(2), 67.

Lorz, W., Craig, R. G., and Cross, W. J., *Erdoel Kohle Erdgas Petrochem.*, 1968, **21**, 610.

Dufallo, J. M., Spence, D. C., and Schwartz, W. A., *Chem. Eng. Prog.*, 1981, **77**(1), 56.

Franck, H.-G. and Stadelhofer, J. W., *Industrial Aromatic Chemistry*, Springer-Verlag, Berlin, 1988, 123.

Llangwell A process for making acetic acid by fermenting the cellulose in corn cobs. Xylose is a co-product. The microorganism was isolated from the gut of the goat. Piloted on a large scale by the Commercial Solvents Company, Terre Haute, IN, from 1928 to 1930.

LM [Lurgi-Mitterberg] A process for extracting copper from chalcopyrite. The ore is subject to "activation grinding" and then dissolved in sulfuric acid under oxygen pressure. The copper dissolves as copper sulfate, leaving a residue of elemental sulfur and gangue. Developed in the early 1970s by Lurgi Chemie and Hüttentecknik, Kupferbergbau Mitterberg, and the Technical University of West Berlin. A demonstration plant operated in Mühlbach, Austria, from 1974 until the mine became exhausted.

Locap A process for removing mercaptans from gasoline by catalytic oxidation to disulfides, using a fixed bed of catalyst that is continuously treated with aqueous sodium sulfide. Commercialized by Petrolite Corporation in 1963.

Hydrocarbon Process., 1964, **43**(9), 210.

O'Brien, G. A., Newell, O., Jr., and DuVon, R. H., Jr., *Hydrocarbon Process.*, 1964, **43**(6), 175.

LO-CAT A process for removing hydrogen sulfide and organic sulfur compounds from petroleum fractions by air oxidation in a cyclic catalytic process similar to the *Stretford process. The aqueous solution contains iron, two proprietary chelating agents, a biocide, and a surfactant; the formulation is known as ARI-310. The sulfur product is removed as a slurry. Developed in 1972 by Air Resources (now ARI Technologies) and first commercialized in 1976. Over 125 units were operating in 1996. An improved version, LO-CAT II, was announced in 1991.

US Patent 4,189,462.

British Patent 1,538,925.

Hardison, L. C., *Oil & Gas J.*, 1984, **82**(23), 60.

Hardison, L. C., in *Acid and Sour Gas Treating Processes,* Newman, S. A., Ed., Gulf Publishing, Houston, TX, 1985, 678.
Dalrymple, D. A., Trofe, T. W., and Evans, J. M., *Chem. Eng. Prog.,* 1989, **85**(3), 43.
Hardison, L. C. and Ramshaw, D. E., *Hydrocarbon Process.,* 1992, **71**(1), 89.
Hydrocarbon Process., 1996, **75**(4), 106.

LO-FIN [Last out-first **in**]   A version of the *PSA process for separating hydrogen from other gases. It includes a unique gas-retaining vessel which preserves the concentration gradient in one stream before using it to repressurize another bed. Developed jointly by Toyo Engineering Corporation and Essex Corporation.

Suzuki, M., in *Adsorption and Ion Exchange: Fundamentals and Applications,* LeVan, M. D., Ed., American Institute of Chemical Engineers, New York, 1998, 122.

Lomax   An outdated name for a hydrocracking process now offered under the name *Unicracking.

LOMI [Low oxidation metal ions]   A process for decontaminating parts of nuclear reactors by washing with aqueous solutions of low-valency transition metal ions. Developed at the Berkley laboratories of the UK Atomic Energy Authority in the early 1980s.

Longmaid-Henderson   A process for recovering copper from the residue from the roasting of pyrites to produce sulfur dioxide for the manufacture of sulfuric acid. The residue was roasted with sodium chloride at 500 to 600°C; the evolved sulfur oxides and hydrochloric acid were scrubbed in water; and the resulting solution was used to leach the copper from the solid residue. Copper was recovered from the leachate by adding scrap iron. The process became obsolete with the general adoption of elemental sulfur as the feedstock for sulfuric acid manufacture.

Lonza (1)   A process for oxidizing isobutene to α-hydroxy-isobutyric acid, a precursor for methacrylic acid, using a solution of dinitrogen tetroxide in acetic acid as the oxidant. *See also* Escambia (1).

Lonza (2)   A process for making malononitrile. Acrylonitrile is reacted continuously with cyanogen chloride, in the vapor phase, in a quartz tube at 900°C. Developed by Lonza, Basle.

German Patents 1,921,662; 1,946,429.

Loop   A continuous process for polymerizing aqueous emulsions of olefinic compounds such as vinyl acetate. Polymerization takes place in a tubular reactor (the loop) with recycle. Invented by Gulf Oil Canada in 1971 and further developed by several United Kingdom paint companies. It is now used for making copolymers of vinyl acetate with ethylene, used in solvent-free paints and adhesives.

Canadian Patent 907,795.
Wilkinson, M. and Geddes, K., *Chem. Br.,* 1933, **29**(12), 1050.

Loprox [Low-pressure wet oxidation]   A process for partially oxidizing waste organic products, to render them digestible in biological waste treatment systems such as the *Activated Sludge process. The oxidant is oxygen, at a pressure of 3 to 20 atm. the temperature 120 to 200°C, and the reaction is catalyzed by quinonoid substances and iron salts. Developed by Bayer in Germany in 1980, for use in its own works, and piloted in several countries from 1991 to 1992. Six units were operating in Europe in 1997. Now engineered and offered by Bertrams, Switzerland.

New Sci., 1992, **133**(1811), 24.
Water Waste Treat., 1998, **41**(5), 22.

LOR [Liquid-phase oxidation reactor]   Not a process but a piece of equipment in which to conduct liquid-phase oxidations (*e.g.* the *Mid-Century process) safely with oxygen rather than with air. The oxygen is introduced into the liquid phase and rapidly dispersed in the form of bubbles 1 to 5 mm. in diameter. Developed by Praxair and ABB Lummus Global in 1996.

Chemical Week, 1996, **158**(15), 28.

Lovacat [**Lo**w **va**lency **cat**alyst]   A catalytic process for making ethylene propylene and EPDM rubbers. Developed by DSM Elastomers in 1996.

Kunstst. en Rubber, 1996, **49**(8), 41.

Lowe   *See* Water gas.

Löwenstein-Riedel   An electrolytic process for making hydrogen peroxide by the electrolysis of a solution of sulfuric acid and ammonium sulfate. Ammonium peroxodisulfate, $(NH_4)_2S_2O_8$, is an intermediate. This, and the *Weissenstein process were made obsolete with the invention of the *AO process.

Schumb, W. C., Satterfield, C. N., and Wentworth, R. L., *Hydrogen Peroxide,* Reinhold Publishing, New York, 1955, 145.

Löwig   Also called Ferrite. A causticization process—the conversion of sodium carbonate to sodium hydroxide. The sodium carbonate is mixed with iron oxide and heated for several hours in a rotating kiln. Carbon dioxide is evolved and sodium ferrite remains:

$$Na_2CO_3 + Fe_2O_3 = 2NaFeO_2 + CO_2$$

The product is agitated with water, producing sodium hydroxide and ferric oxide for re-use:

$$2NaFeO_2 + H_2O = 2NaOH + Fe_2O_3$$

The process was invented in Germany by C. Löwig in 1882 and used at Joseph Crosfield & Sons, Warrington, UK, in the late 19th and early 20th centuries. *See also* causticization.

British Patent 4,364 (1882).

Musson, A. E., *Enterprise in Soap and Chemicals,* Manchester University Press, Manchester, England, 1965, 82,203.

LPMEOH   A process for making methanol. Developed by Air Products & Chemicals and Chem Systems in the late 1970s.

Cybulski, A., *Catal. Revs., Sci. & Eng.,* 1994, **36**(4), 558.

LPO (1)   *See* OXO.

LPO (2)   Also called *Celanese LPO.

LP OXO [**L**ow **p**ressure **OXO**] *See* OXO.

LPG Unibon   An outdated UOP version of the hydrocracking process for simultaneously hydrogenating and cracking a naphtha petroleum fraction to form C_3 and C_4 hydrocarbons. In 1992 the technology was offered under the umbrella of *Unicracking.

LR   *See* Lurgi-Ruhrgas.

LSE [Liquid solvent extraction]   A coal liquifaction process, under development in 1990 by British Coal, at Point of Ayr, North Wales. The coal is dissolved in a coal-derived hydrocarbon solvent and then catalytically hydrocracked.

Chem. Br., 1990, **26**(10), 922.

LT Unibon   A two-stage, catalytic *hydrotreating process for removing deleterious components from naphtha without cracking it. Developed by UOP.

LTC [Low temperature chlorination]   A general term for processes for making 1,2-dichloroethane from ethylene and chlorine by processes operated below the boiling point of the product (83°C). *See also* HTC.

Lucas [Lurgi-Claus-Abgas-Schwefelgewinnung]   A process for removing residual sulfur compounds from the tail gases from the *Claus process. The gases are incinerated and then passed over a bed of hot coke, which converts all the sulfur to sulfur dioxide. This is absorbed in aqueous sodium phosphate, which releases it on heating:

$$Na_2HPO_4 + H_2O + SO_2 \rightleftharpoons NaH_2PO_4 + NaHSO_3$$

The sulfur dioxide is returned to the Claus process for re-use. Developed by Lurgi Mineralöltechnik, Germany.

Doerges, A., Bratzler, K., and Schlauer, J., *Hydrocarbon Process.*, 1976, **55**(10), 110.
Sulphur, 1977, (128), 41.

Luce-Rozan   A variation of the *Pattinson process, in which steam is blown through the molten metal as cold water is sprayed on the surface.

Lurgi [Metallurgischegesellschaft]   Lurgi, previously called Metallurgischegesellschaft, now a subsidiary of Metallgesellschaft, is a large chemical engineering company which has particularly given its name to two coal gasification processes. The first was a fixed-bed gasifier in which a bed of coal particles rested on a rotating hearth through which oxygen and steam were injected. The temperature was kept below the slagging temperature of the ash, which was withdrawn from the base of the bed. The process was first used commercially at Zittau, Germany, in 1936; 65 units had been built by 1979. The second design was a slagging gasifier, of which an experimental model was installed in Solihull, UK, in 1956.

Wilke, G., *Chem. Fabr.,* 1938, **11**, 563.
Chem. Eng. News, 1958, **36**(16), 88.
Dainton, A. D., in *Coal and Modern Coal Processing,* Pitt, G. J., and Millward, G. R., Eds., Academic Press, London, 1979, 135.
Cornils, B., in *Chemicals from Coal: New Processes,* Payne, K. R., Ed., John Wiley & Sons, Chichester, England, 1987, 12.

Lurgi-Ruhrgas   Also known as **LR**. A process originally intended for pyrolizing fine-grained solids such as coal, peat, shale, and tar sands to produce mixed hydrocarbons. The process is based on flash heating in a mixer by means of circulated hot powders, usually obtained from the process. The process was later modified to allow ethylene to be produced from heavy hydrocarbons. Ethylene was first made in this way in Germany in 1958. Developed jointly by the Lurgi and Ruhrgas companies in Germany in 1949 and now offered by Lurgi.

Rammler, R. W. and Weiss, H.-J., in *Handbook of Synfuels Technology,* Meyers, R. A., Ed., McGraw-Hill, New York, 1984, 4-17.

Lurgi Spülgas   *See* Spülgas.

Lyocell   A papermaking process, based on dissolving wood pulp in N-methyl morpholine. Developed by Courtaulds, but the subject of a patent dispute with Lenzing, Austria, since 1993. Courtaulds has a plant in Mobile, AL, and Lenzing has one in Heiligenkreuz, Austria.

MacArthur-Forrest *See* Cyanide.

Macrox A pulp-bleaching process using hydrogen peroxide.

> Troughton, N. A. and Sarot, P., TAPPI Pulping Conference, 1992.

Madaras *See* DR.

Madison An improved version of the *Scholler-Tornesch *saccharification process. Developed at the U.S. Forest Products Laboratory, Madison, WI.

> *Riegel's Handbook of Industrial Chemistry,* 9th ed., Kent, J. A., Ed, Van Nostrand Reinhold, New York, 1992, 256.

Madsenell *See* metal surface treatment.

Magchar A process for extracting gold from solution by adsorption on composites of magnetic particles deposited on grains of activated carbon. Invented by E. Herkenhoff and N. Hedley.

> *Eng. Min. J.,* 1982, Aug, 84.
> Yannopoulos, J. C., *The Extractive Metallurgy of Gold,* Van Nostrand Reinhold, New York, 1991, 224.

MagnaCat A process for selectively removing metal-contaminated catalyst particles from an * FCC reactor by magnetic separation. Developed by the Ashland Petroleum Company, which sold it to MW Kellogg in 1997.

> Hettinger, W. P., Jr., *Catal. Today,* 1992, **13**, 157.
> Andersson, S.-I. and Myrstad, T., *Appl. Catal. A: Gen.,* 1997, **159**(1–2), 291.
> *Eur. Chem. News,* 1997, **68**(1784), 31.
> *Eur. Chem. News,* 1998, **69**(1797), 32.
> *Chem. Eng. (N.Y.),* 1997, **104**(11), 122.

Magnaforming A *catalytic reforming process developed by the Atlantic Richfield Corporation and Englehard Corporation. First announced in 1965, it was commercialized in 1967 and by 1988, 150 units were operating worldwide. Hydrocarbon Research has installed units in Argentina, Algeria, and the USSR.

> Nevison, J. A., Obaditch, C. J., and Dalson, M. H., *Hydrocarbon Process.,* 1974, **53**(6), 111.
> Little, D. M., *Catalytic Reforming,* PennWell Publishing, Tulsa, OK, 1985, 158.
> *Hydrocarbon Process.,* 1988, **67**(9), 80.

Magnefite [**Magne**sium sul**fite**] A process for separating lignosulfonates from the sulfite liquor from papermaking. The lignosulfonates are separated as magnesium salts. Six paper mills in the United States and Canada were using this process as of 1981.

Magnetherm [**Mag**ee magnesium **E**isenberg **therm**al] A process for making magnesium by reducing dolomite with ferrosilicon at 1,600°C in a vertical electric arc furnace:

$$2CaO \cdot MgO + Fe_xSi = 2Mg + Ca_2SiO_4 + xFe$$

A flux containing Ca, Al, and Mg oxides, or bauxite, is used, and the process takes place in a partial vacuum. Invented in 1965 by E. M. Magee and B. Eisenberg at Esso Research and Engineering Company. An improved version of this process, *MAGRAM, is under development at the University of Manchester Institute of Science and Technology, which does not require a vacuum and which uses a mixture of waste products to provide the flux.

U.S. Patent 3,441,402.
New Sci., 1995, **145**(1967), 21.
Trocmé, F., in *Advances in Extractive Metallurgy and Refining,* Jones, M. J., Ed., Institution of Mining and Metallurgy, London, 1971, 517.
Chem. Br., 1996, **32**(4), 12.

Magnex   A process for removing mineral matter from coal by first rendering it magnetic. The coal is treated with iron carbonyl vapor, which deposits a thin skin of magnetic material on the pyrite and other mineral matter, but not on the coal. Conventional magnetic separation is then used. Developed by Hazen Research in 1976.

IEA Coal Research, *The Problems of Sulphur,* Butterworths, London, 1989, 38.

Magnicol   [**Magn**etic **col**umnar]   A process for making Alnico (an iron-based magnetic alloy containing Al, Ni, Co, and Cu) crystallize with a columnar grain structure in order to optimize its magnetic properties. Successive additions of silicon, carbon, and sulfur are made to the initial melt.

Palmer, D. J. and Shaw, S. W. K., *Cobalt,* 1967, **43**, 63 (*Chem. Abstr.,* **71**, 32827).

MAGRAM   A process for extracting magnesium metal from dolomite. Similar to *Magnetherm, but using asbestos as the fluxing material and not requiring the use of vacuum. Developed by an international consortium financed by the European Union at the University of Manchester Institute of Science and Technology. *See* Magnetherm.

Chem. Br., 1996, **32**(4), 12.

MAGSORB   A process for removing carbon dioxide from hot gas streams by reversible absorption on magnesium oxide modified with potassium carbonate. Developed by the Institute of Gas Technology, Chicago, for fuel gas derived from coal.

MAK fining   A petroleum refining process which combines *MAK hydrocracking with a cold-flow improvement process. Developed by Nippon Ketjen and first licensed in 1998.

Jpn. Chem. Week, 1998, **39**(1976), 4,7.

MAK hydrocracking   [**M**obil **A**kzo **K**ellogg]   A process for making high-quality, low-sulfur fuels from a variety of petroleum intermediates. Developed jointly by the three companies named. Two units were operating in 1996.

Hydrocarbon Process., 1996, **75**(11), 126.

Malaprop   A process for removing carbonyl sulfide from gas streams by scrubbing with diethylene glycolamine (DGA).

Moore, T. F., Dingman, J. C., and Johnson, F. L., Jr., in *Acid and Sour Gas Treating Processes,* Newman, S. A., Ed., Gulf Publishing, Houston, TX, 1985, 290,313.

Mallet   A process for separating oxygen from air by selective dissolution in water. Oxygen is more soluble than nitrogen in water, so by contacting water with compressed air, and desorbing the gases at a lower pressure, it is possible to make air enriched in oxygen. Repetition of the process yields progressively purer oxygen. Invented by J. T. A. Mallet in Paris in 1869; supplanted by liquifaction processes at the start of the 20th century.

British Patent 2,137 (1869).

Maloney   A system for drying gases with zeolites. Three beds are used, with one of them being regenerated by hot gas at any one time. Developed by Maloney Steel Company, Calgary, Canada.

Palmer, G. H., *Hydrocarbon Process.*, 1977, **56**(4), 103.
Kohl, A. L. and Riesenfeld, F. C., *Gas Purification*, 4th ed., Gulf Publishing, Houston, TX, 1985, 656.

Manasevit   A process for making electronic devices by depositing thin films of elements or simple compounds such as gallium arsenide on flat substrates by *CVD from volatile compounds such as trimethyl gallium and arsine.

Manasevit, H. M., *Appl. Phys. Lett.*, 1968, **12**, 156.
Manesivit, H. M. and Simpson, W. I., *J. Electrochem. Soc.*, 1969, **116**, 1725.
Manesevit, H. M., *J. Electrochem. Soc.*, 1971, **118**, 647.

Manchester   A variation on the *Ferrox process for removing hydrogen sulfide from industrial gases in which several absorbers are used, and delay stages permit completion of the reaction with the iron oxide absorbent. Developed by the Manchester Corporation Gas Department in the 1940s and installed in several British gasworks.

British Patents 550,272; 611,917.
Kohl, A. L. and Riesenfeld, F. C., *Gas Purification*, 4th ed., Gulf Publishing, Houston, TX, 1985, 494.

Manhès   A metallurgical process for removing sulfur from copper matte by blowing air through the molten material. Invented by P. Manhès in France in 1880.

Manhès, P., *Ber. Dtsch. Chem. Ges.*, 1881, **14**, 2432
German Patent 15,562.

Manley   An early thermal process for cracking petroleum.

Mannheim (1)   A process for making hydrochloric acid by roasting sulfuric acid and sodium chloride together in a closed cast iron furnace equipped with a plough. The by-product sodium sulfate, known as salt cake, may be recrystallized after neutralization and filtration, and used as a detergent ingredient. A potassium variant is used in those locations where native potassium chloride can be found.

Mannheim (2)   An early version of the *Contact process for making sulfuric acid. Two catalysts were used: ferric oxide, followed by platinum. The first Mannheim plant was built in Buffalo, NY, in 1903.

Miles, F. D., *The Manufacture of Sulfuric Acid (Contact Process)*, Gurney & Jackson, London, 1925, Chap. 10.
Levy, S. I., *An Introduction to Industrial Chemistry*, G. Bell & Sons, London, 1926, 218.

Mansfield   A process for extracting copper from sulfide ores by roasting with anthracite or coke and a silicious flux in a special blast furnace.

Mark and Wulff   A process for making styrene from benzene and ethylene. Developed in Germany in the 1930s.

Marqueyrol and Loriette   A process for making nitroguanidine, an explosive. Cyanamide dimer is converted to guanidinium sulfate by heating with sulfuric acid; this is then nitrated with nitric acid. *See also* Welland.

Aubertein, P., *Mem. Poudres*, 1948, **30**, 143 (*Chem. Abstr.*, **45**, 8250).

Maruzen (1)   A process for making terephthalic acid from *p*-xylene. Similar to the *Amoco process but yielding a purer product in one stage. Operated in Japan by Matsuyama Chemical Company.

Raghavendrachar, P. and Ramachandran, S., *Ind. Eng. Chem. Res.*, 1992, **31**, 453.

Maruzen (2)   A process for purifying *p*-xylene by crystallization, using ethylene as the direct coolant. Developed by Maruzen Gas Oil Company, United States. Now probably superseded by the *Parex (1) process.

> Hatanaka, Y. and Nakamura, T., *Oil & Gas J.,* 1972, **70**(47), 60.

MAS [Methanolo alcooli superiori]   A process for making mixtures of methanol with higher alcohols, for use as gasoline extenders, developed by a consortium of Snamprogetti, Haldor Topsoe, and Anic. Piloted in a demonstration plant in Italy.

> Asinger, F., *Methanol—Chemie und Energierhostoff,* Springer-Verlag, Berlin, 1986, 120.

Massener   *See* Hoerde.

Mathieson (1)   A process for making chlorine dioxide gas by passing sulfur dioxide, diluted with air, into aqueous sodium chlorate and sulfuric acid. The product is absorbed in water. Operated in the United States on a large scale for pulp-bleaching.

> Sheltmire, W. H., in *Chlorine, Its Manufacture, Properties and Uses,* Sconce, J. S., Ed., Reinhold Publishing, New York, 1962, 303,539.

Mathieson (2)   A process for making calcium hypochlorite dihydrate by mixing sodium hypochlorite and calcium chloride. Invented by A. George and R. B. MacMullin at the Mathieson Alkali Works, New York, in the 1920s. *See also* Perchloron.

> U.S. Patents 1,713,650; 1,787,048.
> Sheltmire, W. H., in *Chlorine, Its Manufacture, Properties and Uses,* Sconce, J. S., Ed., Reinhold Publishing, New York, 1962, 523.

Matthey   A complex sequence of chemical operations for purifying platinum. Developed by G. Matthey and used in his factory since 1879.

> McDonald, D., *A History of Platinum,* Johnson Matthey, London, 1960, 220.

MAWR [Mobil alkanolamine waste recovery]   A process which reduces the quantity of waste generated by alkanolamine processes, which remove acid gases from oil refinery gas streams. Developed by Mobil Oil, Germany, and used commercially there since 1979.

Maxofin   A dehydrogenation process for converting light hydrocarbons such as propane and isobutane into olefins and hydrogen. Competetive with *Catofin and *Oleflex. Under development by Mobil Research & Development Corporation since 1995.

Mazzoni   A family of continuous soapmaking processes.

> Lanteri, A., *Seifen, Oele, Fette, Wachse,* 1958, **84**, 589.

MBG [MAN Bergbauforschung Gasification]   A coal gasification process, suitable for all grades of coal, especially those difficult to gasify. Under development by MAN Gutehoffnungshütte, in collaboration with Deutsche Montan Technologie, in 1991.

MBR [Mobil benzene reduction]   A catalytic process for reducing the benzene content of gasoline. It combines features of three earlier processes: benzene alkylation with light olefins, olefin equilibration with aromatization, and selective paraffin cracking. The olefins are obtained from *FCC offgas. The catalyst is a modified ZSM-5 zeolite. Developed by Mobil Research & Development Corporation in 1993.

> *Chem. Eng. News,* 1993, **71**(38), 36.
> *Hydrocarbon Process.,* 1994, **73**(11), 90.

M-C   *See* Mid-Century.

McKechnie-Seybolt   A process for making vanadium by reducing vanadium pentoxide with calcium in the presence of iodine. It is conducted in a steel bomb at 700°C.

McKenna *See* Menstruum.

MCRC A variation on the *CBA sulfur recovery process using multiple Claus converters. Developed by the Delta Engineering Corporation in 1983 and used in Canada, China, and Mexico.

> Davis, G. W., *Oil & Gas J.,* 1985, **83**(8), 110.
> Kohl, A. L. and Riesenfeld, F. C., *Gas Purification,* 4th ed., Gulf Publishing, Houston, TX, 1985, 451.
> *Sulphur,* 1994, (231), 50.

MDDW [Mobil Distillate DeWaxing] A process for removing waxes (long-chain normal paraffins) from petroleum fractions by cracking over the zeolite ZSM-5. The waxes are converted to liquid hydrocarbon fuels. Twenty one units were operating in 1990.

> Chen, N. Y., Gorring, R. L., Ireland, H. R., and Stein, T. R., *Oil & Gas J.,* 1977, **75**(23), 165.
> Perry, R. H., Jr., Davis, F. E., and Smith, R. B., *Oil & Gas J.,* 1978, **76**(21), 78.
> Ireland, H. R., Redini, C., Raff, A. S., and Fava, L., *Oil & Gas J.,* 1979, **77**(24), 82.

MDEA [Methyl diethanolamine] A general name for processes using methyl diethanolamine for absorbing hydrogen sulfide and carbon dioxide from other gases. *See also* Activated MDEA.

> Kohl, A. L. and Riesenfeld, F. C., *Gas Purification,* 4th ed., Gulf Publishing, Houston, TX, 1985, 29.

MECER A process for recovering copper from waste streams by extraction with a β-diketone solution. Used in Germany, Sweden, the United Kingdom, and the USSR for treating effluent from the etching of printed circuit boards.

> Cox, M., in *Developments in Solvent Extraction,* Alegret, S., Ed., Ellis Horwood, Chichester, England, 1988, 177.

MEDISORBON An adsorptive process for removing mercury and dioxins from flue-gas. The adsorbent is a dealuminated zeolite Y manufactured by Degussa. For mercury removal, the zeolite is impregnated with sulfur. Developed in 1994 by Lurgi Energie und Umwelt and piloted in Germany and The Netherlands.

> *Chem. Eng., N.Y.,* 1994, **101**(10), 19.

MEGOX A process for increasing the rates of microbiological processes by the use of pure oxygen instead of air.

Mehra (1) [Named after the inventor] A process for extracting particular hydrocarbons from natural or synthetic gas streams using solvent extraction into polyalkylene glycol dialkyl ethers. Invented in 1982 by Y. R. Mehra at the El Paso Hydrocarbons Company, Odessa, TX.

> U.S. Patent 4,421,535.
> Mehra, Y. R., in *Encyclopedia of Chemical Processing and Design,* McKetta, J. J. and Cunningham, W. A., Eds., Marcel Dekker, New York, 1990, **31**, 35.

Mehra (2) [Named after the inventor] A gas separation process utilizing absorption in a solvent at moderate pressures. Developed by Advanced Extraction Technologies and applied to hydrogen recovery, nitrogen rejection, and recovery of natural gas liquids.

> *Hydrocarbon Process.,* 1997, **76**(5), 15.
> Bell, C. J. and Mehra, Y. R., *Oil & Gas J.,* 1997, **95**(39), 86.

Meissner *See* Schmidt.

Mellon A process for re-refining used oil, using solvent extraction and distillation.

> *Oil & Gas J.,* 1994, **92**(22), 87.

Membrane cell   A refinement of the *Diaphragm cell process in which the diaphragm is made from a cation-exchange membrane. *See also* Castner-Kellner.

MEMBREL   A process for making an aqueous solution of ozone by electrolyzing water, using a solid perfluorinated cation-exchange membrane as the electrolyte. The membrane was invented at Brookhaven National Laboratory, originally for use in fuel cells. It was subsequently developed in the 1980s by Asea Brown Boveri, Switzerland, for making ozone for purifying water. The process is now offered by Ozonia International, Switzerland. Ozonia is a joint venture of Degremont with L'Air Liquide. Water purified in this way is used in the electronics and pharmaceutical industries.

> Stucki, S., Theis, G., Kötz, R., Devanaty, H., and Christen, H. J., *J. Electrochem. Soc.,* 1985, **132**(2), 367.
> *Water Waste Treat.,* 1996, Mar, 26.

Menstruum   Also known as the McKenna process. A process for making the carbides of niobium and tantalum from their respective oxides. The oxide is reduced by heating to 2,000°C with aluminum in a graphite vessel. Graphite lumps are then added and the heating continued. After cooling, the excess of aluminum and aluminum carbide is dissolved out with hydrochloric acid. Graphite is removed by flotation, leaving the crystalline carbide. Invented in 1937 by P. M. McKenna.

> U.S. Patents 2,113,353; 2,113,354; 2,113,355; 2,113,356.

Mercapsol   A process for removing mercaptans from petroleum fractions, using aqueous sodium or potassium hydroxide containing cresols and solubility promoters. Developed by the Pure Oil Company, a division of the Union Oil Company of California, and first operated in West Virginia in 1941.

> Unzelman, G. H. and Wolf, C. J., in *Petroleum Processing Handbook,* Bland, W. F. and Davidson, R. L., Eds., McGraw-Hill, New York, 1967, 3-116.

Mercerization   A process for modifying cotton textiles by treatment with alkali. The alkali is cold, conentrated aqueous sodium hydroxide; it is subsequently removed by washing with acetic acid. The process is generally conducted while the textile is held under tension. The product has improved lustre and is easier to dye. Invented by J. Mercer in 1844.

MercOx   A process for removing mercury and sulfur dioxide from flue-gases. Hydrogen peroxide is first sprayed into the gas, converting metallic mercury to mercuric ions in solution. A water spray removes the sulfur dioxide as sulfuric acid. Mercury is removed from the liquor by ion-exchange, and the sulphate is precipitated as gypsum. Developed by Uhde and Gotaverken, with the Institut für Technische Chemie.

> *Chem. Eng. (N.Y.),* 1996, **103**(6), 19.

Mercury cell   *See* Castner-Kellner.

MERICAT   A process for removing mercaptans from petroleum fractions by a combination of catalytic oxidation and extraction with aqueous sodium hydroxide, using a proprietary contactor based on a bundle of hollow fibers. The sulfur products are disulfides, which remain in the hydrocarbon product. Developed by the Merichem Company, Houston, TX, and used in 61 plants as of 1991. Mericat II is a variation which includes a carbon bed too; there were four installations as of 1991. *See also* Thiolex.

> *Hydrocarbon Process.,* 1996, **75**(4), 126.

MERICON   A process for oxidizing and neutralizing spent alkali solutions from oil refining. Developed by the Merichem Company, Houston, TX. Five units were operating as of 1991.

> *Hydrocarbon Process.,* 1992, **71**(4), 120.

MERIFINING A process for extracting aromatic mercaptans and organic acids from cracked hydrocarbon fractions by aqueous alkali, using a bundle of hollow fibers. Developed by the Merichem Company, Houston, TX. Twelve units were operating as of 1991.

> *Hydrocarbon Process.,* 1996, **75**(4), 126.

Merox [**Mer**captan **ox**idation] A process for removing mercaptans from petroleum fractions by extracting them into aqueous sodium hydroxide and then catalytically oxidizing them to disulfides using air. The catalyst is an organometallic compound, either a vanadium phthalocyanine supported on charcoal, or a sulfonated cobalt phthalocyanine. Developed by UOP in 1958 and widely licensed; by 1994, more than 1,500 units had been built, worldwide.

> Unzelman, G. H. and Wolf, C. J., in *Petroleum Processing Handbook,* Bland, W. F. and Davidson, R. L., Eds., McGraw-Hill, New York, 1967, 3-128.
> Basu, B., Satapathy, S., and Bhatnagar, A. K., *Catal. Rev. Sci. & Eng.,* 1993, **35**(4), 571.
> *Hydrocarbon Process.,* 1996, **75**(4), 128.
> Holbrook, D. L., in *Handbook of Petroleum Refining Processes,* Meyers, R. A., Ed., McGraw-Hill, New York, 1997, 11.29.

Merrill-Crowe An improvement on the *cyanide process for extracting gold from rock. The solution of gold cyanide is reduced with zinc dust, thereby precipitating the gold as a fine powder which is filtered off and smelted. Operated in South Africa.

> Yannopoulos, J. C., *The Extractive Metallurgy of Gold,* Van Nostrand Reinhold, New York, 1991, 72, 189.

Merseburg A process for making ammonium sulfate fertilizer from gypsum. The gypsum is slurried with water and ammonium carbonate solution added. Calcium carbonate precipitates and is removed, any excess of ammonium carbonate is neutralized with sulfuric acid, and the solution is concentrated until it crystallizes:

$$CaSO_4 + (NH_4)_2CO_3 = CaCO_3 + (NH_4)_2SO_4$$

Developed by IG Farbenindustrie and first installed at Oppau, Germany, in 1913. Subsequently, widely used worldwide.

> Gopinath, N. D., in *Phosphoric Acid,* Vol. 1, Part 2, Slack, A. V., Ed., Marcel Dekker, New York, 1968, 541.

META-4 [**Meta**thesis–**C4**] A process for converting a mixture of C_4 hydrocarbons to propylene by metathesis in the presence of ethylene. Developed by Institut Français du Pétrole.

> Torck, B., *Chem. Ind. (London),* 1993, (19), 742.

metal surface treatment Many processes have been developed for treating the surfaces of metals in order to protect or decorate them. Some involve chemical reactions. Others are ostensibly physical, although surface chemical reactions doubtless occur in them all. Such processes fall outside the scope of this dictionary, but for convenience the major ones having special names are listed below. Descriptions of most of them may be found in the references below. Aldip, Alplate, Alrak, Alumilite, Alzak, Angus Smith, Atrament, Banox, Barff, Bengough-Stuart, Bethanising, Bonderising, Borchers-Schmidt, Bowers-Barff, Bullard-Dunn, Calorizing, Chromizing, Coslettizing, Footner, Granodizing, Hanson-Van Winkle-Munning, Laxal, Madsenell, MVB (modified Bauer Vogler), Nitralizing, ONERA, Parkerizing, Protal, Pylumin, Sendzimir, Sheppard, Sherardizing, Shimer, Walterization, Zincote.

> Burns, R. M. and Bradley, W. W., *Protective Coatings for Metals,* 3rd ed., Reinhold Publishing, New York, 1967.
> Tottle, C. R., *An Encyclopedia of Metallurgy and Materials,* The Metals Society and Macdonald & Evans, London, 1984.

METC   A coal gasification process based on a stirred, fixed-bed gasifier. Developed for the U.S. Department of Energy.

METEX   [Metal extraction]   A process for extracting heavy metals from industrial waste waters by adsorption on activated sludge under anaerobic conditions. It is operated in an up-flow, cylindrical reactor with a conical separation zone at the top. Developed by Linde, orig-inally for removing dissolved copper from winemaking wastes. First commercialized in 1987.

METLCAP   A process for encapsulating hazardous heavy metal wastes in a proprietary type of cement. Developed and offered by Environmental Remediation Technology, Cleveland, OH.

Metrex   A process for recycling spent hydroprocessing catalysts. Developed in 1993 by Metrex BV.

Met-X   A continuous process for removing traces of metals from cracking catalysts by ion-exchange. Developed by Atlantic Refining Company and first operated in Philadelphia in 1961.

> Leum, L. M. and Connor, J. E., Jr., *Ind. Eng. Chem. Prod. Res. Dev.,* 1962, **1**(3), 145.
> Unzelman, G. H. and Wolf, C. J., in *Petroleum Processing Handbook,* Bland, W. F. and Davidson, R. L., Eds., McGraw-Hill, New York, 1967, 3-15.

Meyers   *See* TRW Meyers.

M-forming   A process for increasing the octane rating of gasoline by cracking and iso-merization, catalyzed by the zeolite ZSM-5. Developed in the 1970s by Mobil Corporation, but not commercialized. A related process, M2-forming, for aromatizing light aliphatic hy-drocarbons over HZSM-5, was not commercialized either. *See also* Cyclar.

> Chen, N. Y. and Yan, T. Y., *Ind. Eng. Chem. Proc. Des. Dev.,* 1986, **25**(1), 151.
> Chen, N. Y., Garwood, W. E., and Heck, R. H. (1987). *Ind. Eng. Chem. Prod. Res. Dev.,* 1987, **26**, 706.
> Chen, N. Y. and Degnan, T. F., *Chem. Eng. Prog.,* 1988, **84**(2), 32.

MGCC   [Mitsubishi Gas-Chemical Company]   Also called JGCC. A process for extract-ing m-xylene from mixed xylene isomers by making the fluoroboric acid complex. All the xylene isomers form such complexes, but that formed by the m-isomer is much more stable than the others. Development started in 1962; by 1979, three plants were operating.

> *Hydrocarbon Process.,* 1969, **48**(11), 254.
> Herrin, G. R. and Martel, E. H., *Chem. Eng. (London),* 1971, (253), 319.
> Masseling, J. J. H., *CHEMTECH,* 1976, **6,** 714.

MHC   [Mitsubishi hydrocracking]   A process for making benzene and other aromatic hy-drocarbons by hydrogenating cracked petroleum fractions.

MHC Unibon   [Mild hydrocracking]   A mild *hydrocracking process for desulfurizing gas oil and converting it to lower molecular weight hydrocarbons, suitable for further pro-cessing by catalytic cracking. Developed by UOP.

MHD   [Mitsubishi hydrodealkylation]   A thermal process for converting toluene to ben-zene. Developed by Mitsubishi Chemical.

> Weissermel, K. and Arpe, H.-J. *Industrial Organic Chemistry,* 3rd ed., VCH Publishers, Weinheim, Germany, 1997, 330.

MHDV   [Mobil Hydrogen Donor Visbreaking]   A modified *visbreaking process in which a hydrogen donor stream from the oil refinery is added to the heavy hydrocarbon stream before thermal cracking. Developed by Mobil Corporation.

MHO [Metallurgie Hoboken-Overpelt]   A process for extracting manganese and other metals from nodules from the sea bed by extraction with hydrochloric acid:

$$MnO_2 + 4HCl = MnCl_2 + 2H_2O + Cl_2$$

The chlorine formed in this stage is re-used in a subsequent stage where it oxidizes the manganese (II) to manganese (IV), which precipitates as MnO_2. Developed by the Belgian company named.

Gupta, C. K. and Mukherjee, T. K., *Hydrometallurgy in Extraction Processes,* Vol. 1, CRC Press, Boca Raton, FL, 1990, 10.

MHTI [Mobil high temperature isomerization]   A process for converting mixed xylene streams to *p*-xylene. The catalyst is the zeolite ZSM-5. Developed by Mobil Research & Development Corporation and first commercialized in 1981. Eleven units were operating as of 1991. *See also* MLPI and MVPI.

Hydrocarbon Process., 1991, **70**(3), 166.

Michigan   *See* Grainer.

Micro-Simplex   *See* MS.

Mid-Century   Also called **M-C.** A process for oxidizing *p*-xylene to terephthalic acid, using oxygen in acetic acid and catalyzed by a mixture of cobalt and manganese bromides. Developed in the 1950s by Halcon International and commercialized by Standard Oil Company (Indiana). The first plant was built at Jolet, IA, in 1938. The *Amoco and *Maruzen processes are improved versions.

U.S. Patent 2,833,816.
Landau, R. and Saffer, A., *Chem. Eng. Prog.,* 1968, **64**(10), 20.
Landau, R., *Chem. Eng. Prog.,* 1988, **84**(7), 31.
Parteinheimer, W., in *Catalysis of Organic Reactions,* Blackburn, D. W., Ed., Marcel Dekker, New York, 1990, 321.
Raghavendrachar, P. and Ramachandran, S., *Ind. Eng. Chem. Res.,* 1992, **31**, 453.

Middox   A process for delignifying wood pulp by the use of oxygen. Developed jointly by Air Products & Chemicals, and Black Clawson Company. The process removes half of the lignin from the pulp, thereby halving the chlorine usage.

Chem. Week, 1981, **129**(17), 17.

Midforming [**Mid**dle-range distillate **forming**]   A process for converting lower olefins to transport fuels. The catalyst is either a ZSM-5–type zeolite in which some of the aluminum has been replaced by iron, or a hetero-poly acid. Developed in the 1980s by the National Chemical Laboratory, Pune, India. To be piloted by Bharat Petrochemical Corporation, Bombay, and Davy Powergas.

Indian Patent Appl. 985/DEL/87.
U.S. Patent 4,950,821.

MIDREX [**Mid**land-Ross **ex**traction]   A process for the direct reduction of iron ore, using a mixture of carbon monoxide and hydrogen obtained by carbon dioxide *reforming. The key to the process is a special shaft furnace. The reformer is fed by a mixture of natural gas and off-gas from the shaft furnace. The reformer catalyst is nickel oxide supported on alumina. Developed by the Midland-Ross Corporation in the 1960s, based on earlier work by the Surface Combustion Company, Toledo, OH. The prototype plant was completed in Portland, OR, in 1969; the first large-scale plant was built in 1971 by the Georgetown Steel Corporation, Georgetown, SC. By 1998, 47 plants were either operating or under construction worldwide. The process is now licensed by Midrex International, Rotterdam, a subsidiary of Kobe Steel, Japan. *See also* DR.

Dayton, S., *Eng. Min. J.,* 1979, **180**(1), 80.

MIDW [Mobil isomerization dewaxing] A petroleum refining process which improves yield and quality by isomerizing and selectively cracking paraffins in waxy oils. The catalyst is a noble metal, supported on a zeolite. Developed by Mobil Corporation from 1991 to 1996.

> *Chem. Eng.*, 1996, **103**(3), 19.

MIGAS [Mitsubishi Gas] A process for making methyl methacrylate. Developed by the Mitsubishi Gas Chemical Company in 1992.

Miller A process for purifying and removing silver from gold by passing chlorine gas through the molten metal, covered with borax. The silver forms silver chloride, which floats to the top. Bismuth, antimony, and arsenic are eliminated as their volatile chlorides. Developed by F. B. Miller at the Sydney Mint in Australia in 1867 and soon in world-wide use.

> Yannopoulos, J. C., *The Extractive Metallurgy of Gold*, Van Nostrand Reinhold, New York, 1991, 242.

MILOX [Milieu oxidative] A wood-pulping and bleaching process. Wood chips are treated with hydrogen peroxide and formic acid in a three-stage process. Developed by the Finnish Pulp and Paper Research Institute (KCL) and Kemira. Piloted at Oulu, Finland in 1991.

> *Eur. Chem. News*, 1990, **55**, 8 Oct, 28.
> *Eur. Chem. News (Finland Suppl)*, 1991, **56**, May, 4.
> *Chem. Br.*, 1991, **27**, 687.
> Sundquist, J. and Poppius-Lerlin, K., in *Environmentally Friendly Technologies for the Pulp and Paper Industries*, Young, R. A. and Akhar, M., Eds., John Wiley & Sons, New York, 1998, 157.

Minemet A hydrometallurgical process for extracting metals from sulfide ores by leaching with ferric chloride solution. Developed by Minemet Recherche, France.

> Gupta, C. K. and Mukherjee, T. K., *Hydrometallurgy in Extraction Processes*, Vol. 1, CRC Press, Boca Raton, FL, 1990, 196.

Minex A process for removing hydrogen sulfide and carbon dioxide from gases and light hydrocarbon streams in oil refineries. Developed by the Merichem Company, Houston, TX.

> *Hydrocarbon Process.*, 1992, **71**(4), 120.

Minifos A process for making mono-ammonium and di-ammonium phosphates by reacting ammonia with phosphoric acid. Offered by Lurgi.

> Ranney, M. W., *Ammonium Phosphates*, Noyes Data Corp. Process Review No. 35, Noyes Publications, Park Ridge, NJ, 1969, 30.

Mitsui-Toatsu A high-pressure process for making urea from ammonia and carbon dioxide. Invented in 1967 by Toyo Koatsu Industries.

> U.S. Patent 3,506,710.
> Zardi, U., *Nitrogen*, 1982, (135), 26

MixAlco [Mixed Alcohol] A fermentation process for making a mixed alcohol fuel from biomass.

> *Fuels and Chemicals from Biomass*, Saha, B. C. and Woodward, J., Eds., American Chemical Society, Washington, DC, 1997.

MLDW [Mobil lube dewaxing] A catalytic process for removing waxes (long-chain linear aliphatic hydrocarbons and alkyl aromatic hydrocarbons) from lubricating oil. Developed by Mobil Research & Development Corporation and operated at Mobil Oil refineries since 1981. Eight units were operating in 1991.

MLPI [Mobil low pressure isomerization]   One of a family of processes developed by Mobil Corporation for isomerizing xylene mixtures, using a zeolite catalyst. This one was developed in 1977. *See also* LTI, MHTI, MVPI.

U.S. Patent 4,101,596.

Mobil   The Mobil Corporation has developed many processes, but in the 1990s the one most associated with its name was the Methanol to Gasoline process, using a zeolite catalyst. *See* MTG.

Mobil/Badger   A process for making ethylbenzene by reacting benzene with ethylene, in the vapor phase, over a ZSM-5 zeolite catalyst containing phosphorus. Diluted ethylene streams, from a variety of industrial sources, may be used. Developed in the 1970s by The Badger Company, using a catalyst developed by Mobil Corporation. First operated on a large scale in 1980 by American Hoechst Company; by 1991, 21 plants had been built. An improved version of the ZSM-5 catalyst, EBUF-1, was developed by Fina Oil & Chemical Company and United Catalysts from 1981 and used at the Cosmar company's plant in Louisiana from 1994.

U.S. Patent 3,962,364.
Hölderich, W. F. and van Bekkum, H., in *Introduction to Zeolite Science and Practice,* van Bekkum, H., Flanigen, E. M., and Jansen, J. C., Eds., Elsevier, Amsterdam, 1991, 664.
Fallon, K. J., Wang, H. K. W., and Venkat, C. R., *Oil & Gas J.,* 1995, **93**(16), 50.

Mobil/Badger cumene   A process for making cumene by reacting benzene with propylene. Developed from the *Mobil/Badger process for making ethylbenzene. First commercialized at the Georgia Gulf plant in Pasadena, TX, in 1996. A variation of this process, announced in 1993, reduces the benzene content of gasoline by reaction with propylene from an *FCC plant. The process now uses zeolite MCM-22 as the catalyst.

Goelzer, A. R., Hernandez-Robinson, A., Ram, S., Chin, A. A., Harandi, M. H., and Smith, C. M., *Oil & Gas J.,* 1993, **91**(937), 63.
Eur. Chem. News, 1997, **68**(1792), 26.

Mobil-Witco-shell   A process for making poly (1-butene) by polymerizing 1-butene with a *Ziegler catalyst in an excess of liquid monomer.

Chem. Week, 1977, 7 Dec, 9.

MOCVD [Metal oxide chemical vapor deposition]   A general name for a group of processes used to make micro-electronic devices by depositing thin films of metal oxides on suitable substrate surfaces by means of chemical vapor deposition. *See* CVD.

MODAR [Named after Modell, the inventor, and his original partners **O'Donnell** and **Rich**]   An application of *Supercritical Water Oxidation (SCWO). A process for destroying organic wastes, especially hydrocarbons, by oxidation under supercritical aqueous conditions. The waste water, and alkali, are heated to 450 to 650°C under a pressure of >200 atm. Hydrocarbons are rapidly oxidized to carbon dioxide; organic compounds of halogens, sulfur, and phosphorus are converted to the respective inorganic acids which are neutralized. Invented in 1980 by M. Modell in Cambridge, MA; developed by MODAR and engineered and commercialized with ABB Lummus Crest in 1989. It was piloted with pharmaceutical wastes at Pfinztal, near Karlsruhe, in 1994. Presently offered by Modell Environmental Corporation (MODEC).

The name is used also by ICI as an acronym for a range of modified acrylic resins.

U.S. Patents 4,113,446; 4,338,199; 4,543,190; 4,822,497.
Chem. Week, 1986, **139**(14), 40.
Modell, M., 1992, *Chem. Eng. News,* **70**(4), 2.

Bettinger, J. A., *Chem. Eng. News,* 1992, **70**(10), 2.
Water Waste Treat., 1994, **37**(9), 32.
Chem. Eng. (Rugby, England), 1994, (568), 24.

Modop [**Mo**bil **O**il **D**irect **O**xidation **P**rocess] A process for removing residual sulfur-containing gases from the tail gas from the *Claus process. The catalyst is titanium dioxide pelletized with calcium sulfate. Developed in the 1980s by Rhône-Poulenc, Procatalyse, and Mobil Oil. Three plants were operating in Germany in 1995 and one in the United States.

European Patents 60,742; 78,690.
Kettner, R. and Liermann, N., *Oil & Gas J.,* 1988, **86**(2), 63.
Hydrocarbon Process., 1992, **71**(4), 122.
Wieckowska, J., *Catal. Today,* 1995, **24**(4), 445.

Moebius An electrolytic process for removing gold and platinum from silver. The crude metal, known as Doré, is used as the anode. The cathodes are of silver or stainless steel. The electrolyte is a diluted solution of silver nitrate and nitric acid. Gold and other metals collect as anode slimes. Invented in Mexico by B. Moebius, first operated there in 1884, and subsequently widely operated in Germany and the United States. *See also* Balbach, Thum, Wohlwill.

British Patent 16,554 (1884).
Dennis, W. H., *A Hundred Years of Metallurgy,* Gerald Duckworth, London, 1963, 288.

Mofex A liquid-liquid extraction process for removing aromatic hydrocarbons from hydrocarbon mixtures. The solvent is a monomethylformamide/water mixture, operated at 20 to 30°C, 0.1 to 0.4 bar. Developed by Leuna-Werke.

Weissermel, K. and Arpe, H.-J., *Industrial Organic Chemistry,* 3rd ed., VCH Publishers, Weinheim, Germany, 1997, 320.

Moffat *See* steelmaking.

MOG [**M**obil **o**lefins to **g**asoline] A process for converting dilute streams of C_2- to C_4-hydrocarbons to gasoline, using a fluidized bed of zeolite ZSM-5 catalyst. Developed by Mobil Research & Development Corporation and piloted in 1990.

Hydrocarbon Process., 1994, **73**(11), 142.

MOGD [**M**obil **o**lefine to **g**asoline and **d**istillate] A process for converting C_2- to C_{10}-olefins to high-octane gasoline and other hydrocarbons. Developed by Mobil Corporation and first used at its refinery at Paulsboro, NJ, in 1982.

Garwood, W. E., in *Intrazeolite Chemistry,* Stucky, G. D. and Dwyer, F. G., Eds., American Chemical Society, Washington, D.C., 1983, 383.

Mohawk A process for recovering used automotive oils for re-use. Developed in Canada. The first plant was to be built by Evergreen Holdings in Newark, CA.

Hydrocarbon Process., 1990, **69**(9), 26.

MOI [**M**obil **o**lefin **i**nterconversion] A process for increasing the yield of propylene from *steam crackers and *fluid catalytic crackers, using a ZSM-type catalyst. Developed in 1998 by Mobil Technology.

Eur. Chem. News, 1998, **69**(1808), 39.

Molex A version of the *Sorbex process, for separating linear aliphatic hydrocarbons from branched-chain and cyclic hydrocarbons in naphtha, kerosene, or gas oil. The process operates in the liquid phase and the adsorbent is a modified 5A zeolite; the pores in this zeolite will admit only the linear hydrocarbons, so the separation factor is very large. First commercialized in 1964; by 1992, 33 plants had been licensed worldwide. *See also* Parex (2).

Carson, D. B. and Broughton, D. B., *Pet. Refin.,* 1959, **38**(4), 130.
Broughton, D. B., *Chem. Eng. Prog.,* 1968, **64**(8), 60.
Sohn, S. W., in *Handbook of Petroleum Refining Processes,* Meyers, R. A., Ed., McGraw-Hill, New York, 1997, 10.75.

MOLPSA-nitrogen [Molecular sieve pressure swing adsorption] A version of the *PSA process for separating nitrogen from air, developed by Kobe Steel. Most PSA processes for nitrogen use molecular sieve carbon as the adsorbent, but this one uses zeolite X. Water and carbon dioxide are first removed in a two-bed PSA system, and then the nitrogen is concentrated and purified in a three-bed system.

Suzuki, M., in *Adsorption and Ion Exchange: Fundamentals and Applications,* LeVan, M. D., Ed., American Institute of Chemical Engineers, New York, 1998, 121.

Molten Carbonate A *flue-gas desulfurization process in which the sulfur dioxide contacts a molten mixture of inorganic carbonates. These are converted to sulfates and sulfides and then reduced to hydrogen sulfide, which is treated in a *Claus kiln. The advantage of this process over most others is that it does not cool the flue-gases. Not commercialized.

Oldenkamp, R. D. and Margolin, E. D., *Chem. Eng. Prog.,* 1969, **65**(11), 73.
Kohl, A. L. and Riesenfeld, F. C., *Gas Purification,* 4th ed., Gulf Publishing, Houston, TX, 1985, 386.

Moltox A process for separating oxygen from air by selective absorption in a molten salt mixture at high temperature. Invented by D. C. Erickson of Energy Concepts, and developed by Air Products and Chemicals. The salts are a mixture of the nitrites and nitrates of sodium and potassium. The reaction is:

$$NO_2^- + \tfrac{1}{2}O_2 \rightarrow NO_3^-.$$

The operating temperature range is 450 to 700°C; either pressure-swing or thermal-swing modes can be used.

U.S. Patents 4,132,766; 4,287,170; 4,340,578.
Erickson, D. C., *Chem. Eng. (N.Y.),* 1983, **90**(21), 28.
Dunbobbin, B. R. and Brown, W. R., *Gas Sep. Purif.,* 1987, **1**, 23.

Mond A process for recovering sulfur from the residues from the *Leblanc process. The sulfur is partially oxidized to thiosulfate and converted to elemental sulfur by adding hydrochloric acid. This process recovers only half the sulfur; it was supplanted by the *Chance process. Invented by L. Mond and operated by the Netham Chemical Company at Bristol from 1868 to 1888.

Holland, R., *Chem. Ind. (London),* 1985, (11), 367.

Mond gas A process for gasifying coal at a relatively low temperature, using a mixture of air and steam. The use of steam increases the yield of ammonia. The process was invented primarily to produce the ammonia needed for the *ammonia-soda process. The gas is of low calorific value but can be used for industrial heating. Developed by L. Mond at Brunner Mond, Winnington, Cheshire, in 1883. Subsequently commercialized by the South Staffordshire Mond Gas Corporation at Dudley Port, near Birmingham, England, which distributed the gas to local industry through the world's first gas grid. The engineering and further commercialization was carried out by the Power Gas Corporation.

British Patent 3,923 (1883).
Hill, W. H., in *Chemistry of Coal Utilization,* Vol. 2, Lowry, H. H., Ed., John Wiley & Sons, New York, 1945, 1028.
Cohen, J. M., *The Life of Ludwig Mond,* Methuen & Co, London, 1956, 176.

Mond nickel A process for extracting nickel from its ores by the intermediary of the volatile nickel tetracarbonyl. Sulfide ores are first roasted to convert sulfides to oxides and then reduced by heating in hydrogen and carbon monoxide (water gas). The crude metal is reacted with carbon monoxide at 50°C, producing $Ni(CO)_4$, which is subsequently decomposed at 180 to 200°C. Invented by L. Mond and C. Langer in 1889, piloted at the works of Henry Wiggin & Company in Smethwick, Scotland in 1892, and subsequently commercialized on a large scale in Swansea, South Wales, where it still operates. A new plant was built in Canada in 1986.

British Patent 12,626 (1890).
Canadian Patents 35,427; 35,428.
U.S. Patents 455,228; 455,229; 455,230.
Mond, L., Langer, C., and Quincke, F., *J. Chem. Soc.,* 1890, **57,** 749.
Cohen, J. M., *The Life of Ludwig Mond,* Methuen, London, 1956, 282.
Abel, E., *Chem. Br.,* 1989, **25,** 1014.

Monell *See* steelmaking.

Monk-Irwin An unsuccessful predecessor of the *Sulfate process for making titanium dioxide pigment from ilmenite. Invented by C. R. Whittemore at McGill University, Montreal, in the early 1920s and subsequently developed by J. Irwin and R. H. Monk in Canada and B. Laporte Limited in Luton, England. Ilmenite from the deposit at Ivry, Quebec, was reduced by heating with coke, leached with ferric chloride solution, and then roasted with a mixture of sulfuric acid and sodium sulfate. The resulting cake, containing titanyl sulfate, was dissolved in water, hydrolyzed, and the titania hydrate calcined. Some of the product was extended with barium sulfate. The project was abandoned in 1928.

U.S. Patent 1,542,350.

Monsanto (1) A process for making adiponitrile, an intermediate in the manufacture of Nylon 66, by the electrolytic hydrodimerization (EHD) of acrylonitrile:

$$2CH_2{=}CHCN + 2H_2O + 2e^- = NC(CH_2)_4CN + 2OH^-$$

The original process used aqueous tetraethylammonium ethylsulfate as the electrolyte, a lead cathode, and a lead-silver alloy anode. The Mark II process, commercialized in the mid-1970s, uses an emulsion of acrylonitrile in aqueous sodium phosphate containing a salt of the hexamethylene-bis-(ethyldibutylammonium) cation. The process was invented in 1959 by M. M. Baizer at Monsanto Corporation, St. Louis, MO. It was commercialized in 1965 and has been continuously improved ever since. The process is also operated in Japan by Asahi Chemical Industry Company. In 1990, the world production of adiponitrile by this process was over 200,000 tonnes per year.

Prescott, J. H., *Chem. Eng. (N.Y.).,* 1965, **72**(23), 238.
Baizer, M. M. and Danly, D. E., *Chem. Ind. (London),* 1979, (435), 439.
Pletcher, D. and Walsh, F. C., *Industrial Electrochemistry,* 2nd. ed., Chapman & Hall, London, 1990, 298.

Monsanto (2) A catalytic process for synthesizing the drug L-DOPA. The catalyst is a chiral diphosphine-rhodium complex. Invented in the early 1970s.

Monsanto acetic acid A process for making acetic acid by carbonylation of methanol, catalyzed by rhodium iodide. Operated by BP.

Haynes, A., Mann, B. E., Morris, G. E., and Maitlis, P. M., *J. Amer. Chem. Soc.,* 1993, **115**(10), 4093.
Parkins, A. W., in *Insights into Speciality Inorganic Chemicals,* Thompson, D., Ed., Royal Society of Chemistry, Cambridge, 1995, 110.

Mon Savon A continuous soapmaking process.

Kirk-Othmer's Encyclopedia of Chemical Technology, 3rd ed., Vol. 21, John Wiley & Sons, New York, 1983, 173.

Mont Cenis   [Named after a coal mine in the Ruhr]   An early ammonia synthesis process, basically similar to the *Haber-Bosch process but using coke-oven gas. Operated by The Royal Dutch Group at Ymuiden, The Netherlands, since 1929.

Scholvien, W. F., *Chem. Met. Eng.,* 1931, **38**(2), 82.

Spitz, P. H., *Petrochemicals, the Rise of an Industry,* John Wiley & Sons, New York, 1988, 84.

Montoro   A process for making styrene and propylene oxide. Named after the eponomous company. The process was to be used in Repsol Quimica's plant in Tarragona.

Morgas   [**Mor**gantown **gas**ification]   A coal gasification process using a stirred, fixed-bed gasifier. Piloted in the 1970s at the Morgantown Research Center of the U.S. Bureau of Mines, West Virginia.

Hebden, D. and Stroud, H. J. F., in *Chemistry of Coal Utilization,* 2nd. Suppl. Vol., Elliott, M. A., Ed., John Wiley & Sons, New York, 1981, 1627.

MORPHYLANE   A process for removing aromatic hydrocarbons from hydrocarbon mixtures by extractive distillation. The added solvent is N-formyl morpholine. The process was developed by Krupp Koppers in the 1960s and by 1994, 22 units had been built. *See also* MORPHYLEX, OCTENAR.

MORPHYLEX   A liquid-liquid extraction process for removing aromatic hydrocarbons from hydrocarbon mixtures. The solvent is N-formyl morpholine, the operating temperature is 180 to 200°C. The process was developed by Krupp Koppers in the 1960s and first commercialized in 1972. Only one plant had been built as of 1994. *See also* MORPHYLANE, OCTENAR.

Franck, H.-G. and Stadelhofer, J. W., *Industrial Aromatic Chemistry,* Springer-Verlag, Berlin, 1988, 110.

Morse   *See* steelmaking.

MOSC   [**M**obil **o**il **s**ludge **c**oking]   A process used in oil refineries which converts aqueous sludges to coke, thereby reducing the quantity of waste discharged. Developed by Mobil Corporation.

MOST   [**M**obil **O**il **SO**$_x$ **T**reatment]   A catalytic process for removing sulfur-containing gases from the tail gases from the *Claus process and other SO$_x$-containing gases. The gases are first combusted with air, converting all the sulfur-containing species to SO$_2$. The SO$_2$ is adsorbed on a solid sorbent/catalyst such as vanadia-promoted magnesia spinel, and then reductively desorbed as a mixture of H$_2$S and SO$_2$ for recycle to the Claus plant.

Buchanan, J. S., Stern, D. L., Nariman, K. E., Teitman, G. J., Sodomin, J. F., and Johnson, D. L., *Ind. Eng. Chem. Res.,* 1996, **35**, 2495.

MOXY   [**M**ead **oxy**gen]   A variation of the *Kraft papermaking process in which the sulfides are oxidized to polysulfides, with some increase in efficiency. Developed by Mead Corporation.

MPC   [**M**itsui **P**etrochemical]   A continuous process for polymerizing propylene, based on the *Ziegler-Natta process, but using a much more active catalyst so that de-ashing (catalyst removal) is not required. The catalyst contains magnesium in addition to titanium; successive versions of it have been known as HY-HS (high yield, high stereospecifity), HY-HS II, and T-catalyst. Developed jointly by Mitsui Petrochemical Industries, Japan, and Montedison SpA, Italy, in 1975, and now licensed in 56 plants worldwide.

MRG   [**M**ethane **r**ich **g**as]   A catalytic steam-reforming system, similar to the classic *syngas reaction of steam with a hydrocarbon mixture, but yielding hydrogen, methane, and carbon monoxide in different proportions. The system is thermodynamically balanced, requiring no heat other than that required to raise the reactants to the operating temperature. Developed by the Japan Gasoline Company.

MRH (1) [Methanol reformer hydrogen] A process for generating hydrogen from methanol, separating it by *PSA. Developed by the Marutani CPE Company.

Suzuki, M., in *Adsorption and Ion Exchange: Fundamentals and Applications*, LeVan, M. D., Ed., American Institute of Chemical Engineers, New York, 1998, 121.

MRH (2) A *hydrocracking process for "difficult" petroleum residues, i. e., those containing high levels of metals, sulfur, and nitrogen compounds. It uses catalytic hydrogenation in a slurry bed. Developed by the MW Kellogg Company.

Marcos, F. and Rosa-Brussin, D., *Catal. Rev. Sci. Eng.*, 1995, **37**(1), 3.

MRU [Methanol recovery unit] A process for removing methanol from the unreacted components from the synthesis of methyl *t*-butyl ether. It uses selective adsorption on multiple beds of a zeolite such as 4A. Developed by Union Carbide Corporation and now licensed by UOP; as of 1992, eight units had been licensed. *See also* ORU.

U.S. Patent 4,740,631.

MS [Micro-Simplex] A *steam reforming process for making town gas from petroleum fractions or LPG. Developed by Gaz de France and Stein & Roubaix.

British Petroleum Co., *Gas Making and Natural Gas*, British Petroleum Co., London, 1972, 91.

MS-2 A molecular sieving processes for separating branched-chain aliphatic hydrocarbons from unbranched ones by selective adsorption on a zeolite. Developed by the British Petroleum Company in the 1970s but not commercialized.

Grebell, J., *Oil & Gas J.*, 1975, **73**(15), 85.

MSCC [Millisecond catalytic cracking] A *fluid catalytic cracking process which uses an ultra-short contact time reaction system. It is claimed that less capital investment and higher liquid yields can be achieved using this process, compared with conventional *FCC units. Developed by Bar-Co and now offered by UOP; it has been operating since 1994.

Eur. Chem. News, 1995, **64**(1682), 28.
Hydrocarbon Process., 1996, **75**(11), 96.

MSDW [Mobil selective dewaxing] A catalytic dewaxing process which uses a catalyst containing a shape-selective molecular sieve and a noble metal.

Oil & Gas J., 1997, **95**(35), 64.

MSP3 [Micro-Suspension Process] A process for making polyvinyl chloride in suspension. Developed by Atochem, which has granted four licenses since 1977.

Chem. Mark. Rep., 1990, 22 Oct, 4.

MS Sorbex A *Sorbex process used in the production of *m*-xylene from C_8 aromatic mixtures. A zeolite is used as the sorbent and toluene is the desorbent.

Eur. Chem. News, 1995, **64**(1687), 32.

MSTDP [Mobil selective toluene disproportionation] A process for converting toluene to benzene and a xylene mixture rich in *p*-xylene. The catalyst is the zeolite ZSM-5, selectively coked to constrict the pores and thus increase the yield of *p*-xylene produced. Developed and licensed by the Mobil Oil Corporation and first commercialized in Sicily in 1988. *See also* MTDP.

Chen, N. Y., Kaeding, W. W., and Dwter, F. G., *J. Am. Chem. Soc.*, 1979, **101**, 6783.
Kaeding, W. W., Chu, C., Young, L. B., and Butter, S. A., *J. Catal.*, 1981, **69**, 392.

Hydrocarbon Process., 1989, **68**(11), 93.
Hydrocarbon Process., 1991, **70**(3), 140.

MTA [Methanol to aromatics]   A common abbreviation for any process which achieves this conversion, notably a Mobil process.

MT-chlor [Mitsui Toatsu Chlorine]   A process for recovering chlorine from hydrogen chloride. The hydrogen chloride is mixed with oxygen and passed through a fluidized bed of chromia/silica catalyst. Developed by Mitsui Toatsu and first operated in Japan in 1988. *See also* Deacon, Kel-Chlor.

Tozuka, Y., in *Science and Technology in Catalysis,* Izumi, Y., Aral, H., and Iwamoto, M., Eds., Elsevier, Amsterdam, 1994, 41.

MTDP [Mobil toluene disproportionation]   A catalytic process which converts 2 moles of toluene to 1 mole of mixed xylenes and 1 mole of benzene. The catalyst is the zeolite ZSM-5. Developed by Mobil Research & Development Corporation and first commercialized in 1975. Supersed by *MSTDP.

MTE   A process for recovering sulfur from acid gases, based on the *Claus process, but using a circulating, powdered catalyst instead of the usual fixed catalyst bed. Developed in 1987 but not yet commercialized.

U.S. Patent 4,801,443.
Simek, I. O., *Hydrocarbon Process.,* 1991, **70**(4), 45.

MTG [Methanol to gasoline]   A common abbreviation for any process achieving this conversion, notably the Mobil process. This uses as a catalyst the synthetic zeolite ZSM-5, invented at the Mobil Research Laboratory in 1972. The process was first disclosed in 1976 and commercialized in 1985 by New Zealand Synfuels, a joint venture of Mobil Corporation and Petrocorp. In 1990, this process was providing one third of New Zealand's gasoline requirements.

Kam, A. Y., Schreiner, M., and Yurchak, S., in *Handbook of Synfuels Technology,* Meyers, R. A., Ed., McGraw-Hill, New York, 1984, 2-75.
Weissermel, K. and Arpe, H.-J., *Industrial Organic Chemistry,* 3rd ed., VCH Publishers, Weinheim, Germany, 1997, 32.

MTH [Mittel-Temperatur-Hydrierung; German, meaning medium-temperature hydrogenation]   A version of the *TTH process, using different processing conditions, by which a larger proportion of transport fuels could be produced.

Weisser, O. and Landa, S., *Sulphide Catalysts, Their Properties and Applications,* Pergamon Press, Oxford, 1973, 333.

MTO [Methanol to olefins]   A catalytic process for converting methanol to olefins, mainly propylenes and butenes. Developed by Mobil Research & Development Corporation and first demonstrated in 1985. Another version of this process was developed by UOP and Norsk Hydro and has been run at a demonstration unit at Porsgrunn, Norway, since June 1995. It is based on fluidized bed technology using a SAPO molecular sieve catalyst. It converts 80 percent of the carbon in the feed to ethylene and propylene.

Chem. Eng. (N.Y.), 1996, **103**(1), 17.
Picciotti, M., *Oil & Gas J.,* 1997, **95**(26), 72.

MTPX [Mobil toluene to *p*-xylene]   A catalytic process for making *p*-xylene from toluene, developed by Mobil Corporation in 1995. First installed at Mobil's plant in Fairfax, VA.

Ind. Eng. Chem., 1995, **73**(35), 12.

Mülheim   *See* Alfol.

Müller-Kühne   A process for recovering sulfuric acid from phosphogypsum, the waste product from the manufacture of phosphoric acid. The process is economic only if the lime co-product is converted to cement. Based on the work of W. S. Müller and H. H. Kühne at Bayer, Leverkusen, from 1915 to 1918. Further developed in Germany in the 1950s and still in operation in Germany and Austria in 1989.

> Hull, W. Q., Schon, F., and Zirngibl, H., *Ind. Eng. Chem.*, 1957, **49**(8), 1204.
> Becker, P., *Phosphates and Phosphoric Acid,* 2nd. ed., Marcel Dekker, New York, 1989, 560.

Munich   An integrated process for making chlorine dioxide from hydrochloric acid. Sodium chlorate is made electrochemically from sodium chloride, and this is reduced with hydrochloric acid. Developed from the *Kesting process by H. Fröhler and E. Rossberger at Elektrochemische Werke München, and first commercialized in 1974. The essential improvement over the Kesting process is the use of titanium electrodes coated with ruthenium oxide for the electrolytic cell, and the use of titanium metal for the construction generally. The hydrogen produced in the electrolysis is burnt with the chlorine from the reduction stage to produce hydrochloric acid. The process is licensed by Fröhler A.G. and has been engineered by Uhde. As of 1990, 18 plants had been built, with capacities of between 8 and 40 tonnes per day.

> German Patents 2,407,312; 2,645,121.

Murso   *See* Musro.

Musro   [**Mu**rphy ores, CSIRO]   Also written **Murso.** A process for beneficiating ilmenite by a combination of oxidation, reduction, and pressure leaching with hydrochloric acid. Invented in Australia in 1967 and developed jointly by Murphyores Pty and the Commonwealth Scientific and Industrial Research Organization, but not commercialized then. Further developed in 1992 by Pivot Mining NL, Queensland.

> British Patent 1,225,826.
> Gupta, C. K. and Mukherjee, T. K., *Hydrometallurgy in Extraction Processes,* Vol. 1, CRC Press, Boca Raton, FL, 1990, 111.

MVB   *See* metal surface treatment.

MVPI   [**M**obil **v**apor **p**hase **i**somerization]   A process for converting mixed xylene streams to *p*-xylene, catalyzed by the zeolite ZSM-5. Invented by Mobil Corporation in 1973, later superseded by *MHTI. *See also* LTI, MHTI, MLPI.

> U.S. Patents 3,856,871; 3,856,872; 3,856,873; 3,856,874.

MWB   *See* Sulzer MWB.

N

Nahcolite   A *flue-gas desulfurization process. Nahcolite is a mineral containing 70 to 90 percent sodium bicarbonate, which is found in Colorado. In this process, the powdered nahcolite is injected into the baghouse and the following reaction occurs:

$$SO_2 + 2NaHCO_3 + \tfrac{1}{2}O_2 = Na_2SO_4 + 2CO_2 + H_2O$$

The solid product is dumped. Piloted by Battelle Columbus Laboratories.

Genco, J. M., Rosenberg, H. S., Anastis, M. Y., Rosar, E. C., and Dulin, J. M., *J. Air Pollut. Control Assoc.,* 1975, **25**(12), 1244.

Genco, J. M. and Rosenberg, H. S., *J. Air Pollut. Control Assoc.,* 1976, **26**(10), 989.

Nalfining   A process for purifying petroleum fractions by extraction with aqueous sodium hydroxide.

Unzelman, G. H. and Wolf, C. J., in *Petroleum Processing Handbook,* Bland, W. F. and Davidson, R. L., Eds., McGraw-Hill, New York, 1967, 3-20.

NAPFINING   A process for removing naphthenic acids from petroleum fractions by extracting with aqueous alkali, using a bundle of hollow fibers. Developed by the Merichem Company, Houston, TX, and used in 19 plants in 1991.

Hydrocarbon Process., 1996, **75**(4), 126.

Naphtachimie   A gas-phase process for making high-density polyethylene in a fluidized bed. Invented by Naphtachimie in 1973, and operated by that company at Lavera, France, since 1975.

U.S. Patent 3,922,322.

Natta   A process for polymerizing propylene and other higher olefins, catalyzed by crystalline titanium trichloride and an alkyl aluminum compound such as triethyl aluminum. The polymer can exhibit various types of stereoregularity, depending on the catalyst and the conditions. Invented in 1954 by G. Natta at the Istituto de Chimica Industriale del Politecnico di Milano, Italy, and commercialized in 1957. Now used widely, worldwide. *See also* Ziegler, Ziegler-Natta.

U.S. Patents 3,112,300; 3,112,301.

Natta, G., Pino, P., Corradini, P., Danusso, F., Mantica, E., Mazzanti, G., and Moraglio, G., *J. Am. Chem. Soc.,* 1955, **77,** 1708.

Natta, G., *J. Polymer Sci.,* 1955, **16,** 143.

Natta, G., *Angew. Chem.,* 1956, **68,** 393.

Natta, G., *Inaugural Lecture, 16th. Internat. Conf. Pure Appl. Chem., Paris,* Birkhauser Verlag, Basel, 1957, 21.

Raff, R. A. V., in *Ethylene and Its Industrial Derivatives,* Miller, S. A., Ed., Ernest Benn, London, 1969, 335.

NCB   *See* carbonization.

NDA   [Natural detergent alcohols]   A process for making long-chain alcohols, for use in detergent synthesis, from fatty acids from vegetable oils. The fatty acids are esterified with methanol and the resulting methyl esters are catalytically hydrogenated. Developed by Kvaerner Process Technology in 1988; the first commercial plant began operation in the Philippines in 1998.

Eur. Chem. News, 1998, **69**(1807), 19.

NEC [Nitrogen Engineering Corporation] A modification of the *Haber process for making ammonia.

> Vancini, C. A., *Synthesis of Ammonia,* translated by L. Pirt, Macmillan, Basingstoke, England, 1971, 237.

Nedol [New Energy Development Organization liquifaction] A coal liquifaction process in development in Japan by the New Energy and Industrial Technology Development Organization (NEDO), Tokyo. Crushed coal is mixed with a pyrite catalyst and slurried in a hydrogenated heavy oil. Liquifaction takes place at 450°C, 170 bar. The overall oil yield is 59 percent.

> *Chem. Eng. (N.Y.),* 1998, **105**(2), 29.

NEOCHROME A process for making colored acrylonitrile fibers by dyeing during the spinning process. Developed by Courtaulds.

> *Eur. Chem. News CHEMSCOPE,* 1995, May, 8.

NEUTREC A *flue-gas desulfurization process, intended for treating the waste gases from incinerators for municipal, hospital, and industrial wastes. Sodium bicarbonate, optionally mixed with active carbon, is injected into the gases after the usual bag filter, and the solid products are removed in a second bag filter. Sodium compounds can be recovered from the product for reuse, and any toxic compounds disposed of separately. Developed by Solvay and operated in Europe since 1991.

> *Chem. Ind. (London),* 1997, (19), 762.

NExETHERS A process for converting C_5, C_6, and C_7 olefins to ethers for blending into gasoline to increase its octane rating. The process resembles *Nextame but uses additional methanol to increase the yield of ethers.

> *Oil & Gas J.,* 1997, **95**(1), 44.

NExSELECT A catalytic, selective hydrogenation process developed by Neste Oy. Operated in Porvoo, Finland, since 1996.

> *Oil & Gas J.,* 1997, **95**(1), 45.

NExTAME [Neste tertiary amyl methyl ether] A catalytic process for converting C_5, C_6, and C_7 tertiary olefins to ethers for blending into gasoline to increase its octane rating. Developed by Neste Oy in 1994 and commercialized at Porvoo, Finland, in 1995.

> *Hydrocarbon Process.,* 1996, **75**(11), 110.
> *Oil & Gas J.,* 1997, **95**(1), 44.

NGOP [Natural gas oxypyrolysis] *See* IFP Oxypyrolysis.

Neostar A process for destroying waste organic chlorides (*e.g.*, polychlorinated biphenyls) by heating with steam and hydrogen at over 1,000°C. The products are methane, ethane, other chlorine-free hydrocarbons, and hydrochloric acid. Developed by Cerchar, France.

Nesbitt *See* steelmaking.

Netto An early process for extracting aluminum from cryolite by reducing it with sodium. Operated in the 1890s.

Neuberg A process for increasing the yield of glycerol from the fermentation of glucose by adding sodium sulfite.

> Neuberg, C. and Reinfurth, E., *Biochem. Z.,* 1918, **89**, 365 (*Chem. Abstr., 13, 328*).
> Baldwin, E., *Dynamic Aspects of Biochemistry,* 5th ed., Cambridge University Press, Cambridge, 1967, 347.

Neutralysis A process for converting municipal waste into lightweight aggregate. Developed by Neutralysis, Austria, and offered by Davy Corporation.

neutralization A family of processes for making sodium cyanide by neutralizing anhydrous hydrocyanic acid with aqueous sodium hydroxide. These replaced the *Castner (2) process in the 1960s.

Newcell A process for making reconstituted cellulose fibers by dissolving cellulose in N-methyl morpholine N-oxide and injecting the solution into water. Invented in 1977 by Akzona, NC.

U.S. Patents 4,142,913; 4,144,080; 4,145,532.

New Jersey A continuous process for extracting zinc from zinc oxide, made by roasting zinc sulfide ore, by reduction with carbon in a vertical retort. First operated by the New Jersey Zinc Company in Palmerton, PA, in 1929, and introduced into the Avonmouth, UK, works of the Imperial Smelting Company in 1934.

Newton Chambers A process for purifying benzene by fractional solidification; cooling is accomplished by mixing it with refrigerated brine. The process does not remove thiophene.

Nicaro [named after the Cuban town] A process for extracting nickel from low-grade ores. The ore is reduced by heating with *producer gas and is then leached with aqueous ammonia.

Niers A process for treating the aqueous effluent from dyeworks by a combination of chemical precipitation and biological purification.

Nippon Steel *See* steelmaking.

NIPR *See* carbonization.

N-ISELF A process for separating linear hydrocarbons from light naphtha by selective adsorption on a zeolite. Developed by Société Nationale Elf-Aquitaine, France.

Bernard, J. R., Gourlia, J.-P., and Guttierrez, M. J., *Chem. Eng. (N.Y.)*, 1981, **88**(10), 92.

NITECH A cryogenic process for removing nitrogen from natural gas, mainly methane. The high-pressure gas is liquified by expansion and then fractionated. The essential feature is the use of an internal reflux condenser within the fractionating column. Developed by BCCK Engineering and demonstrated on a full-scale plant in Oregon in 1994.

U.S. Patent 5,375,422.
Butts, R. C., Chou, K., and Slaton, B., *Oil & Gas J.*, 1995, **93**(11), 92.

Nitralizing *See* metal surface treatment.

NitRem [Nitrate Removal] A process for removing nitrate from water supplies by electrodyalysis through a selective membrane. Developed in the 1980s by OTTO Oeko-Tech & Company, Germany.

NITREX A process for removing nitrogen from natural gas by *PSA. Developed by UOP.

Hydrocarbon Process., 1996, **75**(4), 128.

NitroGEN [Nitrogen generator] A version of the *VPSA process for separating nitrogen from air by vacuum pressure swing adsorption. Developed by the Linde Division of the Union Carbide Corporation. The name has been used also for two membrane systems for extracting pure nitrogen from air.

Eur. Chem. News, 1989, **53**(1391), 31.
Chem. Mark. Rep., 1990, 29 Oct. 5.

Nitro Nobel A process for making nitrate esters such as nitroglycerol. A special injector is used to mix the liquid polyol with the nitrating acid.

Nittetu A process for destroying waste organic chlorides by submerged combustion.

> Santoleri, J. J., *Chem. Eng. Prog.,* 1973, **69**(1), 68.

Nixan [Nitrocyclohexane] A process for making cyclohexane oxime (an intermediate in the manufacture of nylon) from benzene by liquid phase nitration, followed by hydrogenation of the nitrobenzene. Invented by Du Pont and operated from 1963 to 1967.

> Weissermel, K. and Arpe, H.-J., *Industrial Organic Chemistry,* 3rd ed., VCH Publishers, Weinheim, Germany, 1997, 254.

Noguchi A catalytic process for hydrogenating lignin to a mixture of monophenols. Invented in 1952 at the Noguchi Institute of Japan, but not commercialized because the yields were uneconomic.

> Goheen, D. W., in *Lignin Structure and Reactions,* American Chemical Society, Washington, D. C., 1966, 205 (*Chem. Abstr.,* **43,** 21091).

NoNO$_x$ A two-stage combustion system which does not produce oxides of nitrogen. The first stage operates under reducing conditions, at a controlled, low partial-pressure of oxygen. The second stage uses a small excess of oxygen. Developed by Boliden Contech, Sweden, originally for the roasting of arsenical pyrite, but now used principally for incinerating waste.

Noranda A continuous copper smelting process, developed in Canada by Noranda Mines, from 1964.

> Canadian Patent 758,020.
> Themelis, N. J. and McKerrow, G. C., in *Advances in Extractive Metallurgy and Refining,* Jones, M. J., Ed., Institution of Mining and Metallurgy, London, 1971, 3.

Nordac A process for concentrating *wet-process phosphoric acid by submerged combustion. Operated in Europe since 1947. *See also* Ozark Mahoning.

> Forster, J. H., in *Phosphoric Acid,* Vol. 1, Part 2, Slack, A. V., Ed., Marcel Dekker, New York, 1968, 594.

Nordac-Aman *See* Woodall-Duckham.

Normann Also called Sabatier-Normann, after P. Sabatier, one of the inventors of catalytic hydrogenation. The first commercial process for hardening fats by catalytic hydrogenation over nickel. Invented in 1902 by K. P. W. T. Normann at the Herforder Maschinenfett-und Ölfabrik, Germany, and initially licensed to Joseph Crosfield & Sons, United Kingdom. The first large-scale plant was built at Warrington in 1909. Competing processes were developed by E. Erdmann, C. Paal, N. Testrup, and M. Wilbuschewitsch and much patent litigation ensued. Normann's patent was eventually declared invalid in 1913 because of incomplete disclosure.

> German Patents 139,457; 141,029.
> British Patent 1,515 (1903).
> Musson, A. E., *Enterprise in Soap and Chemicals,* Manchester University Press, Manchester, 1965, 165.
> Wilson, C., *The History of Unilever,* Cassell, London, 1954, 110.

Norsk-Hydro This large Norwegian company has given its name to a number of processes based on hydro-electric power. One such process, offered by Lurgi, is for producing ammonium phosphates. Another is for making magnesium by electrolyzing molten magnesium chloride, derived indirectly from seawater; this has been in operation at Porsgrun, Norway, since 1951.

> Höy-Petersen, N., *J. Met.,* 1969, **21**(4), 43.

NORSOLOR A continuous process for making polystyrene. Licensed by Badger Company and operated in France and South Korea.

Hydrocarbon Process., 1989, **68**(11), 110.

North Thames Gas Board A process for removing organic sulfur compounds from coal gas by catalytic oxidation over nickel sulfide at 380°C. The sulfur dioxide produced is removed by scrubbing with dilute aqueous sodium hydroxide. Operated by the gasworks of the North Thames Gas Board, London, between 1937 and 1953.

Plant, J. H. G. and Newling, W. B. S., *Trans. Inst. Gas Eng.,* 1948, **98**, 308.

Norzinc *See* Boliden/Norzinc.

NoTICE [**No Tie In Claus Expansion**] A process for oxidizing sulfur for the manufacture of sulfuric acid. Oxygen is introduced below the surface of a pool of molten sulfur. This permits easy temperature control. Developed by Brown & Root Braun and first used at Port Newches, TX, in 1989.

Schendel, R. L., *Oil & Gas J.,* 1993, **91**(39), 63.

Novacon An adsorptive process for removing oxides of sulfur and nitrogen, and carbon monoxide, from combustion gases. The adsorbent is an active form of natural marble.

Novalfer *See* DR.

Novolen A process for making polypropylene in the gas-phase, using a vertical stirred-bed reactor. Developed by BASF and engineered by Uhde. Eight plants had been licensed as of 1985. A metallocene-catalyzed version was introduced in 1996. The name is used also for the product.

NOVOX Not a process, but a trademark used by BOC to designate its *PSA process for separating oxygen from air.

NO$_x$Out A process for removing oxides of nitrogen from flue-gases by reaction with urea:

$$CO(NH_2)_2 + 2NO + \tfrac{1}{2}O_2 = 2N_2 + CO_2 + 2H_2O$$

The problem with this reaction is that it takes place over a narrow temperature range, between about 930 and 1,030°C. Below this range, ammonia is formed; above it, more nitrogen oxides are formed. In the NO$_x$Out process, proprietary additives are used to widen the usable temperature range. Developed by the Electric Power Research Institute, Palo Alto, CA, from 1976 to 1980, and then further developed by Fuel Tech. It was first commercialized in Germany in 1988. In February 1990, Fuel Tech formed a joint venture with Nalco Chemical Company – Nalco Fuel Tech – to further develop and promote the process. By September 1990, 16 systems had been sold in the United States and 2 in Europe. In November 1991, an improved version – NO$_x$Out Plus – was announced.

U.S. Patents 4,208,386; 4,325,924; 4,719,092.
Chem. Mark. Rep., 1990, **238**(12), 31.
Hydrocarbon Process., 1993, **72**(8), 80.
Lin, M. L., Comparato, J. R., and Sun, W. H., in *Reduction of Nitrogen Oxide Emissions,* Ozkan, U.S., Agarwal, S. K., and Marcelin, G., Eds., American Chemical Society, Washington, D.C., 1995, Chap. 17.

NOXSO A process for simultaneously removing sulfur oxides and nitrogen oxides from flue-gases. The sorbent is a regenerable mixture of sodium carbonate with alumina in a hot fluidized bed. The gases are retained as sodium sulfate, nitrate, and nitrite. Regeneration is carried out by heating first with air and then with a reducing gas such as hydrogen or natural gas. Developed and piloted by the Noxso Corporation with MK-Ferguson Company and W. R. Grace and Company. A demonstration plant was installed in the Ohio Edison power plant in Toronto, in 1991, funded by the U.S. Department of Energy and the Ohio Coal Development Office. Noxso Corporation was declared bankrupt in June 1997, following a dispute with Olin Corporation.

U.S. Patent 4,755,499.
Chem. Eng. (N.Y.), 1989, **96**(6), 21.
Chem. Eng. News, 1990, **68**(38), 35.
Neal, L. G., Woods, M. C., and Bolli, R. E., in *Processing and Utilization of High-sulfur Coals, IV,* Dugan, P. R., Quigley, D. R., and Attia, Y. A., Eds., Elsevier, Amsterdam, 1991, 651.
Chem. Eng. News, 1997, **75**(23), 15.

NOXSORB   A process for removing oxides of sulfur and nitrogen from "dirty" gases. Developed by Trimer Corporation in 1995.

Amer. Ceram. Soc. Bul., 1995, **74**(8), 86.

NRS   [New Regeneration System]   A process for regenerating the ion-exchange resin used for removing calcium from sugar solution. If sodium chloride were used, the waste calcium chloride solution would have to be disposed of; if sodium hydroxide were used, calcium hydroxide would be precipitated in the resin. The NRS process uses sodium hydroxide in the presence of sucrose, which retains the calcium in solution as calcium saccharate. Developed by the IMACTI Division of Duolite International, The Netherlands.

NSC   An obsolete direct reduction ironmaking process, operated in Japan and Malaysia. *See* DR.

NSM   A Dutch process for making ammonium nitrate, offered by Uhde. Not to be confused with another NSM (New Smoking Material), a tobacco substitute developed by ICI in the 1970s but later abandoned.

NSSC [Neutral sulfite semichemical pulping]   A papermaking process in which wood chips are digested in an aqueous solution of sodium sulfite and sodium carbonate at 140 to 170°C for several hours.

Nu-Iron   *See* DR.

Nulite   A catalytic, photochemical process for oxidizing toxic organic compounds in water. The catalyst is titanium dioxide supported on a mesh; the light is sunlight. Intended for treating groundwater containing not more than 500 ppm of toxic organic materials. Developed and offered by Nutech Environmental, London, Ontario.

Nurex   A process for extracting $C_8 - C_{30}$ linear hydrocarbons from petroleum fractions, using their ability to form urea inclusion complexes. Branched-chain hydrocarbons do not form such complexes. Developed by the Nippon Mining Company, Japan, and operated until 1979.

Scholten, G. G., in *Petroleum Processing Handbook,* McKetta, J. J., Ed., Marcel Dekker, New York, 1992, 587.
Weissermel, K. and Arpe, H.-J., *Industrial Organic Chemistry,* 3rd ed., VCH Publishers, Weinheim, Germany, 1997, 78.

Nutriox   A process for eliminating the odor and septicity of liquid effluent. Developed by Norsk Hydro in 1996.

Nuvalon   A development of the *Aloton process for extracting aluminum from clay. As in the Aloton process, clay is first heated with ammonium hydrogen sulfate. In the Nuvalon version, the product from this reaction is digested under pressure with 30 percent nitric acid, producing a solution of basic aluminum nitrate. Iron is removed by hydrolysis or by cooling.

Normal aluminum nitrate is crystallized out and calcined to alumina. The process was piloted in Germany in 1951 but not commercialized.

Gewecke, F., *Chem. Fabr.*, 1934, **21/22**, 6 June, 199.

O'Connor, D. J., *Alumina Extraction from Non-bauxitic Materials*, Aluminium-Verlag, Düsseldorf, 1988, 163.

O

Oberphos A version of the superphosphate process for making a fertilizer by treating phosphate rock with sulfuric acid, which yields a granular product. Used in the United States and Canada, but superseded in the United States by the *Davison process.

Gray, A. N., *Phosphates and Superphosphate*, Vol. 1, Interscience Publishers, New York, 1947, 124.

OBM *See* steelmaking.

OCET [Opti-Crude Enhancement Technology] A process for converting residual refinery oil into petroleum distillates and a coal substitute. A pulsed electric field is applied to the oil. Developed by SGI International in 1996 and expected to be commercialized by 1998.

Hydrocarbon Process., 1996, **75**(1), 42.

OCM [Oxidative Coupling of Methane] *See* oxidative coupling.

OCP [Oxygène Chaux Pulverisée] A steelmaking process in which powdered lime is blown into the furnace through the oxygen stream in order to combine with the phosphorus. Developed in the 1950s by CNRM, a Belgian metallurgical research organization. Similar to the *OLP process. *See also* LD/AC.

OCR A process for upgrading petroleum residues by catalytic hydrogenation.

Hydrocarbon Process., 1997, **76**(2), 45.

OCT [Olefins conversion technology] A process for making propylene from mixed petrochemical feedstocks. Developed by Phillips and acquired by ABB Lummus Global in 1997. First installed at the Karlsruhe oil refinery of Mineraloelraffinerie Oberrhein (Miro) for startup in 2000.

Eur. Chem. News, 1997, **68**(1792), 45.

Octafining A process for isomerizing *m*-xylene to *o*- and *p*-xylene, developed by the Atlantic Richfield Company in 1960. The catalyst was originally platinum on an aluminum silicate base; now a zeolite base is used. The reaction takes place in a hydrogen atmosphere. Hydrocarbon Research installed units in Argentina and the USSR.

Hydrocarbon Process., 1963, **42**(11), 206.

Uhlig, H. F. and Pfefferle, W. C., in *Refining Petroleum for Chemicals*, Spillane, L. J. and Leftin, H.P., Eds., American Chemical Society, Washington, D.C., 1970, 204.

Octamix [Octane mixture] A process for converting *syngas to a mixture of methanol with higher alcohols by reducing the CO/H_2 ratio below that required for the usual process for making methanol. The process is operated at 270 to 300°C, 50 to 100 bar, in the presence of a copper-based catalyst. The name is also a trade name used by Lurgi to denote a mixture of methanol and higher alcohols made by this process, suitable for blending with gasoline to increase its octane number.

> Weissermel, K. and Arpe, H.-J., *Industrial Organic Chemistry,* 3rd ed., VCH Publishers, Weinheim, Germany, 1997, 32.

Octgain   A *hydrofinishing process which reduces the sulfur and olefin content of gasoline without reducing its octane number. A zeolite catalyst is used. Developed by Mobil in 1994.

> *Chem. Eng. (N.Y.),* 1994, **101**(7), 25.

OCTENAR   [Octane enhancement by removing aromatics]   A process for removing aromatic hydrocarbons from petroleum reformate by extractive distillation with N-formyl morphylane. The product can be blended with gasoline to increase its octane number — hence the name. A paraffin mixture is obtained as a side-product. Developed by Krupp Koppers from its *MORPHYLANE and *MORPHYLEX processes.

Octol   A process for making mixed linear octenes by the catalytic dimerization of mixed butenes. A proprietary heterogeneous catalyst is used. Developed jointly by Hüls and UOP, and now offered for license by UOP. First operated in 1983 in the Hüls refinery in Marl, Germany. Another installation began production in 1986 at the General Sekiyu Refineries in Japan.

> Friedlander, R. G., Ward, D. J., Obenaus, F., Nierlich, F., and Neumeister, J., *Hydrocarbon Process.,* 1986, **65**(2), 31.
> Nierlich, F., *Erdoel, Erdgas, Kohle,* 1987, **103**(11), 486.
> *Hydrocarbon Process.,* 1991, **70**(3), 166.

Odda   A process for making a fertilizer by treating phosphate rock with nitric acid. Developed by Odda Smelteverk, Norway, in the early 1900s and still in use in 1988. Licensed by BASF and offered by Uhde.

> Piepers, R. J., in *Phosphoric Acid,* Vol. 1, Part 2, Slack, A. V., Ed., Marcel Dekker, New York, 1968, 916.

ODORGARD   A process for removing odorous gases from air streams by scrubbing with an aqueous solution of sodium hypochlorite in the presence of a proprietary heterogeneous catalyst. The catalyst contains nickel and is based on the *HYDECAT catalyst. Developed by ICI Katalco and F. H. H. Valentin. Nine units had been installed in the United Kingdom by 1995.

> World Patent WO 94/11091.
> *Chem. Eng. Prog.,* 1995, **91**(6), 19.
> Hancock, F. E., King, F., Flavell, W. R., and Islam, M. S., *Catal. Today,* 1998, **40**(4), 289.

ODS   *See* oxydesulfurization (2).

OFS   [Oil From Sludge]   A generic term for processes for converting sewage sludge into fuel oils. *See* Enersludge.

OGR   [Off-gas recovery]   A solvent-based technology for recovering olefins and/or hydrogen from *FCC or coker off-gases. Developed by KTI.

oil gasification   Processes which convert liquid petroleum fractions into gaseous fuels. Such processes with special names which are described in this dictionary are: CRG, HTR, Petrogas, Pintsch, Recatro, SEGAS, SSC.

Olefining   [Olefin refining]   A process for converting *syngas or methanol to a mixture of ethylene, propylene, and butenes. The catalyst is a ZSM-5–type zeolite in which some of the aluminum has been replaced by iron. Developed in 1984 by the National Chemical

Laboratory, Pune, India; the process was to be piloted by Bharat Petrochemical Corporation, Bombay, in 1992.

Indian Patents 159,164; 160,038; 160,212.
European Patent 161,360.

OlefinSiv A process for isolating isobutene from a mixture of C_4-hydrocarbons by chromatography over a zeolite molecular sieve. Developed by the Linde Division of the Union Carbide Corporation, as one of its *IsoSiv family of processes.

Adler, M. S. and Johnson, D. R., *Chem. Eng. Prog.,* 1979, **75**(1), 77.

Oleflex [Olefin flexibility] A process for converting normal hydrocarbons into the corresponding olefins (*e.g.,* propane to propylene, or isobutane to isobutene) by catalytic dehydrogenation. Similar to the *Pacol process but incorporating a continuous catalyst regeneration unit. So-called because of its flexibility in the production of a range of mono-olefins from a range of $C_2 - C_5$ hydrocarbons. Developed and offered for license by UOP. The first commercial plant began operation in Thailand in 1989. In 1997, two units for propane, four units for isobutane, and one unit for mixed propane/isobutane were operating. A third unit for propane was scheduled to start up in 1997. UOP and Packinox have developed a temperature-controlled reactor for this process, expected to be operating in a demonstration plant in 1998.

Vora, B. V. and Imai, T., *Hydrocarbon Process.,* 1982, **61**(4), 171.
Pujado, P. R. and Vora, B. V., *Hydrocarbon Process.,* 1990, **69**(3), 65.
Gregor, J., in *Handbook of Petroleum Refining Processes,* Meyers, R. A., Ed., McGraw-Hill, New York, 1997, 5.3.

Olex A version of the *Sorbex process for separating olefins from paraffins in wide-boiling mixtures. It can be used for hydrocarbons in the range $C_6 - C_{20}$. Based on the selective adsorption of olefins in a zeolite and their subsequent recovery by displacement with a liquid at a different boiling point. Mainly used for extracting $C_{11} - C_{14}$ olefins from the *Pacol process. As of 1990, six plants had been licensed.

Broughton, D. B. and Berg, R. C., *Hydrocarbon Process.,* 1969, **48**(6), 115.
Sohn, S. W., in *Handbook of Petroleum Refining Processes,* Meyers, R. A., Ed., McGraw-Hill, New York, 1997, 10.79.

Olin Raschig A refinement of the basic *Raschig (1) process for making hydroxylamine. *See* Raschig (1).

OLP [Oxygène Lance Poudre] A steelmaking process in which powdered lime is blown into the furnace through the oxygen stream in order to combine with the phosphorus. Developed in the late 1950s by l'Institut des Recherches de la Siderurgie. *See also* LD/AC.

Jackson, A., *Oxygen Steelmaking for Steelmakers,* Newnes-Butterworths, London, 1969, 165.

ONERA *See* metal surface treatment.

One-shot *See* Siroc.

Onia-Gegi [Offica National Industriel de l'Azota, and Gaz a l'Eau et Gas Industriel] A cyclic catalytic process for producing either town gas by *steam reforming, or *syngas from a variety of hydrocarbon feeds, by reaction with oxygen. Developed by the companies named, engineered by Humphreys & Glasgow, and used in England since the 1950s.

Claxton, G., *Benzoles, Production and Uses,* National Benzole & Allied Products Association, London, 1961, 96.
British Petroleum Co., *Gas Making and Natural Gas,* British Petroleum Co., London, 1972, 85.

Open Hearth   Also called the Siemens process, and the pig and scrap process. A steel-making process in which a mixture of pig iron, iron ore, scrap iron, and limestone was heated in a special reverberatory furnace using regenerative heating. It differed from the earlier Siemens-Martin process in that additional iron ore was added to the slag. The regenerative principle was invented by Frederic Siemens (who changed his name to Charles William Siemens) in England in 1856 and applied to steelmaking by him and E. Martin and P. Martin at Sireuil, France in 1864. *See also* Acid Open Hearth, Basic Open Hearth, Thomas.

> British Patent 2,861 (1856).
> Barraclough, K. C., *Steelmaking Before Bessemer, Vol. 2, Crucible Steel,* The Metals Society, London, 1984, 106.
> Barraclough, K. C., *Steelmaking 1850–1900,* The Institute of Metals, London, 1990, 137.

OptiCAT-plus   A process for regenerating *hydrotreating catalysts. It uses a fluidized-bed pretreatment to control the initial exotherms found in regeneration, followed by a moderate heat soak to eliminate sulfur and carbon. Developed in the 1990s by the Criterion Catalyst Company.

> *Eur. Chem. News,* 1996, **66**(1738), 25.

Optisol   A process for removing acid gases and sulfur compounds from gas streams by scrubbing with an aqueous solution of an amine and a proprietary physical solvent.

> *Hydrocarbon Process.,* 1986, **65**(4), 82.

ORC   [Occidental Research Corporation]   Also called the Garrett process. A coal gasification process using flash-pyrolysis at approximately 900°C in the absence of oxygen. Piloted by the Occidental Research Corporation in the 1970s.

> Sass, A., *Chem. Eng. Prog.,* 1974, **70**(1), 72.
> McMath, H. G., Lumpkin, R. E., Longanbach, J. R., and Sass, A., *Chem. Eng. Prog.,* 1974, **70**(6), 72.

Orcarb   *See* DR.

ORF   *See* DR.

Orford   An obsolete metallurgical process for separating copper and nickel. Ores containing these metals were smelted in a *Bessemer converter, forming matte. This was melted with sodium sulfate and coke, which yielded copper and nickel sulfides, which are immiscible and easily separated on cooling.

> Morgan, G. T. and Pratt, D. D., *British Chemical Industry,* Edward Arnold & Co., London, 1938, 107.

Organocell   A combined process for delignifying and bleaching wood pulp. Developed by Technocell, Düsseldorf.

> *Chem. Eng. (N.Y.),* 1991, **98**(1), 39.

Organosolv   A papermaking process which achieves separation of the lignin from the cellulose by dissolving the lignin in an organic solvent. The first pilot plant, built in Munich in 1984 with support from the European Commission, used aqueous methanol containing a small concentration of anthraquinone. Several variations on this process were tried in 1989 but none had been commercialized. A mixture of acetic and hydrochloric acids has been investigated. *See also* ASAM.

> Phillips, G. O., *Chem. Br.,* 1989, **25**, 1007.
> Parajo, J. C., Alonso, J. L., and Santos, V., *Ind. Eng. Chem. Res.,* 1995, **34**, 4333.
> Hergert, H. L., in *Environmentally Friendly Technologies for the Pulp and Paper Industries,* Young, R. A. and Akhar, M., Eds., John Wiley & Sons, New York, 1998, 5.

Orkla   A complex process for recovering sulfur from pyrite. The ore was smelted with coke, limestone, and quartz, with very little air at 1,600°C, and the iron was removed as a slag. The copper and other nonferrous metals formed a "matte" with the sulfur. Pyrolysis of this matte removed half of the sulfur. An air blast removed the other half without oxidizing it. Developed by the Orkla Mining Company, Norway, between 1919 and 1927. First commercialized at Thamshavn, Norway, in 1931; but the plant closed in 1962. The process was used for many years in Spain, Portugal, and Hungary.

> Haynes, W., *The Stone that Burns,* D. Van Nostrand, New York, 1942, 271.
> Katz, M. and Cole, R. J., *Ind. Eng. Chem.,* 1950, **42**, 2266.
> Kaier, T., *Eng. Min. J.,* 1954, **155**(7), 88.

Orthoflow   A fluidized-bed *catalytic cracking process in which the reactor and regenerator are combined in a single vessel. Designed by the MW Kellogg Company and widely used in the 1950s. First operated in 1951 by the British American Oil Company at Edmonton, Alberta. By 1994, more than 120 units had been built.

> Reidel, J. C., *Oil & Gas J.,* 1952, **50**(46), 200.
> Unzelman, G. H. and Wolf, C. J., in *Petroleum Processing Handbook,* Bland, W. F. and Davidson, R. L., Eds., McGraw-Hill, New York, 1967, 3-4.
> *Hydrocarbon Process.,* 1994, **73**(11), 114.

Orthoforming   A fluidized-bed *catalytic reforming process. Developed by the MW Kellogg Company in 1953 and first operated commercially in 1955.

> *Oil & Gas J.,* 1955, **53**(46), 162.
> Little, D. M., *Catalytic Reforming,* PennWell Publishing, Tulsa, OK, 1985, xv.

ORU   [Oxygenate removal unit]   A fixed-bed adsorption system for removing oxygenated hydrocarbons (*e.g.,* methanol, methyl *t*-butyl ether) from light hydrocarbon liquid streams. Used particularly for removing all alcohols and ethers from streams resulting from the manufacture of methyl *t*-butyl ether. Developed by UOP in the mid-1980s. By 1992, more than 25 process licenses had been granted. *See also* MRU.

> U.S. Patents 4,575,566; 4,575,567.

OSAG   A process for making ammonium sulfate from gypsum. Developed by Österreichische Stickstoff-Werke, Linz, Austria, and offered by Power-Gas. *See also* Merseburg.

> Gopinath, N. D., in *Phosphoric Acid,* Vol. 1, Part 2, Slack, A. V., Ed., Marcel Dekker, New York, 1968, 545.

OSIL   A *DR ironmaking process. Two plants were operating in India in 1997.

Ostromislenski   A process for making butadiene by condensing ethanol with acetaldehyde over an oxide catalyst at 360 to 440°C. Invented by I. I. Ostromislensky in Russia in 1915.

> Ostromislensky, I. I., *J. Russ. Phys. Chem.,* 1915, **47**, 1472.
> *Riegel's Handbook of Industrial Chemistry,* 7th ed., Kent, J. A., Ed., Van Nostrand Reinhold, New York, 1974, 232.

Ostwald   The basis of the modern family of processes for making nitric acid by the catalytic oxidation of ammonia.

OSW   [Österreichische Stickstoff-Werke]   An *ammoxidation process for making acrylonitrile from propylene. Operated in Austria by the named company.

> Dumas, T. and Bulani, W., *Oxidation of Petrochemicals: Chemistry and Technology,* Applied Science Publishers, London, 1974, 155.

OSW/Krupp   A process for making sulfuric acid and cement from gypsum, developed by Krupp Koppers.

Otto Aqua-Tech HCR [High capacity reactor] A high-intensity biological treatment process for purifying effluents from food processing, certain industrial processes, and landfill leachate. The waste is circulated rapidly through a vertical loop reactor and air is injected at the top. Invented at the Technical University of Clausthal-Zellerfeld; developed and commercialized by Otto Oeko-Tech. Eleven plants had been installed in Germany and Italy by 1991. *See also* Deep Shaft, Biobor HSR.

Otto-Rummel A coal-gasification process based on a double shaft furnace, developed in Germany by Dr. C. Otto & Company.

Outokumpu [Named after a hill in Finland, near Kuusjärvi] A flash-smelting process for sulfide ores. It is an energy-efficient process (also called an autogenous process), using mainly the heat of combustion of the contained sulfur to sulfur dioxide, rather than any external source of heat. Developed P. Bryk and J. Ryselin at the Harjavalta works of Outokumpu Oy, Finland, in 1946. Used mainly for copper ores, but also for nickel, iron, and lead; by 1988, 40 plants were using the process worldwide.

  Kuisma, M., *A History of Outokumpu,* Gummerus Kirjapaino Oy, Jyväskylä, Finland, 1989.

Oxco [Oxidative coupling] A process for converting natural gas to transport fuels and chemicals, based on the oxidative coupling of methane to ethane in a fluidized-bed reactor. Developed in Australia by the Division of Coal and Energy Technology, CSIRO, and BHP. *See also* IFP Oxypyrolysis.

  Edwards, J. H., Do, K. T., and Tyler, R. J., *Catal. Today,* 1990, **6**, 435.
  Hutchings, G. H. and Joyner, R. W., *Chem. Ind. (London),* 1991, 575.
  Edwards, J. H., Do, K. T., and Tyler, R. J., in *Natural Gas Conversion,* Holmen, A., Jens, K.-J., and Kolboe, S., Eds., Elsevier, Amsterdam, 1991, 489.

OXD *See* oxidative dehydrogenation.

O-X-D [Oxidative dehydrogenation] A process for converting *n*-butane to butadiene by selective atmospheric oxidation over a catalyst. Developed by the Phillips Petroleum Company and used by that company in Texas from 1971 to 1976. *See also* Oxo-D.

  Husen, P. C., Deel, K. R., and Peters, W. D., *Oil & Gas J.,* 1971, **69**(31), 60.
  Weissermel, K. and Arpe, H.-J., *Industrial Organic Chemistry,* 3rd ed., VCH Publishers, Weinheim, Germany, 1997, 111.

oxidative coupling A general term for processes which convert methane to ethane, ethylene, and higher hydrocarbons by heterogeneous catalytic oxidation. If the feed is methane, the process is also called OCM. *See* IFP Oxypyrolysis, Oxco.

  Srivastava, R. D., Zhou, P., Stiegel, V. U. S. Rao, and Cinquegrane, G., in *Catalysis,* Vol. 9, Spivey, J. J., Ed., Royal Society of Chemistry, Cambridge, 1992, 191.

oxidative dehydrogenation Also called OXD. A general term for processes which convert mono-alkenes to di-alkenes, or alkanes to alkenes, by partial oxidation. *O-X-D is a commercialized example. The conversion of ethane to ethylene by such a process has been studied by several companies in the 1970s but is not yet commercial.

  Eastman, A. D., Kolts, J. H., and Kimble, J. B., in *Novel Production Methods for Ethylene, Light Hydrocarbons, and Aromatics,* Albright, L. F., Crynes, B. L., and Nowak, S., Eds., Marcel Dekker, New York, 1992, 21.

Oxirane A general process for oxidizing olefins to olefin oxides by using an organic hydroperoxide, made by autoxidation of a hydrocarbon. Two versions are commercial. The first to be developed oxidizes propylene to propylene oxide, using as the oxidant *t*-butyl hydroperoxide made by the atmospheric oxidation of isobutane. Molybdenum naphthenate is used as a homogéneous catalyst. The *t*-butanol co-product may be reconverted to isobutane, or sold. The second uses ethylbenzene hydroperoxide as the oxidant; the co-product 2-phenylethanol is converted to styrene. The process was developed by Halcon Corporation and the first plant began operation in 1968. Halcon formed a joint venture with Atlantic Richfield Corporation (ARCO) under the name Oxirane Chemical Corporation, so the process came to be called the Oxirane process. It is operated on a large scale in Texas, The Netherlands, and France.

Gait, A. J., in *Propylene and Its Industrial Derivatives,* Hancock, E. G., Ed., Ernest Benn, London, 1973, 282.

Inform. Chim., 1979, (188), 175.

Braithwaite, E. R., in *Speciality Inorganic Chemicals,* Thompson, R., Ed., Royal Society of Chemistry, London, 1981, 359.

Weissermel, K. and Arpe, H.-J., *Industrial Organic Chemistry,* 3rd ed., VCH Publishers, Weinheim, Germany, 1997, 268.

Oxispec   A catalytic process for removing trace impurities from effluent gas streams. Developed by ICI Katalco in the 1990s.

Eur. Chem. News Suppl., 1996, Dec, 13.

OXITRON   A municipal sewage treatment process designed by Dorr Oliver. Starfish Industries, UK, has adapted it for coastal towns. The reactor is a biological fluidized bed installed on the seabed.

Water Waste Treat., 1995, **38**(6), 42.

OXO   [From **Ox**ierung, German, meaning ketonization]   Also called hydroformylation and Oxoation. A process for converting olefins to aldehydes containing an additional carbon atom, provided by carbon monoxide:

$$R\text{--}CH\text{=}CH_2 + CO + H_2 \rightarrow xR\text{--}CH_2\text{--}CH_2\text{--}CHO + (1-x)R\text{--}CH\text{--}CHO$$
$$\underset{\displaystyle CH_3}{\big|}$$

In recent years the name has come to include the production of downstream products (e.g. alcohols and acids) from the aldehydes too. Invented by O. Roelen in 1938 at the Chemische Verwertungsgesellschaft Oberhausen. Further developed by Ruhr Chemie and IG Farbenindustrie in Germany during World War II. It was first commercialized in 1948. Originally the process operated at high pressure, and dicobalt octacarbonyl, $Co_2(CO)_8$, was used as a homogeneous catalyst. The present process, known as the LP OXO process, also the LPO process, developed in the 1970s jointly by Union Carbide Corporation, Johnson Matthey, and Davy Corporation, operates at a lower pressure and uses the Wilkinson catalyst (a complex of rhodium carbonyl hydride with triphenyl phosphine). The resulting carbonyl compound can be hydrogenated *in situ* to the corresponding alcohol if an excess of hydrogen is used, or oxidized in a subsequent operation to the corresponding carboxylic acid. In 1990, 11 plants had been licensed worldwide. More than two million tons of chemicals were made by this family of processes in the United States in the 1980s. *See also* RCH/RP.

German Patent 849,548.

U.S. Patent 2,327,066.

Kirch, L. and Orchin, M., *J. Am. Chem. Soc.,* 1959, **81**, 3597.

Storch, H. H., Golumbic, N., and Anderson, R. B., *The Fischer-Tropsch and Related Syntheses,* John Wiley & Sons, New York, 1951, 441.

Allen, P. W., Pruett, R. L., and Wickson, E. J., in *Encyclopedia of Chemical Processing and Design,* McKetta, J. J. and Cunningham, W. A., Eds., Marcel Dekker, New York, 1990, **33**, 46.

Chem. Eng. (Rugby, England), 1990, Mar, 65.

Cornils, B., Herrmann, W. A., and Rasch, M., *Angew. Chem., Internat. Edn.,* 1994, **33**, 2144.

Weissermel, K. and Arpe, H.-J., *Industrial Organic Chemistry*, 3rd ed., VCH Publishers, Weinheim, Germany, 1997, 133.

Oxoation *See* OXO.

Oxo-D A process for converting *n*-butene to butadiene by selective atmospheric oxidation over a catalyst in a fixed-bed reactor. Developed by Petro-Tex Chemical Corporation and operated by that company in Texas since 1965. *See also* O-X-D.

Welch, L. M., Croce, L. J., and Christmann, H. F., *Hydrocarbon Process.*, 1978, **57**(11), 131.

Oxorbon A process for removing sulfur compounds from *syngas by adsorption on activated carbon. Offered by Lurgi. *See also* Desorex.

Oxycat A catalytic oxidation process for removing combustible vapors from air and industrial exhaust gases. The catalyst is platinum on alumina, supported inside a porcelain tube.

Houdry, J. H. and Hayes, C. T., *J. Air Pollut. Control Assoc.*, 1957, 7(3), 182.
Resen, L., *Oil & Gas J.*, 1958, **56**(1), 110.

oxychlorination An adaptation of the *Deacon process, used for converting ethylene to 1, 2-dichloroethane. A mixture of ethylene, air, and hydrogen chloride is passed over a catalyst of cupric chloride on potassium chloride. For the manufacture of vinyl chloride, the dichloroethane is cracked and the hydrogen chloride recycled. Several companies developed and commercialized this process in the 1960s.

Spitz, P. H., *Petrochemicals, the Rise of an Industry*, John Wiley & Sons, New York, 1988, 403.

OxyClaus A variation of the *Claus process, using combustion with oxygen to convert a fraction of the sulfur compounds to sulfur dioxide before reaction. Developed by Lurgi Oel Gas Chemie and Pritchard Corporation.

oxydehydrogenation A general name for the conversion of saturated aliphatic hydrocarbons to olefins, using atmospheric oxygen; exemplified by *Oxo-D, *O-X-D.

OXYDEP An *Activated Sludge process using pure oxygen. Developed by Air Products and Chemicals.

oxydesulfurization (1) A general name for processes which remove sulfur from coal by oxidation. *See also* IGT, PETC, oxydesulfurization (2).

oxydesulfurization (2) Also known as **ODS**. A process for extracting elemental sulfur from carbon dioxide or natural gas, developed by the MW Kellogg Company. *See also* oxydesulfurization (1).

OxyGEN [**Oxygen generator**] A version of the *VPSA process for separating oxygen from air by vacuum pressure swing adsorption. Developed by the Linde Division of the Union Carbide Corporation. Economical for quantities between 10,000 and 150,000 cf/h, at purities between 90 and 95 percent. Intended primarily for providing air enriched in oxygen for the steel and pulp and paper industries. First operated in 1989 at the Manitoba Rolling Mills in Selkirk, Manitoba.

Eur. Chem. News, 1989, **53**(1391), 31.
Chem. Mark. Rep., 1989, 11 Sept, 3.

Oxyhydrochlorination A two-stage process for making gasoline from lower paraffinic hydrocarbons, especially methane. The methane, mixed with oxygen and hydrogen chloride, is passed over a supported copper chloride catalyst, yielding a mixture of chloromethanes:

$$CH_4 + \tfrac{1}{2}O_2 + HCl = CH_3Cl + H_2O$$

These chloromethanes are converted to gasoline using a zeolite catalyst, and the hydrogen chloride co-product is recycled. Developed from 1987 by Allied Chemical Corporation and the Pittsburgh Energy Technology Center of the U.S. Department of Energy.

U.S. Patent 4,769,504.

Srivastava, R. D., Zhou, P., Stiegel, V. U. S. Rao, and Cinquegrane, G., in *Catalysis,* Vol. 9, Spivey, J. J., Ed., Royal Society of Chemistry, Cambridge, 1992, 205.

Oxypro (1) A process for making di-isopropyl ether (DOPE) from a propane/propylene stream from *FCC. The catalyst system is superior to other acid catalysts such as zeolites because of its greater activity at low temperatures. The Oxypro catalyst functions at below 175°C, whereas zeolites require temperatures closer to 260°C. DOPE is used as a gasoline additive. Developed by UOP in 1994; first licensed in Chile in 1996 for completion in 1997.

Hydrocarbon Process., 1995, **74**(7), 15.
Hydrocarbon Process., 1995, **74**(8), 42.
Davis, S., in *Handbook of Petroleum Refining Processes,* Meyers, R. A., Ed., McGraw-Hill, New York, 1997, 13.19.

Oxypro (2) A family of pulp-bleaching processes developed by Air Products and Chemicals. Piloted in 1995 at Pittsfield, MA. Oxypro O_R is for processing mixed office waste papers; it uses molecular oxygen, optionally hydrogen peroxide, sodium hydroxide, and a stabilizer. First commercialized in Scotland in 1994.

U.S. Patent 5,211,809.
Chem. Eng. (N.Y.), 1995, **102**(12), 66.

Oxypyrolysis *See* IFP Oxypyrolysis.

Oxy-Rich [Oxygen en**rich**ment] A version of the *PSA process for producing air enriched with oxygen in the range 23 to 50 percent. It is for use where higher degrees of enrichment are unnecessary — for enhanced combustion, and for improved oxidation in chemical and biochemical reactors. It is more energy-efficient than the standard PSA process because only some of the air is compressed to the highest level of the cycle. It was announced publicly in 1987 by Air Products & Chemicals.

U.S. Patent 4,685,939.

Sircar, S. and Kratz, W. C., in *Adsorption and Ion Exchange: Fundamentals and Applications,* LeVan, M. D., Ed., American Institute of Chemical Engineers, New York, 1988, 141.

Oxysulfreen A multi-stage variation of the *Sulfreen process for removing sulfur compounds from the off-gases from the *Claus process. In the first stage, the sulfur in all the organic sulfur compounds is converted to hydrogen sulfide by *hydrodesulfurization over a cobalt/molybdenum/alumina catalyst. After removal of water, some of this hydrogen sulfide is oxidized to sulfur dioxide over a titania catalyst. The final stage is the Sulfreen version of the Claus process, in which the remaining hydrogen sulfide is reacted with the sulfur dioxide to produce elemental sulfur. *See also* Hydrosulfreen, Sulfreen.

OXYWELL A process for separating oxygen from air by *PSA, using vacuum desorption from a zeolite. Used for medical oxygen generators, hence the name. Developed by Nippon Sanso.

Suzuki, M., in *Adsorption and Ion Exchange: Fundamentals and Applications,* LeVan, M. D., Ed., American Institute of Chemical Engineers, New York, 1998, 120.

Ozalid *See* reprography.

Ozark Mahoning A process for concentrating *Wet process phosphoric acid by submerged combustion evaporation. Developed by the Ozark Mahoning Company and first installed in Pasadena, TX, in 1945. *See also* Nordac.

Forster, J. H., *Phosphoric Acid,* Vol. 1 (Part 2), Slack, A. V., Ed., Marcel Dekker, New York, 1968, 599.

OZIOLE   *See* carbonization.

Ozocarb   A process for purifying municipal drinking water by treatment with ozone, hydrogen peroxide, and activated carbon. Developed by Trailigaz, a subsidiary of Compagnie Général des Eaux.

OZOFLOT   A process for treating raw water with ozone. The design of the treatment vessel encourages bubbles of ozonized air to become attached to algae and particulate solids and float to the surface where they can be skimmed off. Developed in France by OTV.

P

Paal   *See* Normann.

Pacol   [**Pa**raffin **c**onversion, linear]   A process for converting $C_{10} - C_{20}$ linear aliphatic hydrocarbons to the corresponding olefins by catalytic dehydrogenation. Used mainly for the production of linear intermediates for detergents. Used also, in combination with *Olex, for making mono-olefins. The catalyst is a platinum metal on an alumina support; several generations of the catalyst have been developed; the latest is known as DeH-7. Developed by UOP in the 1960s and first commercialized in 1968. By 1990, UOP had licensed 29 plants worldwide. See also DeFine, Detergent Alkylate.

Bloch, M. S., *Eur. Chem. News,* 1966, **10**(254), 46.
Broughton, D. B. and Berg, C. R., *Chem. Eng. (N.Y.),* 1970, **77**(2), 86.
Vora, B., Pujado, P. R., Imai, T., and Fritsch, T., *Chem. Ind. (London),* 1990, 6 Mar, 187.
Pujado, P. R., in *Handbook of Petroleum Refining Processes,* Meyers, R. A., Ed., McGraw-Hill, New York, 1997, 5.11.

PACT   [**P**owdered **a**ctivated **c**arbon **t**reatment]   A wastewater treatment process which combines activated carbon treatment with biological treatment, providing a single-stage treatment of toxic liquid wastes. Developed by DuPont in the 1970s at its Chambers Works, Deepwater, NJ, and now licensed by U.S. Filter/Zimpro. More than 50 units were operating in 1990.

Lankford, P. W., in *Toxicity Reduction in Industrial Effluents,* Lankford, P. W. and Eckenfelder, W. W., Jr., Eds., Van Nostrand Reinhold, New York, 1990, 229.
Chem. Eng. (N.Y.), 1990, **97**(2), 44.
Hutton, D. G., Meidl, J. A., and O'Brien, G. J., in *Environmental Chemistry of Dyes and Pigments,* Reife, A. and Freeman, H. S., Eds., John Wiley & Sons, New York, 1996, 105–164.
McIntyre, D. C., *loc. cit.,* 165–190.

PAL   [**p**eroxide **a**ssisted **l**each]   An improved *cyanide process for extracting gold from its ores. Addition of hydrogen peroxide to the system improves the yield of gold and reduces the usage of cyanide. First operated in South Africa in 1987.

Loroesch, J., *Randol Gold Forum,* Squaw Valley, CO, 1990, 215.

PAMCO   [**P**ittsburgh **a**nd **M**idway **Co**al Mining Company]   *See* SRC.

PAMELA [Pilotanlage (originally Prototypanlage) Mol zur Erzeugung lagerfähiger Abfälle] A continuous process for immobilizing nuclear waste in a borosilicate glass. Developed by the Deutsche Gesellschaft für Wiederaufarbeitung von Kernbrennstoffen, and Eurochemic, initially in Germany and later in Mol, Belgium. The plant was first operated with radioactive materials in 1985. *See also* VERA.

Lutze, W., in *Radioactive Waste Forms for the Future,* Lutze, W. and Ewing, R. C., Eds., North-Holland, Amsterdam, 1988, 7,612.

Panindco An early entrained-flow coal gasification process.

Paragon A two-stage *hydrocracking process, based on the zeolite ZSM-5, claimed to increase the yield and quality of the gasoline produced. Developed by Chevron Research Company, but not commercialized by 1991.

O'Rear, D. J., *Ind. Eng. Chem. Res.,* 1987, **26,** 2337.
Maxwell, I. E. and Stork, W. H. J., in *Introduction to Zeolite Science and Practice,* van Bekkum, H., Flanigen, E. M., and Jansen, J. C., Eds., Elsevier, Amsterdam, 1991, 610.

Paraho [**Para ho**mem, Portuguese, "for mankind"] A process for making oil and gas from oil shale. Development began in 1971 by the Paraho Development Corporation at Grand Junction, CO. Since then, in conjunction with a variety of American companies and agencies, a number of pilot plants have been operated and plants designed.

Jones, J. B., Jr. and Glassett, J. M., in *Handbook of Synfuels Technology,* Meyers, R. A., Ed., McGraw-Hill, New York, 1984, 4-63.

Paralene [**para**-xylene] Also called Gorham and also spelled parylene. A process for coating articles with poly-*p*-xylene. The vapor of di-*p*-xylylene is pyrolyzed at 550°C, yielding *p*-xylyl free radicals, $\cdot CH_2-C_6H_4-CH_2\cdot$, which deposit and polymerize on cooled surfaces. Developed by W. F. Gorham at Union Carbide Corporation.

Gorham, W. F., *J. Polymer Sci., A-1,* 1966, **4,** 3027.

PARC A process for making ammonia, developed by KTI.

Parex (1) [**Par**a extraction] A version of the *Sorbex process, for selectively extracting *p*-xylene from mixtures of xylene isomers, ethylbenzene, and aliphatic hydrocarbons. The feedstock is usually a C_8 stream from a catalytic reformer, mixed with a xylene stream from a xylene isomerization unit. The process is operated at 177°C; the desorbent is usually *p*-diethylbenzene. The first commercial plant began operation in Germany in 1971; by 1992, 453 plants had been licensed worldwide. Not to be confused with Parex (2).

U.S. Patents 3,524,895; 3,626,020; 3,734,974.
Broughton, D. B., Neuzil, R. W., Pharis, J. M., and Brearley, C. S., *Chem. Eng. Prog.,* 1970, **66**(9), 70.
Ruthven, D. M., *Principles of Adsorption and Adsorption Processes,* John Wiley, New York, 1984, 400.
Seidel, R. and Staudte, B., *Zeolites,* 1993, **13,** 348.
Jenneret, J. J., in *Handbook of Petroleum Refining Processes,* Meyers, R. A., Ed., McGraw-Hill, New York, 1997, 2.45.

Parex (2) [**Par**affin extraction] A process for separating linear aliphatic hydrocarbons from branched-chain and cyclic hydrocarbons by means of a zeolite 5A adsorbent. The products are desorbed with a mixture of steam and ammonia. Developed in the mid-1960s by Luena-Werke and Schwedt in East Germany and operated in East Germany, Bulgaria and the USSR. Broadly similar to *Molex and not to be confused with *Parex (1).

East German Patents 49,962; 64,766.
Wehner, K., Welker, J., and Seidel, G., *Chem. Tech. (Leipzig),* 1969, **21,** 548.

Seidel, G., Welker, J., Ermischer, W., and Wehner, K., *Chem. Tech. (Leipzig),* 1979, **31**(8), 405.

Schirmer, W., Fiedler, K., Stach, H., and Suckow, M., in *Zeolites as Catalysts, Sorbents and Detergent Builders,* Karge, H. G. and Weitkamp, J., Eds., Elsevier, Amsterdam, 1989, 439.

Seidel, R. and Staudte, B., *Zeolites,* 1993, **13**, 348.

Par-Isom [**Par**affin **isom**erization] A process for isomerizing light naphtha in order to improve the octane number. The proprietary catalyst was developed by Cosmo Oil Company and Mitsubishi Heavy Industries, and the process was developed by UOP. The oxide catalyst is claimed to be more efficient than zeolite catalysts currently used for this process.

Parkerizing *See* metal surface treatment.

Parkes A process for removing silver from lead, based on the use of zinc, which forms intermetallic compounds of lower melting point. Developed by A. Parkes in Birmingham, England, in the 1850s. Parkes also invented the first plastic (Parkesine), used for making billiard balls.

British Patents 13,675 (1850); 13,997 (1852).

Barrett, K. R. and Knight, R. P., *Silver—Exploration, Mining and Treatment,* Institute of Mining and Metallurgy, London, 1988.

parting A general name for the separation of silver, gold, and platinum from each other, practised since antiquity. Early processes involved dissolution in nitric acid, but only electrochemical processes are used now. *See* Balbach, Moebius, Thum, Wohlwill.

Patera A process for extracting silver from its ores, invented in 1858.

Patio [Spanish, a courtyard] A medieval process for extracting silver from argentite, Ag_2S. The ore was mixed with salt, mercury, and roasted pyrites, which contains cupric sulfate. This mixture was crushed by stones dragged by mules walking on the paved floor of a courtyard. The overall reactions are:

$$Ag_2S + 2NaCl + 2Hg + CuSO_4 = 2Ag + Hg_2Cl_2 + CuS + Na_2SO_4$$

$$2Ag + 2Hg = Hg_2Ag_2$$

Invented by Bartolomé de Medina, a Spanish trader, in Mexico in 1554, and used there until the end of the 19th century. The invention changed the course of economic history in all Hispanic America; 40 percent of all the silver recorded to have been produced in the world before 1900 was extracted by this process. *See also* Cazo, Washoe.

Probert, A., *J. West,* 1969, **8**, 90 (118 refs).

Jacobsen, R. H. and Murphy, J. W., *Silver—Exploration, Mining and Treatment,* Institute of Mining & Metallurgy, London, 1988, 283.

Nriago, J. O., *Chem. Ind. (London),* 1994, **30**(8), 650.

Pattinson (1) A process for extracting silver from lead by selective crystallization. When molten lead is cooled, the first crystals of lead contain less silver than the residual melt. Repetition of this process a number of times yields a silver concentrate which is further purified by *cupellation. Invented in 1833 by H. L. Pattinson. Largely superseded by the *Parkes process, except for metals containing bismuth for which the Pattinson is the preferred process. *See also* Luce-Rozan.

Dennis, W. H., *A Hundred Years of Metallurgy,* Gerald Duckworth, London, 1963, 194.

Pattinson (2) A process for making pure magnesium compounds from calcined dolomite, using the high solubility in water of magnesium hydrogen carbonate, $Mg(HCO_3)_2$. Invented by H. L. Pattinson in Gateshead, England, in 1841.

British Patent 9,102 (1841).

Pauling-Plinke A process for concentrating and purifying waste sulfuric acid by distillation and addition of nitric acid. It was obsolete by 1994.

Büchner, W., Schliebs, R., Winter, G., and Büchel, K. H., *Industrial Inorganic Chemistry,* VCH Publishers, Weinheim, Germany, 1989, 117.

PCA [Precipitation with a compressed anti-solvent] A process for making a solid with unusual morphology by spraying a solution of it into a supercritical fluid. The process resembles spray drying into a supercritical fluid. Used for making microspheres, microporous fibers, and hollow microporous fibers.

Brennecke, J. F., *Chem. Ind. (London),* 1996, (21), 831.

PCC [Partial combustion cracking] Not to be confused with precipitated calcium carbonate. A process for *cracking crude petroleum or heavy oil to a mixture of olefins and aromatic hydrocarbons. The heat carrier is steam, produced by the partial combustion of the feed. Developed by Dow Chemical Company. It was piloted in 1979 and a larger plant was built in Freeport, TX, in 1984.

Kirk, R. O., *Chem. Eng. Prog.,* 1983, **79**(2), 78.
Hu, Y. C., in *Chemical Processing Handbook,* Marcel Dekker, New York, 1993, 780.

Peachy A process for vulcanizing rubber by successive exposure to hydrogen sulfide and sulfur dioxide. Not commercialized.

peak shaving A term used in the gas industry for gas-producing processes which can be started quickly, to satisfy sudden increases in demand. One such process is *Hytanol.

PEATGAS A process for converting peat to gaseous fuels. Developed from 1974 to 1980 by the Institute of Gas Technology, Chicago, and the Minnesota Gas Company, to use the peat deposits in Northern Minnesota.

Pechiney (1) A process for making aluminum by electrolyzing a molten mixture of the chlorides and fluorides of Al, Ba, Ca, Na, and alumina. Developed in 1922.

Pechiney (2) A process for making urea from ammonia and carbon dioxide. The ammonium carbamate intermediate is handled and heated as an oil slurry.

Lowenheim, F. A. and Moran, M. M., *Faith, Keyes, and Clark's Industrial Chemicals,* 4th ed., John Wiley & Sons, New York, 1975, 856.

Pechiney H⁺ A process for extracting aluminum from clays and other aluminous ores and wastes by hydrochloric acid. The ore is first attacked by sulfuric acid and a hydrated aluminum chloride sulfate is isolated. Sparging a solution of this with hydrogen chloride precipitates aluminum trichloride hexahydrate, which is pyrohydrolyzed in two stages. Invented in 1977 by J. Cohen and A. Adjemian at Aluminium Pechiney, France, and subsequently developed in association with Alcan. Piloted in France but not yet commercialized.

French Patent 1,558,347.
European Patents 5,679; 6,070.
U.S. Patent 4,124,680.
Cohen, J. and Mercier, H., *Light Met. Met. Ind.,* 1976, **2**, 3.
O'Connor, D. J., *Alumina Extraction from Non-bauxitic Materials,* Aluminium-Verlag, Düsseldorf, 1988, 87,112,145.

Pechini A process for making mixed oxide ceramics from organic precursors. It is based on the ability of certain α-hydroxy-carboxylic acids, such as citric acid, to form polybasic acid chelates with metal ions. The chelates undergo polyesterification when heated with a polyol such as ethylene glycol. Further heating produces a resin that is then calcined. This method yields very homogeneous products. It was originally developed for making alkaline earth and lead titanates and has since been used for making niobates, ferrites, nickelates, and even ceramic superconductors. Invented in 1963 by M. P. Pechini at the Sprague Electric Company, MA.

U.S. Patent 3,330,697.
Falter, L. M., Payne, D. A., Friedmann, T. A., Wright, W. H., and Ginsburg, *Br. Ceram. Proc.,* 1989, (41), 261.

Pecor   *See* Woodall-Duckham.

Pedersen   A process for extracting aluminum from bauxite, which also yields metallic iron. The ore is first smelted in an electric furnace with limestone, iron ore, and coke at 1,350 to 1,400°C to produce a calcium aluminate slag and metallic iron. Aluminium is leached from the slag by sodium carbonate solution, and alumina is then precipitated from the leachate by carbon dioxide. The process requires cheap electricity and a market for the iron. It was invented by H. Pedersen in 1924 and operated at Hoyanger, Norway, from 1928 until the mid-1960s.

British Patent 232,930.
Miller, J. and Irgens, A., *Light Met. Met. Ind.,* 1974, **3**, 789.
O'Connor, D. J., *Alumina Extraction from Non-bauxitic Materials,* Aluminium-Verlag, Düsseldorf, 1988, 233.

Pekilo   [Paecilomyces]   A process for making single-cell protein from waste sulfite liquor from the paper industry. The organism is *Paecilomyces variotti.* Used in Finland.

Romantschuk, H. and Lehtomaki, M., *Proc. Biochem.,* 1978, **13**(3), 16.

Penex   [Pentane and hexane isomerization]   A process for converting *n*-pentane and *n*-hexane and their mixtures into branched-chain pentanes and hexanes of higher octane number by catalytic isomerization. The catalyst is similar to the *Butamer catalyst. The product is used in high-octane gasoline. First commercialized by UOP in 1958. More than 75 units were operating as of 1996.

Unzelman, G. H. and Wolf, C. J., in *Petroleum Processing Handbook,* Bland, W. F. and Davidson, R. L., Eds., McGraw-Hill, New York, 1967, 3-48.
Schmidt, R. J., Weiszmann, J. A., and Johnson, J. A., *Oil & Gas J.,* 1985, **83**(21), 80.
Hydrocarbon Process., 1996, **75**(11), 142.
Cusher, N. A., in *Handbook of Petroleum Refining Processes,* Meyers, R. A., Ed., McGraw-Hill, New York, 1997, 9.15.

Penex-Plus   A petroleum refining process that combines the *Penex process with a process for hydrogenating benzene to cyclohexane. Developed by UOP for reducing the benzene content of gasoline; first offered for license in 1991.

Peniakoff   A process for extracting aluminum from bauxite or other aluminous ores. The ore is roasted with coke and sodium sulfate in a rotary kiln at 1,200 to 1,400°C; this converts the aluminum to sodium aluminate, which is leached out with dilute sodium hydroxide solution. The basic reactions are:

$$Na_2SO_4 + 4C = Na_2S + 4CO$$

$$Na_2SO_4 + 2C = Na_2S + 2CO_2$$

$$4Al_2O_3 + Na_2S + 3Na_2SO_4 = 8NaAlO_2 + 4SO_2$$

The sulfur dioxide is recovered as sulfuric acid and reconverted to sodium sulfate. Alumina hydrate is precipitated from the sodium aluminate by carbon dioxide. The process has not become widely accepted because the product is contaminated by silica, but it was used in Belgium before and after World War I and in Germany in the 1920s and 1940s.

O'Connor, D. J., *Alumina Extraction from Non-bauxitic Materials,* Aluminium-Verlag, Düsseldorf, 1988, 306.

Penna *See* Woodall-Duckham.

Penniman A process for oxidizing petroleum, or its distillates, to mixtures of acids, phenols, and aldehydes. Powdered coal or coke is added to the petroleum, and air is passed through under high temperature and pressure. Invented by W. B. D. Penniman in 1925.

U.S. Patent 252,327.

Ellis, C., *The Chemistry of Petroleum Derivatives,* The Chemical Catalog Co., New York, 1934, 839.

Penniman-Zoph A process for making a yellow iron oxide pigment. Hydrated ferric oxide seed is added to a solution of ferrous sulfate and the suspension circulated over scrap iron, with air being passed through. Hydrated ferric oxide deposits on the seed crystals, giving a finely divided, yellow pigment:

$$4FeSO_4 + 6H_2O + O_2 = 4FeO \cdot OH + 4H_2SO_4$$

The pigment can be used in this form, or calcined to a red ferric oxide pigment.

Kirk-Othmer's Encyclopedia of Chemical Technology, 4th ed., Vol. 19, John Wiley & Sons, New York, 1991–1998, 24.

Pennsalt *See* Pennwalt.

Pennwalt Formerly called Pennsalt. A process for making calcium hypochlorite by passing chlorine into an aqueous suspension of calcium hypochlorite and calcium hydroxide. Developed by the Pennsylvania Salt Manufacturing Corporation in 1948.

U.S. Patent 2,441,337.

Pentafining A process for isomerizing pentane in a hydrogen atmosphere, using a platinum catalyst supported on silica-alumina. Developed by the Atlantic Richfield Company.

Unzelman, G. H. and Wolf, C. J., in *Petroleum Processing Handbook,* Bland, W. F. and Davidson, R. L., Eds., McGraw-Hill, New York, 1967, 3-50.

Pentesom *See* C$_5$ Pentesom.

Pep Set A process for making foundry molds developed by Ashland Chemical. In 1990 it was announced that a pilot plant was to be built in cooperation with the USSR and that the process had been licensed in China. *See also* Isocure.

Peracidox A process for removing sulfur dioxide from the tail gases from sulfuric acid plants by absorption in peroxomonosulfuric acid (Caro's acid). The peroxomonosulfuric acid is generated on-site by the electrolytic oxidation of sulfuric acid. Developed by Lurgi and Süd-Chemie and first operated in 1972.

Perc A process for making a heavy fuel oil by reacting a slurry of biomass in aqueous sodium carbonate solution with carbon monoxide. Under development in the United States in 1980.

Perchloron A process for making calcium hypochlorite, similar to the *Mathieson (2) process but using more chlorine and yielding a more readily filterable material. Developed by the Pennsylvania Salt Manufacturing Company. The name is also used in Germany as a product name for calcium hypochlorite.

Sheltmire, W. H., in *Chlorine, Its Manufacture, Properties, and Uses,* Sconce, J. S., Ed., Reinhold Publishing, New York, 1962, 523.

Perco An early petroleum sweetening process, using an adsorbent bed containing copper sulfate and sodium chloride.

PERCOS A process for removing sulfur dioxide from waste gases by scrubbing with aqueous hydrogen peroxide. The product is a commercial grade of 30 to 60 percent sulfuric acid. Developed by Adolph Plinke Söhne and Degussa.

Pernert A process for making perchloric acid by reacting sodium perchlorate with hydrochloric acid. Invented by J. C. Pernert in 1946 and operated by the Hooker Electrochemical Company at Niagara Falls.

> U.S. Patent 2,392,861.
> Schumacher, J. C., *Perchlorates, Their Properties, Manufacture, and Uses,* Reinhold Publishing, New York, 1960, 72.

Perosa A process for extracting beryllium from beryl.

Perox A process for removing hydrogen sulfide from coal gas. The gas is passed through aqueous ammonia containing hydroquinone. Atmospheric oxidation of the resulting solution gives elemental sulfur. Developed in Germany after World War II and still in use, being offered by Krupp Koppers.

> Pippig, H., *Gas Wasserfach.,* 1953, **94**, 62 (*Chem. Abstr.,* **47**, 5096).
> Kohl, A. L. and Riesenfeld, F. C., *Chem. Eng. (N.Y.),* 1959, **66**, 153.
> Kohl, A. L. and Riesenfeld, F. C., *Gas Purification,* 4th ed., Gulf Publishing, Houston, TX, 1985, 520.

perox-pure A process for oxidizing organic contaminants in water by the combined use of hydrogen peroxide and UV radiation. Developed originally for treating contaminated groundwater, it is now used also for purifying recycled water used in semiconductor manufacture and in many chemical manufacturing processes. A variant is known as Rayox. Developed in 1986 by Peroxidation Systems, Tucson, AZ. Calgon Carbon acquired the business from Vulcan Chemicals in December of 1995. By 1996, over 250 installations had been made, worldwide.

> Masten, S. J. and Davies, S. H. R., in *Environmental Oxidants,* Nriagu, J. O. and Simmons, M. S., Eds., John Wiley & Sons, New York, 1994, 534.
> James, S. C., Kovalik, W. W., Jr., and Bassin, J., *Chem. Ind. (London),* 1995, (13), 492.

Perrin A modification of the *Bessemer process which accomplishes the removal of phosphorus from iron by treating the initial molten metal with a molten mixture of lime, alumina, and fluorspar.

Persson A process for making chlorine dioxide by reducing sodium chlorate with chromium (III) in the presence of sulfuric acid. The chromium (III) becomes oxidized to chromium (VI) and is then reduced back to chromium (III) with sulfur dioxide. This cyclic redox process with chromium avoids complications that would occur if sulfur dioxide itself were used as the reductant. Installed at the Stora Kopparbergs paper mill, Sweden, in 1946.

> Sheltmire, W. H., in *Chlorine, its Manufacture, Properties and Uses,* Sconce, J. S., Ed., Reinhold Publishing, New York, 1962, 275,538.

Pertraction A process for removing organic pollutants from industrial wastewater. The water is contacted with an organic solvent via a hollow-fiber membrane. Developed in 1994 by TNO Institute for Environmental and Energy Technology, in collaboration with Tauw Environmental Consultancy and Hoechst.

> *Eureka,* 1994, **14**(11), 16.

PETC [Pittsburgh Energy Technology Center] A chemical coal-cleaning process based on *oxydesulfurization in which the oxidant is air and lime is used to fix the sulfur. Developed by the Pittsburgh Energy Technology Center, funded by the U.S. Department of Energy from 1970 to 1981. *See also* Ames (2).

IEA Coal Research, *The Problems of Sulphur,* Butterworths, London, 1989, 14.

Petit A process for removing hydrogen sulfide and hydrogen cyanide from gas streams by scrubbing with an alkali carbonate solution and regenerating the liquor with carbon dioxide. Invented by T. P. L. Petit.

German Patent 396,353.

Petrifix A process for solidifying aqueous wastes, converting them to a solid form suitable for landfill. Cementitious additives are used, based on the compositions used by the Romans for making Pozzolanic cements. Developed by Pec-Engineering, Paris, France. In 1979 it had been used in France and Germany.

Pichat, P., Broadsky, M., and Le Bourg, M., in *Toxic and Hazardous Waste Disposal,* Vol. 1, Pojasek, R. B., Ed., Ann Arbor Science, Ann Arbor, MI, 1979, Chap. 9.

PetroFlux A refrigeration process for removing liquid hydrocarbons from natural gas. Developed by Costain Engineering. Twenty three plants had been operating by 1992.

Hydrocarbon Process., 1996, **75**(4), 132.

Petrogas A thermal cracking process for converting heavy petroleum fractions to fuel gas. Developed by Gasco.

Kirk-Othmer's Encyclopedia of Chemical Technology, 3rd ed., Vol. 11, John Wiley & Sons, New York, 1980, 428.

Petrosix [Named after the oil company **Petro**bus and the oil shale company **Superintendecia da Industrializacao da Xisto**] A method for extracting oil and gas from shale. A large demonstration plant was operated in Brazil in the 1970s.

Smith, J. W., in *Handbook of Synfuels Technology,* Meyers, R. A., Ed., McGraw-Hill, New York, 1984, 4-185.

Petro-Tex A process for oxidizing butenes to maleic anhydride. Developed by the Petro-Tex Chemical Corporation and used at its plant in Houston, TX.

Skinner, W. A. and Tieszin, D., *Ind. Eng. Chem.,* 1961, **53**(7), 557.

PETROX An *ammoxidation process for making acrylonitrile from propane or propylene. Developed by BOC Group and partially piloted in New Jersey.

Chem. Eng. (Rugby, England), 1991, (489), 14.

PFH [Pressurized fluidized-bed hydroretorting] A process for making fuel gas from oil shale. Developed and piloted by the Institute of Gas Technology, Chicago.

Phenolsolvan A process for extracting phenols from coke-oven liquor and tar acids from tar by selective solvent extraction with di-isopropyl ether (formerly with *n*-butyl acetate). Developed by Lurgi in 1937.

Wurm, H.-J., *Chem. Ing. Tech.,* 1976, **48**, 840.

Phenoraffin A process for recovering phenols from carbonizer tar and coke-oven tar. The tar is dissolved in aqueous sodium phenolate and extracted with isopropyl ether.

Weissermel, K. and Arpe, H.-J., *Industrial Organic Chemistry,* 3rd ed., VCH Publishers, Weinheim, Germany, 1997, 347.

Phillips (1) A process for polymerizing ethylene and other linear olefins and di-olefins to make linear polymers. This is a liquid-phase process, operated in a hydrocarbon solvent at an intermediate pressure, using a heterogeneous oxide catalyst such as chromia on silica/alumina. Developed in the 1950s by the Phillips Petroleum Company, Bartlesville, OK, and first commercialized at its plant in Pasadena, TX. In 1991, 77 reaction lines were either operating or under construction worldwide, accounting for 34 percent of worldwide capacity for linear polyethylene.

Belgian Patent 530,617.
Chem. Eng. (N.Y.), 1955, **62**(6), 103.
Clark, A. C., Hogan, J. P., Banks, R. L., and Lanning, W. C., *Ind. Eng. Chem.*, 1956, **48**, 1152.
Clark, A. C. and Hogan, J. P., in *Polythylene*, 2nd. ed., Renfrew, A. and Morgan, P., Eds., Iliffe & Sons, London, 1960, 29.
Hydrocarbon Process., 1991, **70**(3), 170.

Phillips (2)   A fractional crystallization process used to freeze-concentrate beer and fruit juices. Formerly used in the production of *p*-xylene.

McKay, D. L., in *Fractional Solidification*, Zief, M. and Wilcox, W. R., Eds., Marcel Dekker, New York, 1967, Chap. 16.

Phillips (3)   A two-stage process for dehydrogenating butane to butadiene.

Phoredox   A modification of the *Activated Sludge sewage treatment process, designed for the separate removal of phosphate. *See also* Phostrip.

Horan, N. J., *Water Waste Treat.*, 1992, **35**(2), 16.

Phorex   [**Pho**sphoric acid **ex**traction]   A process for purifying phosphoric acid by solvent extraction with *n*-butyl or *n*-amyl alcohol. Developed by Azote et Produits Chimiques, France.

Bergdorf, J. and Fischer, R., *Chem. Eng. Prog.*, 1978, **74**(11), 41.

PHOSAM   [**Am**monium **phos**phate]   Also called Phosam-W and USS Phosam. A method for removing ammonia from coke-oven gas by scrubbing with a solution of ammonium phosphate. The ammonia is recovered by heating the solution:

$$NH_4H_2PO_4 + NH_3 \rightleftharpoons (NH_4)_2HPO_4$$

Developed by P. D. Rice and others at the U.S. Steel Corporation in the 1960s, and first used at a coke plant at Clairton, PA. In 1984 it was in operation in 20 coke plants in the United States, Canada, and Japan. The process can also be used in oil refineries and synthetic gas plants, but none was operating in 1984.

U.S. Patents 3,024,090; 3,186,795.
Kohl, A. L. and Riesenfeld, F. C., *Gas Purification*, 4th ed., Gulf Publishing, Houston, TX, 1985, 562.
Rice, R. D. and Busa, J. V., in *Acid and Sour Gas Treating Processes*, Newman, S. A., Ed., Gulf Publishing, Houston, TX, 1985, 786.

Phostrip   A modification of the *Activated Sludge sewage treatment process, designed for the separate removal of phosphate. *See also* Phoredox.

Horan, N. J., *Biological Wastewater Treatment Systems*, John Wiley & Sons, Chichester, England, 1990, 233.

PHOTHO   [**Pho**sphate Glass solidification of **tho**rium-bearing reprocessing waste]   A process for immobilizing radioactive waste products from the thorium fuel cycle in a phosphate glass for long-term storage. Developed at Kernforschungsanlage Jülich, Germany, from 1968 until it was abandoned in 1977 in favor of *PAMELA.

Lutze, W., in *Radioactive Waste Forms for the Future*, Lutze, W. and Ewing, R. C., Eds., North-Holland, Amsterdam, 1988, 8.

Photographic processes   These processes, involving chemical reactions initiated by light, come within the declared scope of this work. However, the history of photography is well documented and does not warrant repetition here. Instead, the named processes that have been, or are being, used commercially are listed in the following text and the reader is referred to the bibliography for references to consult for their details. Reprographic processes for line drawings (e.g. Blueprint) are given individual entries—*see* reprography.

Photographic processes include: Agfacolor, Ambrotype, Anscocolor, Autochrome, Autotype, bichromate, Bromoil, Calotype, carbon, Carbro, Chromatone, Chrysotype, Cibachrome, collodion, Cyanotype, Daguerreotype, Dufay, Dufaycolor, Duplex, Duxochrome, Dyebro, Ektachrome, Ektacolor, Energiatype, Ferraniacolor, Ferrotype, Finlay, Flexichrome, Gasparcolor, Gevacolor, Gum, Hicro, hydrotype, Ilfochrome, Ilford, Itek, Joly, Jos-Pe, KDB, Kodachrome, Kodacolor, Kotovachrome, Lignose, Lippman, Lumière, Melainotype, Minicolor, Omnicolore, Ozobrome, Ozotype, Paget, phototype, Pinatype, Platinotype, Polacolor, Polaroid-Land, Polychrome, Printon, Raydex, Sakuracolor, Tannin, Thomas, Tintype, Utocolor, Uvachrome, Vivex, Warner-Powrie, wet collodion, Woodbury type, Zincotype.

Newhall, B., *The History of Photography from 1839 to the Present Day,* Museum of Modern Art, New York, 1964.
Mees, C. E. K. and James, T. H., *The Theory of the Photographic Process,* Macmillan, New York, 1966.
Friedman, J. S., *History of Color Photography,* Focal Press, London, 1968 (re-issue of 1944 ed.).
Gernsheim, H. and Gernsheim, A., *The History of Photography,* Thames & Hudson, London, 1969.
Coe, B., *Colour Photography; the First Hundred Years, 1840–1940,* Ash & Grant, London, 1978.

Physical Vapor Deposition   Often abbreviated to **PVD**. A process for applying a coating of one material to the surface of another, essentially by sublimation. To be distinguished from *Chemical Vapor Deposition.

Hocking, M. G., Vasantasree, V., and Sidky, P. S., *Metallic and Ceramic Coatings,* Longman, Harlow, Essex, 1989.

Pidgeon   A process for making magnesium metal by reducing dolomite with ferrosilicon at 1,200°C in a vacuum retort:

$$2CaO \cdot MgO + Fe_xSi = 2Mg + Ca_2SiO_4 + xFe$$

Used in World War II.

Hughes, W. T., Ransley, C. E., and Emley, E. F., in *Advances in Extractive Metallurgy,* Institute of Mining and Metallurgy, London, 1968, 429.

Pier-Mittasch   A high-pressure, catalytic process for making methanol from carbon monoxide and hydrogen. Developed by M. Pier and A. Mittasch at BASF in the 1920s.

Spitz, P. H., *Petrochemicals, the Rise of an Industry,* John Wiley & Sons, New York, 1988, 36.

Pieters   *See* Staatsmijnen-Otto.

Pietzsch and Adolph   An electrolytic process for making hydrogen peroxide, operated in Germany in 1910 by Elektrochemische Werke München. Its plants at Munich, Bad Lauterberg, and Rhumspringe were used to make the hydrogen peroxide which was used to launch their V1 weapons and to guide their V2 weapons during World War II.

Schumb, W. C., Satterfield, C. N., and Wentworth, R. L., *Hydrogen Peroxide,* Reinhold Publishing, New York, 1955, 136.

Pig and ore   A colloquial name for the *Siemens Open Hearth process.

Pig and scrap   A colloquial name for the *Siemens-Martin Open Hearth process.

Pintsch   The first commercial oil-gasification process; developed in Germany in the 19th century.

Pintsch-Hillebrand   An early, two-stage coal gasification process.

Stief, F. *Gas Wasserfach,* 1932, **75**, 581 (*Chem. Abstr.,* **26**, 5402).
Müller, H., *Gas Wasserfach,* 1935, **78**, 431 (*Chem. Abstr.,* **29**, 6397).

van der Hoeven, B. J. C., in *Chemistry of Coal Utilization,* Vol. 2, Lowry, H. H., Ed., John Wiley & Sons, New York, 1945, 1668.

PIVER   A French process for vitrifying nuclear waste.

Plasmared   [Plasma reduction]   A direct reduction ironmaking process, using natural gas as the reductant and heated by an electric plasma. Operated in Sweden by SKF from 1981 to 1984. *See* DR.

Platfining   The original name for the *LT Unibon process.

Platforming   [Platinum re-**forming]**   A process for converting aliphatic hydrocarbons (acyclic and cyclic) into aromatic hydrocarbons and hydrogen. Methyl cyclopentane can thus be converted to benzene. The catalyst typically incorporates platinum and another metal on an alumina support. Originally the reactors were side by side and the catalyst had to be regenerated annually. In 1970, continuous catalyst regeneration (CCR) was introduced and the three or four reactors were stacked vertically in series (hence another derivation of the name). Invented by V. Haensel at Universal Oil Products, now UOP. First commercialized by that company at Muskegon, MI, in 1949, and now widely licensed. By 1988, 700 units had been licensed, of which 500 were in operation.

Unzelman, G. H. and Wolf, C. J., in *Petroleum Processing Handbook,* Bland, W. F. and Davidson, R. L., Eds., McGraw-Hill, New York, 1967, 3-29.
Little, D.M., *Catalytic Reforming,* PennWell Publishing, Tulsa, OK, 1985, 160.
Peer, R. L., Bennett, R. W., Felch, D. E., and von Schmidt, E., *Catal. Today,* 1993, **18**(4), 473.
Hydrocarbon Process., 1994, **73**(11), 94.
Dachas, N., Kelly, A., Felch, D. E., and Reis, E., in *Handbook of Petroleum Refining Processes,* Meyers, R. A., Ed., McGraw-Hill, New York, 1997, 4.3.

Platinum Reforming   *See* Sovaforming.

Platreating   [Platinum hydro**treating]**   A *hydrotreating process used in oil refining, using a platinum catalyst.

Unzelman, G. H. and Wolf, C. J., in *Petroleum Processing Handbook,* Bland, W. F. and Davidson, R. L., Eds., McGraw-Hill, New York, 1967, 3-44.

Plattner   An early process for extracting gold from auriferous pyrites by chlorination. The resulting gold chloride is extracted by water and reduced with ferrous sulfate:

$$AuCl_3 + 3FeSO_4 = Au + Fe_2(SO_4)_3 + FeCl_3$$

Developed by C. F. Plattner in Germany in 1853.

Mellor, J. W., *Comprehensive Treatise on Inorganic and Theoretical Chemistry,* Vol. 3, Longmans, Green & Co., London, 1923.

PNC   [Photo**n**itros**ation of c**yclohexane]   A photochemical process for making caprolactam (a precursor for nylon) from cyclohexane, nitrosyl chloride, and hydrogen chloride. The first photochemical product is cyclohexanone oxime:

$$C_6H_{12} + NOCl + HCl \xrightarrow{\text{UV}} C_6H_{10}O{=}NOH{\cdot}2HCl.$$

This, on treatment with sulfuric acid, then undergoes the Beckmann rearrangement to caprolactam. The nitrosyl chloride is made by reacting nitrosyl sulfuric acid (made from oxides of nitrogen and sulfuric acid) with hydrogen chloride gas:

$$NOHSO_4 + HCl = NOCl + H_2SO_4$$

The process was developed by Toyo Rayon Company (now Toray Industries), Japan in the 1960s, and is now operated by that company in Nagoya and Tokai.

Aikawa, K., *Hydrocarbon Process.,* 1964, **43**(11), 157.

Horspool, B., in *Light, Chemical Change and Life,* Coyle, J. D., Hill, R. R., and Roberts, D. R., Eds., Open University Press, Milton Keynes, 1982, 276.

Hydrocarbon Process., 1989, **68**(11), 97.

Pneumatic   *See* Bessemer.

Polyad   [**Poly**mer **ad**sorbent]   Also written POLYAD. A family of processes for removing volatile organic compounds from air streams by continuous adsorption on an adsorbent and desorption with hot air. Several adsorbents are used, depending on the sorbate, including a macroporous polymer ("Bonopore"). Used for control of emissions and for solvent recovery. COMBI-AD is a variant for simultaneously removing several solvents, using two different adsorbents. Developed and offered by Chematur Engineering, Sweden. Twelve units had been installed, in several countries, by 1995.

Polybed   A version of the *PSA process for separating and purifying gases by selective adsorption, using a number of adsorbent beds and a complex valving system in order to produce a gas of high purity. Used particularly for purifying hydrogen. Developed by the Union Carbide Corporation in the mid-1970s. *See also* Hydrogen Polybed PSA.

Corr, F., Dropp, F., and Rudelstorfer, E., *Hydrocarbon Process.,* 1979, **58**(3), 119.

Kohl, A. L. and Riesenfeld, F. C., *Gas Purification,* 4th ed., Gulf Publishing, Houston, TX, 1985, 681.

Yang, R. T., *Gas Separation by Adsorption Processes,* Butterworths, Guildford, England, 1987, 244.

Hydrocarbon Process., 1996, **75**(4), 120.

Polyco   A process for converting propylene and butene to liquid fuels, using copper pyrophosphate as the catalyst. The name has also been used as a trade name for a type of polyvinyl acetate made by the Borden Chemical Company.

Polyforming   [**Poly**merization **re**forming]   An early process for converting gas–oil to gasoline. It combined thermal cracking with polymerization.

Asinger, F., *Mono-olefins: Chemistry and Technology,* translated by B. J., Hazzard, Pergamon Press, Oxford, 1968, 362.

polymerization   Those polymerization processes having special names which are described in this dictionary are: Alfin, Alfol, Alphabutol, Borstar, CP, CX, Dimersol, Exxpol, Gorham, GRS, Hexall, Innovene, Insite, LIPP-SHAC, Mobil-Witco-Shell, MOGD, MPC, MSP3, Naphtachimie, Natta, NORSOLOR, Novolen, Octol, Paralene, Phillips (1), Polyco, SDS, Sclair, Sclairtech, Selectopol, SHOP, SPGK, Spheripol, Standard Oil, UNIPOL, Ziegler (1), Ziegler-Natta.

Polynaphta Essence   A process for making a linear olefin fraction for making methyl *t*-butyl ether to use as a fuel additive. Developed by IFP in 1996; replacing UOP's *Catpoly process.

Inform. Chim. Hebdo, 1997, (1294), 13.

Polzeniusz-Krauss   A process for making calcium cyanamide from calcium carbide by heating it in a nitrogen atmosphere in a channel kiln:

$$CaC_2 + N_2 = CaCN_2 + C$$

This was an early process for fixing nitrogen for use as a fertilizer.

Porter-Clark   The original name for the cold lime-soda process. A water-softening process using sodium carbonate and calcium hydroxide. It removes the non-carbonate, as well as the bicarbonate, hardness. Developed by J. H. Porter. *See* Clark.

Porteus   A process for conditioning sewage sludge by heating under pressure to 180 to 220°C for approximately one hour. The solid product is easy to de-water and is sterile. Developed and used in Germany in the 1980s.

POSTech   A process for making copolymers of styrene with polyols. A special stabilizer is used, as well as an organic peroxide initiator.

> *Eur. Chem. News,* 1996, **65**(1715), 35.

Pott-Broche   A coal liquifaction process in which coal is dissolved in a mixture of tetrahydronaphthalene and cresols, and then hydrogenated. Invented by A. Pott and H. Broche at IG Farbenindustrie, Germany in 1927; used by the Ruhrol Company in Germany between 1938 and 1944. *See also* Exxon Donor Solvent.

> British Patent 293,808.
> French Patent 841,201.
> Pott, A. and Broche, H., *Fuel,* 1943, **13**, 91,125,154.

Powerclaus   A *flue-gas desulfurization system which applies the *Aquaclaus process to power station effluent gases.

Powerforming   A *catalytic reforming process, based on a platinum catalyst. Developed by Esso Research & Engineering Company. First commercialized in Baltimore in 1955, and now widely licensed.

> Unzelman, G. H. and Wolf, C. J., in *Petroleum Processing Handbook,* Bland, W. F. and Davidson, R. L., Eds., McGraw-Hill, New York, 1967, 3-32.
> Asinger, F., *Mono-olefins: Chemistry and Technology,* translated by B. J., Hazzard, Pergamon Press, Oxford, 1968, 391.
> Little, D. M., *Catalytic Reforming,* PennWell Publishing, Tulsa, OK, 1985, 162.
> *Hydrocarbon Process.,* 1988, **67**(9), 80.

POX   [Partial oxidation]   A general term for processes which convert natural gas to *syngas or methanol by partial oxidation. Shell International Petroleum and Lurgi developed one such methanol process in the 1980s whch was used at the Mider refinery in Leuna, Germany. The University of Orléans, France, developed non-catalyzed plasma-assisted POX processes for making syngas in the 1980s and 1990s.

> *Oil & Gas J.,* 1997, **95**(11), 49.
> Leseur, H., Czernichowski, J., and Chapelle, J., *Int. J. Hydrogen Energy,* 1994, **19**(2), 139.

POZONE   A process for making ozone. Elemental phosphorus is emulsified in water at above 45°C and air passed through. This generates ozone and orthophosphoric acid; developed at the Lawrence Berkeley Labortatory, Berkeley, CA.

$$P_4 + 5O_2 = P_4O_{10}$$

$$P_4O_{10} + 6H_2O = 4H_3PO_4$$

Proposed as a source of ozone for removing NO_x and SO_2 from flue-gas, and for pulp-bleaching.

> *Chem. Eng. (N.Y.),* 1994, **101**(11), 25.
> Chang, S.-G., Keyuan, H., and Yizhong, W., *J. Environ. Sci.,* 1994, **6**(1), 1.
> Wang, H., Shi, Y., Le, L., Wang, S.-M., Wei, J., and Chang, S.-G., *Ind. Eng. Chem. Res.,* 1997, **36**(9), 3656.

Poz-O-Tec   A *flue-gas desulfurization process which produces a fibrous form of gypsum, convenient for disposal. In a demonstration project, an artificial reef was built from the product in the Atlantic Ocean near Fire Island, New York. Developed by IU Conversion Systems, Philadelphia, PA.

IEA Coal Research, *The Problems of Sulphur,* Butterworths, London, 1989, 127.

PPG   [Pittsburgh Plate Glass Company]   A process for making calcium hypochlorite. Hypochlorous acid and chlorine monoxide, generated by reacting chlorine and carbon dioxide with sodium carbonate monohydrate, are passed into lime slurry. Invented in 1938 by I. E. Muskatt and G. H. Cady at the Pittsburgh Plate Glass Company.

U.S. Patent 2,240,344.

PR   *See* Sovaforming.

Prayon   One of the *Wet processes for making phosphoric acid by reacting phosphate rock with sulfuric acid. The byproduct is gypsum, calcium sulfate dihydrate. It uses a compartmentalized, multi-section, lined, concrete reactor, with finishing tanks in which the gypsum crystals mature. In 1990 one third of the wet process phosphoric acid made in the Western World was made in this way. The process was developed in 1977 by the Société de Prayon, Belgium. Variations are known as PH2, PH11, and PH12. One variation uses solvent extraction with isopropyl ether and tri-*n*-butyl phosphate.

U.S. Patent 4,188,366.
Forster, J. J., in *Phosphoric Acid,* Vol. 1, Part 2, Slack, A. V., Ed., Marcel Dekker, New York, 1968, 585.
Becker, P., *Phosphates and Phosphoric Acid,* 2nd. ed., Marcel Dekker, New York, 1989, 347.
Gard, D. R., in *Encyclopedia of Chemical Processing and Design,* McKetta, J. J. and Cunningham, W. A., Eds., Marcel Dekker, New York, 1990, **35,** 453.

Precht   *See* Engel-Precht.

PRENFLO   [Pressurized entrained-flow]   A coal gasification process using an entrained-flow gasifier. A mixture of coal dust and oxygen is fed horizontally into a gasifier operating at > 2,000°C and 25 atm. It is similar to the *Koppers-Totzek process, but differs from it mainly in the use of elevated pressure. The solid waste is mostly molten and is collected as a granular slag in a water-bath beneath the gasifier; it can be used as a filler in the construction industry. The process can be used to produce *syngas, but the main application forseen is as a part of a combined-cycle electric power plant. Developed by Krupp Koppers with funding from the Ministry for Research and Technology, FRG, and the Commission of the EEC; the first demonstration plant began operation in Hamburg in 1979 and a second one started up in Fürstenhausen in 1986. The first commercial plant was built at Puertollano, in 1996.

Hydrocarbon Process., 1986, **65**(4), 100.
Chem. Eng. (N.Y.), 1991, **98**(9), 31.

Prenox   A pulp-bleaching process using nitrogen dioxide and oxygen instead of chlorine. Developed by AGA, Sweden.

Pressure swing adsorption   *See* PSA.

PRIAM   An electrochemical process for recovering heavy metals. Announced in 1992 by Électricité de France.

Primex   [Pressureless infiltration by metal]   *See* Lanxide.

PRIMOX   A process for injecting oxygen into sewers ("rising mains") in order to prevent the formation of hydrogen sulfide. Developed by BOC.

PROABD   A crystallization process developed by BEFS PROKEM, France. In 1994, more than 80 plants were using PROABD distillation and crystallization processes. *See* Ab der-Halden.

   Hassene, M., *Asia Pacific Chem.*, 1993, Oct, 30.

PROABD MSC   [melt static **crystallization**]   A process for purifying *p*-xylene by crystallization. Used in conjunction with *MSTDP. Piloted in France from 1994 to 1996 and proposed for installation in India in 1997 and in Bulgaria in 1998.

   Chem. Eng. (N.Y.), 1996, **103**(9), 23.

producer   A generic name for processes which completely convert solid fuels to gaseous fuels, and for the products. Coal or coke is the usual solid fuel. The oxidant was originally air but is now more commonly oxygen. The principle combustible component of the product is carbon monoxide. First developed in the early 19th century, these processes became very important. With the development of the natural gas and petroleum industries in the early 20th century, their importance declined; although there was a revival of interest after the oil crisis of 1973. A common feature of producer gas processes is partial combustion of the solid fuel to provide heat for the reactions. These processes evolved into the many *coal gasification processes.

   van der Hoeven, B. J. C., in *Chemistry of Coal Utilization*, Vol. 2, Lowry, H. H., Ed., John Wiley & Sons, New York, 1945, Chap. 36.

Progil   One of the thermal processes for making phosphoric acid. The phosphorus pentoxide, produced by burning elemental phosphorus, is absorbed in a solution of sodium phosphate; the heat of combustion is partially used in concentrating this solution. Invented by, and named after, Progil SA.

Propylane   An extractive distillation process for removing aromatic hydrocarbons from hydrogenated crude benzene, using propylene carbonate. Developed by Koppers.

Propylox   [**Propyl**ene oxidation]   A process for epoxidizing propylene to propylene oxide (1,2-epoxypropane),

$$CH_3-CH-CH_2$$
$$\diagdown \diagup$$
$$O$$

using a peroxycarboxylic acid in an organic solvent. The peroxy-acid is generated in water and immediately extracted into an organic solvent using an "extractor reactor." Invented in 1975 by A. M. Hilden and P. F. Greenhalgh at Laporte Industries, UK, and developed by Interox Chemicals, a joint company of Laporte Industries and Solvay. Piloted in Widnes, England, in the 1970s but not commercialized. Similar processes, without special names, have been developed by Bayer and by Olin Corporation. These processes would be economic only on a large scale, in conjunction with dedicated hydrogen peroxide plants.

   Belgian Patent 838,068.
   U.S. Patents 4,071,541; 4,177,196.

Protal   *See* metal surface treatment.

Provesteen   A microbiological process for making single-cell protein from methanol, ethanol, or whey, developed in the 1980s by Provesta Corporation, a subsidiary of Phillips Petroleum Company. The basis of the process is a special "high cell-density fermenter," which simplifies the isolation of the product from the water. The organism is the torula yeast; the intended products are: speciality flavor enhancers, a high fiber food bar, a food supplement for horses, and various aquaculture products for shrimp and fish. A plant with an annual capacity of 1,500 tonnes began operation in Bartlesville, OK, in 1989.

> McNaughton, K. T., *Chem. Bus.,* 1989, **11**(5), 18.

Proximol   A process for making hydrogen by reforming methanol. Offered by Lurgi.

Pruteen   A microbiological process for making single-cell protein from methane or methanol, developed by ICI. The organism is *Methylophilus methylotropus,* found in the sewers of Naples. A large pilot plant was built in Billingham, England, in the 1970s. The process was never commercialized, but some of the technology was used later in the *Deep Shaft process.

> Weissermel, K. and Arpe, H.-J., *Industrial Organic Chemistry,* 3rd ed., VCH Publishers, Weinheim, Germany, 1997, 35.

PSA   [Pressure swing adsorption]   A general method for separating gases by cyclic adsorption and desorption from a selective adsorbent, at alternating pressures. Invented by C. W. Skarstrom at the Esso Research and Engineering Company in 1958 and subsequently engineered by the Union Carbide Corporation. The Société de L'Air Liquide, France, made a similar development at that time. For separating nitrogen from air, carbon molecular sieve is the preferred adsorbent. For separating oxygen from air, a zeolite is used. Other zeolites have been used for other separations. Many variations on this basic process have been developed for specific gas mixtures and are known by special names; those described in this dictionary are: AUTO-PUREX G, Bergbau-Forschung, COPISA, COPSA, DWN, DWO, Generon, HYSEC, KURASEP, LO-FIN, MOLPSA-nitrogen, Moltox, MRH, NitroGEN, OxyGEN, Oxy-Rich, OXYWELL, Polybed, PSPP, Remet, RPSA, Sumitomo-BF. *See also* TSA.

> U.S. Patents 2,944,627; 3,155,468.
> Skarstrom, C. W., *Ann. N.Y. Acad. Sci.,* 1959, **72**, 751.
> Yang, R. T., *Gas Separation by Adsorption Processes,* Butterworths, Guildford, England, 1987, 237.
> Suzuki, M., in *Adsorption and Ion Exchange: Fundamentals and Applications,* LeVan, M. D., Ed., American Institute of Chemical Engineers, New York, 1998, 119.
> White, D. H., Jr. and Barkley, P. G., *Chem. Eng. Prog.,* 1989, **85**(1), 25.
> Ruthven, D. M., Farook, S., and Knaebel, K. S., *Pressure Swing Adsorption,* VCH Publishers, Weinheim, Germany, 1993.

PS Claus   A process for recovering sulfur from waste gases by a combination of the *Pressure swing process and the *Claus process.

> *Eur. Chem. News,* 1994, **61**(1611), 28.

PSPP   [Pressure swing parametric pumping]   A version of the *PSA process for separating gases by selective adsorption. It operates by rapidly reversing the gas flows through the absorber bed; the pressures are different for each direction of flow. The main use is for generating oxygen-enriched air for medical use. Invented by the Union Carbide Corporation in 1978.

> U.S. Patents 4,194,891; 4,194,892.
> Keller, G.E., II, in *Industrial Gas Separations,* Whyte, T. E., Yon, C. M., and Wagener, E. H., Eds., American Chemical Society, Washington, D.C., 1983.

Puddling   A process for making wrought iron from pig iron, based on the partial decarburization of pig iron in a special furnace. Invented by H. Cort in Titchfield, Southampton, England, in 1784 and widely used in the United Kingdom and Europe until the end of the 19th century. In 1873 there were 8,000 puddling furnaces in the United Kingdom alone.

> British Patent 1,420 (1784).

Barraclough, K. C., *Steelmaking Before Bessemer, Vol. 2, Crucible Steel,* The Metals Society, London, 1984, 91,303.

Pumpherston [Named after the town near Edinburgh, Scotland, where the process was operated] A process for extracting fuel oil from oil shale. The heart of the process was the Pumpherston retort (also called the Bryson retort), down which the shale fell by gravity and up which air and steam were passed. Ammonia was collected as a by-product. Invented by J. Young in 1850 and operated in Scotland between 1883 and 1962.

British Patent 13,292 (1850).

Smith, J. W., in *Handbook of Synfuels Technology,* Meyers, R. A., Ed., McGraw-Hill, New York, 1984, 4-149.

Russell, P. L., *Oil Shales of the World,* Pergamon Press, Oxford, 1990, 712.

PuraSiv Hg An adsorptive process for removing mercury vapor from gaseous effluents from the *Castner-Kellner process by *TSA. The adsorbent is a zeolite molecular sieve containing silver. Developed by UOP.

U.S. Patent 4,874,525.

PuraSiv HR A process for removing solvent vapors from air by adsorption on beaded activated carbon contained in a combined fluidized moving bed. For water-soluble solvents, the gas used for desorption is nitrogen and the process is known as PuraSiv HR, Type N (not to be confused with PuraSiv N); for chlorinated hydrocarbons, steam stripping is used and the process is known as PuraSiv HR, Type S. Developed by Kureha Chemical Company and now marketed by the Union Carbide Corporation. The process was originally known as GASTAK because it was developed by the Taiyo Kaken Company, subsequently acquired by Kureha Chemical Company. It is also marketed by Daikin Industries under the name Soldacs.

Chem. Eng. (N.Y.), 1977, **84**(18), 39.

Keller, G.E., II, *Industrial Gas Separations,* Whyte, T. E., Jr., Yon, C. M., and Wagener, E. H., Eds., American Chemical Society, Washington, D.C. 1983.

Kohl, A. L. and Riesenfeld, F. C., *Gas Purification,* 4th ed., Gulf Publishing, Houston, TX, 1985, 704.

Yang, R. T., *Gas Separation by Adsorption Processes,* Butterworths, Guildford, England, 1987, 217.

PuraSiv N A process for removing nitrogen oxides from the tail gases from nitric acid plants, using an acid-resistant zeolite molecular sieve. Developed by the Union Carbide Corporation in 1971. Not to be confused with PuraSiv HR, Type N (see previous entry).

Kohl, A. L. and Riesenfeld, F. C., *Gas Purification,* 4th ed., Gulf Publishing, Houston, TX, 1985, 674.

PuraSiv S A process for removing sulfur dioxide from the tail gases from sulfuric acid manufacture by adsorption on a special zeolite. Not to be confused with PuraSiv HR, Type S.

Anderson, R. A., in *Molecular Sieves II,* Katzer, J. R., Ed., American Chemical Society, Washington, D.C., 1977, 637.

PURASPEC A process for purifying gaseous and liquid hydrocarbons by the use of fixed beds of catalysts and adsorbents which remove impurities by chemical reaction. Developed in 1990 by ICI Katalco to enable natural gas and natural gas liquids to meet pipeline specifications. Installed in approximately 60 plants worldwide in 1996.

Hydrocarbon Process., 1996, **75**(4), 133.

Purex [Plutonium and uranium recovery by extraction] A process for the solvent extraction of plutonium from solutions of uranium and fission products, obtained by dissolving spent nuclear fuel elements in nitric acid. The solvent is tri-*n*-butyl phosphate (TBP) in kerosene. First operated by the U.S. Atomic Energy Commission at its Savannah River plant in 1954 and at Hanford in 1956. Now in operation, with modifications, in several countries. These include Barnwell (United States), Cap de la Hague (France), Marcoule (France), Dounreay (Scotland), Sellafield (England), Karlsruhe (Germany), and Trombay (India). *See also* Recuplex.

Siddall, T. H., III, in *Chemical Processing of Reactor Fuels,* Flagg, J. F., Ed., Academic Press, New York, 1961, 199.

Nuclear Fuel Reprocessing Technology, British Nuclear Fuels PLC, Risley, England, 1985.

Büchner, W., Schliebs, R., Winter, G., and Büchel, K. H., *Industrial Inorganic Chemistry,* VCH Publishers, Weinheim, Germany, 1989, 586.

Purifier   An ammonia synthesis process, developed and sold by C. F. Braun, CA.

Purisol   A process for removing hydrogen sulfide from gases by selective absorption in N-methyl-2-pyrrolidone (NMP). Developed and licensed by Lurgi, particularly for desulfurizing waste gases from *IGCC plants. Seven units were in operation or under construction in 1996.

Hochgesand, G., *Chem. Ing. Tech.,* 1968, **40**(9/10), 432.

Hydrocarbon Process., 1975, **54**(4), 92.

Kohl, A. L. and Riesenfeld, F. C., *Gas Purification,* 4th ed., Gulf Publishing, Houston, TX, 1985, 851.

Hydrocarbon Process., 1996, **75**(4), 133.

Purlex   An improved version of the *Bufflex process for extracting uranium from its ores. Operated in South Africa.

Eccles, H. and Naylor, A., *Chem. Ind. (London),* 1987, (6), 174.

Purofer   A direct reduction ironmaking process, using gas as the reductant. First in operation in Germany in 1970, but now used only in Iran. *See* DR.

Purox   A process for partially combusting organic wastes in a shaft furnace with oxygen, thereby producing a fuel gas and a molten slag. The gas, not diluted with nitrogen, is suitable for use as a chemical feedstock. Developed by Union Carbide Corporation in 1974 and piloted in Charleston, WV.

Masuda, T. and Fisher, T. F., in *Thermal Conversion of Solid Wastes and Biomass,* Jones, J. L. and Radding, S. B., Eds., American Chemical Society, Washington, D.C., 1980, 573.

Probstein, R. F. and Hicks, R. E., *Synthetic Fuels,* McGraw-Hill, New York, 1982, 408.

PVD   *See* Physical Vapor Deposition.

PX-Plus   A process for disproportionating toluene to *p*-xylene and benzene. Developed by UOP in the 1990s. Competing technologies are Mobil's MSTDP and MTPX. Not commercialized as of 1997.

Eur. Chem. News, 1997, **67**(1753), 23.

Eur. Chem. News Proc. Rev., 1997, May, 26.

Pylumin   *See* metal surface treatment.

Pyral   A process for destroying toxic waste organochlorine compounds. The wastes are mixed with carbon and sodium carbonate and injected into a graphite-lined arc furnace. Metallic sodium, formed by reduction of the sodium carbonate by the carbon, attacks the chlorinated organic compounds, forming sodium chloride. Developed by Hydro-Quebec in the late 1980s but not yet commercialized.

PYROCAT   A steam cracking process for converting petroleum into light olefins in which a catalyst is deposited on the walls of the heat-exchanger coils in the cracking furnace. The catalyst is a proprietary promoter on an alumina/calcia base. Based on the *THERMOCAT process, it was developed jointly by Veba Oel and Linde from 1996 but has not yet been commercialized.

Chem. Eng. (Rugby, England), 1997, (638), 24.

Pyrohydrolysis   Also called spray-roasting. A process for evaporating and calcining metal salt solutions in one step. Originally developed for processing pickle liquors from the steel industry, containing ferrous chloride in hydrochloric acid, but now used for making ceramic raw materials. Recent development of this process has been by the Ruthner Division of Maschinenfabrik Andritz, Austria.

Kladnig, W. F. and Karner, W., *Am. Ceram. Bull.,* 1990, **69,** 814.

Pyron   A process for making iron powder by reducing mill-scale, obtained from steelworks, with hydrogen. In operation in the United States in places where cheap hydrogen is available near steelworks.

Pyroplasma   A high-temperature process for destroying toxic liquid wastes such as polychlorinated biphenyls. The liquid is passed through a d.c. electric arc heater and the exit gases pass into a refractory-lined chamber where further reactions occur. Developed by Westinghouse Pyrolysis Systems, United States, and the Kingston Royal Military College, Ontario, in the late 1980s but not commercialized.

Kolak, N. P., Barton, T. G., Lee, C. C., and Peduto, E. F., *Nucl. Chem. Waste Manage.,* 1987, **7,** 37.

Pyrotol   A process for making benzene from pyrolysis gasoline by hydrocracking. Developed by Houdry Process and Chemical Company. In 1987, 13 units were operating worldwide.

Lorz, W., Craig, R. G., and Cross, W. J., *Erdoel Kohle Erdgas Petrochem.* 1968, **21,** 610.
Hydrocarbon Process., 1970, **49**(9), 223.

QC   *See* TRC.

Q-Max   A process for making cumene from benzene and propylene by catalytic alkylation using a proprietary regenerable zeolite catalyst. Developed by UOP and first installed in 1996 by JLM Chemicals in Illinois.

Eur. Chem. News, 1996, **66**(1737), 41.
Eur. Chem. News, 1997, **67**(1755), 16.
Bentham, M. F., in *Handbook of Petroleum Refining Processes,* Meyers, R. A., Ed., McGraw-Hill, New York, 1997, 1.67.

QQ-BOP   *See* steelmaking.

Q-S   *See* QSL.

QSL   [Queneau-Schumann-Lurgi]   A submerged smelting process for extracting lead from its ores and secondary sources. Pellets of sulfide ore concentrate are fed into a bath of molten slag held in a rotating cylindrical furnace. Oxygen is fed into the bath below the surface of the slag, forming sulfur dioxide and generating heat. Powdered coal is added further along the pool, reducing the lead oxide to metal. Invented by P. E. Queneau and R. Schumann, Jr. and now offered by Lurgi. Operated in China, Canada, Germany, and Korea.

Morgan, S. W. K., *Zinc and Its Alloys and Compounds,* Ellis Horwood, Chichester, England, 1985, 96.

Chem. Eng. (N.Y.), 1990, **97**(4), 55.

Quentin   A process for regenerating the ion exchange resin used in sugar refining, using magnesium chloride solution.

Quentin, G., *Zucker,* 1957, **10**, 408 (*Chem. Abstr.,* **52**, 766).

Landi, S. and Mantovani, G., *Sugar Technol. Rev.,* 1975, **3**(1), 67.

Quick Contact   *See* TRC.

R

R-2   [Rapson]   A process for making chlorine dioxide by reacting sodium chlorate with sodium chloride and sulfuric acid:

$$2NaClO_3 + 2NaCl + 2H_2SO_4 = 2ClO_2 + Cl_2 + 2Na_2SO_4 + 2H_2O$$

The product gas, mixed with by-product chlorine, is stripped from the solution by a current of air and passed into water. The chlorine dioxide dissolves and most of the chlorine does not; the latter is absorbed in a second column containing alkali. Developed by W. H. Rapson at Hooker Chemical Corporation and operated at Springhill, LA, since 1961.

U.S. Patent 2,863,722.

Rapson, W. H., *Tappi,* 1958, **41**(4), 181.

Rapson, W. H. and Partridge, H. de V., *Tappi,* 1961, **44**(10), 698.

Partridge, H. de V., in *Chlorine, Its Manufacture, Properties, and Uses,* Sconce, J. S., Ed., Reinhold Publishing, New York, 1962, 306.

Raecke   *See* Henkel.

Radenthein   *See* Hansgirg.

Radiance   A process for removing organic contaminants from the surfaces of semiconductors by irradiation with deep ultraviolet light while simultaneously passing an inert gas over the surface in laminar flow. Invented by A. Englesberg in 1987 and developed by Radiance Services Company, Bethesda, MD.

Kaplan, H., *Photonics Spectra,* 1996, **30**(9), 48.

Raney   Not a process, but a nickel catalyst widely used for hydrogenating organic compounds. It is made from a 50/50 nickel/aluminum alloy by leaching out the aluminum with concentrated aqueous sodium hydroxide. The product has a spongy texture and is highly active. Invented by M. Raney in 1926. The business was acquired by W. R. Grace in 1963.

U.S. Patent 1,628,190.

Bond, G. C., *Catalysis by Metals,* Academic Press, London, 1982, 34.

RAPRENO$_x$ [**Rap**id **red**uction of **NO$_x$**] A process for removing NO$_x$ from flue-gases by reaction with cyanuric acid (HOCN)$_3$. The acid decomposes to HOCN, which generates NH$_2$ radicals, which in turn reduce NO to molecular nitrogen.

Miller, J. A., Branch, M. C., and Kee, R. J., *Combust. Flame*, 1981, **43**, 81.
Perry, R. A. and Siebers, D. L., *Nature*, 1986, **324**, 657.
Gmelin Handbook of Inorganic Chemistry & Organometallic Chemistry, 8th ed., Nitrogen, Suppl. B1, Springer-Verlag, Berlin, 1993, 233.

RAR [**R**ecycle **A**bsorbtion **R**egeneration] A process for extracting traces of sulfur compounds from the effluent gases from the *Claus process by use of a selective amine absorbent. Developed by KTI.

Raschen A process for making sodium cyanide by reacting ammonia with carbon disulfide. Invented by J. B. Raschen at the United Alkali Company in Widnes, England in the early 1900s.

Raschig (1) A process for making hydrazine by oxidizing ammonia with sodium hypochlorite in the presence of gelatine:

$$NH_3 + NaOCl = NaOH + H_2NCl$$

$$H_2NCl + NH_3 = N_2H_4 + HCl$$

Invented by F. Raschig at Ludwigshafen, Germany, in 1906 and commercialized by Raschig in 1907. The Olin Raschig version has a complex flow chart and does not use gelatine.

German Patents 192,783; 198,307.
Raschig, F., *Ber. Dtsch. Chem. Ges.*, 1907, **40**, 4587.
Reed, R. A., *Hydrazine and Its Derivatives*, Royal Institute of Chemistry, London, 1957, 2.

Raschig (2) Also called Raschig-Hooker. A two-stage regenerative process for making phenol from benzene. The benzene is first chlorinated with hydrochloric acid in the presence of air, at 200 to 260°C, over a copper catalyst on an alumina base:

$$C_6H_6 + HCl + \tfrac{1}{2}O_2 = C_6H_5Cl + H_2O$$

The resulting chlorobenzene is then hydrolyzed with steam, over an apatite catalyst at about 480°C:

$$C_6H_5Cl + H_2O = C_6H_5OH + HCl$$

The hydrochloric acid is recycled. The process was developed by Raschig at Ludwigshafen, Germany, in the 1930s, based on the work of L. Dusart and Ch. Bardy in 1872. A variation, known as the Hooker-Raschig process, which uses a different catalyst, makes use of the by-product dichlorobenzenes and thus increases the overall yield. There are several commercial routes from benzene to phenol; the Raschig (2) route is now economic only for very large plants in special locations.

French Patent 698,341.
U.S. Patents 1,963,761; 2,009,023; 2,035,917.
Mathes, W., *Angew. Chem.*, 1939, **52**, 591 (*Chem. Abstr.*, **34**, 394).
Crawford, R. M., *Chem. Eng. (N.Y.)*, 1950, **46**, 483.
Prahl, W. H., Williams, W. H., and Widiger, A. H., in *Chlorine, Its Manufacture, Properties, and Uses*, Sconce, J. S., Ed., Reinhold Publishing, New York, 1962, 438.
Weissermel, K. and Arpe, H.-J., *Industrial Organic Chemistry*, 3rd ed., VCH Publishers, Weinheim, Germany, 1997, 350.

Raschig (3) A process for making hydroxylamine. Invented by Raschig AG.

rayon Not a process but the generic name for regenerated cellulose fibers made by the Viscose and related processes. *See* Cross-Bevan-Beadle, Cuprammonium, Viscose.

RCA [**R**adio **C**orporation of **A**merica] Also called RCA-2 and HPM. A process for cleaning silicon wafers used in electronics. They are washed successively by three solutions.

The first is an alkaline solution of hydrogen peroxide, which oxidizes organic matter. The second is an acid fluoride solution which removes silica. The third is an acid solution of hydrogen peroxide which removes transition metals. The process was developed by RCA Corporation in 1970 and widely used thereafter by the electronics industry.

Kern, W. and Poutinen, D. A., *RCA Rev.,* 1970, **31,** 187.

Christenson, K. K., Smith, S. M., and Werho, D., *Microcontamination,* 1994, **12**(6), 47.

RCC [Reduced crude oil conversion] A process for converting reduced crude oil (a petroleum fraction), and other petroleum residues, into high-octane gasoline and other lighter fuels. Based on the *FCC process, but adapted to accommodate higher levels of metal contaminants which can harm the catalyst. Developed by Ashland Oil Company and UOP and commercialized in 1983.

RCD Isomax [Reduced crude desulfurization] An obsolete process for desulfurizing high-sulfur residual oils. Developed by UOP, later replaced by *RCD Unibon.

RCD Unibon [Reduced crude desulfurization] Also known as the Black oil conversion process (BOC). A process for removing organic sulfur-, nitrogen-, and metal-compounds from heavy petroleum fractions. Different catalysts are used for different oils. Developed and licensed by UOP.

Cabrera, C. N., in *Handbook of Petroleum Refining Processes,* Meyers, R. A., Ed., McGraw-Hill, New York, 1986, 6–2.

Marcos, F. and Rosa-Brussin, D., *Catal. Rev. Sci. Eng.,* 1995, **37**(1), 3.

Thompson, G. J., in *Handbook of Petroleum Refining Processes,* Meyers, R. A., Ed., McGraw-Hill, New York, 1997, 8.39.

RCE A *flue-gas desulfurization process in which the sulfur dioxide is absorbed in aqueous magnesium hydroxide. The product is reacted with calcium chloride to produce gypsum, and the magnesium hydroxide is regenerated by treatment with dolomite. Developed by Refractories Consulting & Engineering, Germany, and piloted in Austria.

RCH [Ruhrchemie] A process for increasing the octane rating of gasolines by catalytic isomerization of the olefin fraction, the double bonds migrating from the terminal positions. Developed by Ruhr Chemie in the 1940s.

Asinger, F., *Mono-olefins: Chemistry and Technology,* translated by B. J. Hazzard, Pergamon Press, Oxford, 1968, 1096.

RCH/RP [Ruhrchemie/Rhône Poulenc] A variation of the *OXO process in which the triphenyl phosphine (part of the Wilkinson catalyst) is sulfonated, in order to render the catalyst soluble in water for easier recovery. First commercialized in 1984 for the manufacture of butyraldehyde.

Bach, H., Gick, W., Konkol, W., and Wiebus, E., in *Proc. 9th Internat. Conf. Catal.,* Phillips, M. J. and Ternan, M., Eds., Chemical Institute of Canada, 1988.

Chem. Eng. News, 1994, **72**(41), 28.

Beller, M., Cornils, B., Frohning, C. D., and Kohlpaintner, C. W., *J. Mol. Catal.,* 1995, **A104**(1), 32,48.

RDS Isomax [Residuum desulphurization] A *hydrodesulfurization process for removing sulfur compounds from petroleum residues, while converting the residues to fuel oil. Developed by Chevron Research Company in the early 1970s. Ten units were operating in 1988. *See also* VGO Isomax, VRDS Isomax.

Scott, J. W., Bridge, A. G., Christensen, R. I., and Gould, G. D., *Oil & Gas J.,* 1970, **68**(22), 72.

Speight, J. G., *The Desulfurization of Heavy Oils and Residua,* Marcel Dekker, New York, 1981, 194.

Hydrocarbon Process., 1996, **75**(11), 132.

Readman *See* Furnace.

Recatro A process for making gas from liquid fuels and other gaseous hydrocarbons by catalytic conversion into "rich gas," followed by catalytic steam reforming. Developed by BASF and Lurgi.

Recoflo An ion-exchange process based on short beds and small beads. Developed by the University of Toronto in the 1960s and commercialized by Eco-Tec, Canada. Used for wastewater recovery and removal of metals from various metallurgical waste streams. In 1988, 500 units had been installed in 27 countries.

> Brown, C. J. and Fletcher, C. J., in *Ion Exchange for Industry,* Streat, M., Ed., Ellis Horwood, Chichester, England, 1988, 392.

Recrystallizer A process for recrystallizing sodium chloride from brine. Rock salt is dissolved in brine heated with direct steam. The solution is then partially evaporated under reduced pressure. Invented in 1945 by C. M. Hopper and R. B. Richards at the International Salt Company, Scranton, PA. *See also* Alberger.

> U.S. Patents 2,555,340; 2,876,182.

Rectiflow A multi-stage, liquid–liquid extraction process for removing non-paraffinic components from lubricating oils. Furfural has been used as the solvent. Developed and used by the Shell Petroleum Company in the 1940s, subsequently abandoned.

> *The Petroleum Handbook,* 3rd ed., Shell Petroleum Co., London, 1948, 188.

Rectisol A process for removing sulfur compounds from gas mixtures resulting from the partial oxidation of hydrocarbons, based on physical absorption in methanol at low temperatures. Originally developed in 1951 by Lurgi Gesellschaft für Warmetechnik for the SASOL coal gasification plant in South Africa, but now used also for removing sulfur compounds, CO_2, H_2S, HCN, C_6H_6 and gum-forming hydrocarbons from *syngas and fuel gas. Further developed and now offered by Linde. In 1990, over 70 units were in operation or under construction.

> Herbert, W., *Erdoel Kohle,* 1956, **9**(2), 77.
> *Hydrocarbon Process.,* 1975, **54**(4), 93.
> Kohl, A. L. and Riesenfeld, F. C., *Gas Purification,* 4th ed., Gulf Publishing, Houston, TX, 1985, 821.
> Weiss, H., *Gas Sep. Purif.,* 1988, **2**, 171.
> *Hydrocarbon Process.,* 1992, **71**(4), 125.
> *Hydrocarbon Process.,* 1996, **75**(4), 134.

Recuplex A variant of the *Purex process for extracting plutonium, in which the tributyl phosphate is dissolved in carbon tetrachloride in order to make the organic phase denser than the aqueous phase.

Redex [Recycle extract dual extraction] A process for improving the cetane rating of diesel fuel by removing heavy aromatic hydrocarbons by solvent extraction.

> French Patents 792,281; 1,424,225; 1,424,226.
> Benham, A. L., Plummer, M. A., and Robinson, K. W., *Hydrocarbon Process.,* 1967, **46**(9), 134.

Redox [Reduction oxidation] A process for separating the components of used nuclear fuel by solvent extraction. It was the first process to be used and was brought into operation at Hanford, United States, in 1951, but was superseded in 1954 by the *Purex process. The key to the process was the alternate reduction and oxidation of the plutonium, hence the name. The solvent was Hexone (4-methyl-2-pentanone, methyl isobutyl ketone), so the process was also known as the Hexone process. The aqueous phase contained a high concentration of aluminum nitrate to salt out the uranium and plutonium nitrates into the organic phase. The presence of this aluminum nitrate in the wastes from the process, which made them bulky, was the main reason for the abandonment of the process. *See also* Butex.

Taube, M., *Plutonium,* Macmillan, New York, 1964, 130.
Nuclear Fuel Reprocessing Technology, British Nuclear Fuels, Risley, UK, 1985.

reforming A general name for the reaction of a hydrocarbon, such as methane, with water and/or carbon dioxide, to produce a mixture of carbon monoxide and hydrogen. If water is used, it is called steam reforming or steam cracking. The reactions are endothermic and require a catalyst:

$$CH_4 + H_2O = CO + 3H_2$$

The usual catalyst is nickel on an oxide support. A second reaction also takes place, the shift reaction, also known as the water gas shift reaction:

$$CO + H_2O = CO_2 + H_2$$

The shift reaction can be conducted in a second reactor, catalyzed by a mixture of iron and chromium oxides. The product of reforming is known as synthesis gas, or *syngas, and is mostly used in the manufacture of ammonia and methanol. One of the earliest steam reforming processes was developed in Germany by I.G. Farbenindustrie in 1926. *See also* catalytic reforming.

U.S. Patent 1,934,237.
Ridler, D. E. and Twigg, M. V., in *The Catalyst Handbook,* 2nd. ed., Twigg, M. V., Ed., Wolfe Publishing, London, 1989, 225.

REGEN A process for removing mercaptans from hydrocarbon fractions by catalytic oxidation and extraction with aqueous alkali, using a bundle of hollow fibers. Developed by the Merichem Company, Houston, TX, and used in 34 plants as of 1991.

Hydrocarbon Process., 1992, **71**(4), 120.
Hydrocarbon Process., 1996, **75**(4), 126.

REGENOX A catalytic process for oxidizing organic compounds in gaseous effluents. A modified version oxidizes chlorinated and brominated hydrocarbons at 350 to 450°C without forming dioxins. Developed by Haldor Topsoe and first operated by Broomchemie in The Netherlands in 1995. *See* CATOX.

Chem. Eng. (N.Y.), 1995, **102**(9), 17.

Reich (1) A process for purifying carbon dioxide obtained by fermentation. It is first scrubbed by aqueous ethanol, then by aqueous potassium dichromate to oxidize organic compounds, and finally with concentrated sulfuric acid to dry it. Developed in the 1920s by G. T. Reich.

U.S. Patents 1,519,932; 2,225,131.

Reich (2) A complex process for recovering potassium from sugar processing residues.

Thorpe's Dictionary of Applied Chemistry, 4th ed., Longmans, Green & Co., London, 1950, **10,** 139.

Reinluft A *flue-gas desulfurization process using coke. The carbon acts as a catalyst for the oxidation of the sulfur dioxide to sulfur trioxide in the presence of water, and the sulfur trioxide is retained on the coke. The coke is regenerated in another vessel by heating with a hot gas stream, which reduces the sulfur trioxide back to sulfur dioxide and expels it for use in sulfuric acid manufacture. The key to this process is the inexpensive adsorbent. Developed by Reinluft GmbH and Chemiebau Dr. A. Zieren GmbH, and marketed as the Reinluft (Clean Air) Process. Four plants had been built by 1985.

Bienstock, D., Field, J. H., Katell, S., and Plants, K. D., *J. Air Pollut. Control Assoc.,* 1965, **15,** 459.

Chem. Eng. (N.Y.), 1967, **74**(22), 94.

Kohl, A. L. and Riesenfeld, F. C., *Gas Purification,* 4th ed., Gulf Publishing, Houston, TX, 1985, 407.

Relube   A process for removing sulfur and chlorine compounds from waste oils, particularly those contaminated by polychlorinated biphenyls. Developed by Kinetics Technology International, The Netherlands, and operated first in Greece.

Remet   [**R**eforming **met**hanol]   A process for making high-purity methanol by a combination of *steam reforming and *PSA. Licensed by Tokyo Gas Company and Tokyo Cryogenic Industries Company. Two units were operating in 1990.

Hydrocarbon Process., 1990, **69**(4), 82.

RENUGAS   A thermal gasification process for biomass, under development by the Institute of Gas Technology, Chicago, in the 1980s.

Hydrocarbon Process., 1986, **65**(4), 100.

Chem. Eng. (N.Y.), 1996, **102**(3), 39.

Reppe   A family of processes for making a range of aliphatic compounds from acetylene, developed by W. Reppe in IG Farbenindustrie, Germany, before and during World War II. In one of the processes, acetylene is reacted with carbon monoxide to yield acrylic acid:

$$CH{\equiv}CH + CO + H_2O \rightarrow CH_2{=}CH{-}COOH$$

Acrylic esters are formed if alcohols are used instead of water:

$$CH{\equiv}CH + CO + ROH \rightarrow CH_2{=}CH{-}COOR$$

Nickel carbonyl is the catalyst for these reactions.

In another Reppe process, acetylene is reacted with formaldehyde to yield butyndiol, which can be converted to butadiene for the manufacture of the synthetic rubber *"Buna"; the catalyst is nickel cyanide:

$$CH{\equiv}CH + 2CH_2O \rightarrow HOCH_2{-}C{\equiv}C{-}CH_2OH$$

When, in the 1950s, ethylene became the preferred feedstock for making petrochemicals, most of these acetylene-based processes became obsolete.

German Patents 725,326; 728,466.

U.S. Patents 2,806,040; 2,809,976; 2,925,436; 3,023,237.

Reppe, W., *Acetylene Chemistry,* translated, Charles A. Meyer & Co., New York, 1949.

Reppe, W., *Experentia,* 1949, **5,** 93.

Miller, S. A., *Acetylene: Its Properties, Manufacture and Uses,* Vol. 1, Ernest Benn, London, 1965.

Morris, P. J. T., *Chem. Ind. (London),* 1983, (18), 713.

Eur. Chem. News, Process Rev. Suppl., 1988, Oct, 10.

reprography   The processes listed in the following text, described elsewhere in this dictionary, are mostly for reproducing line drawings, rather than pictures. There is, however, some overlap with *photography. These processes are: Blueprint, Diazo, Dual-Spectrum, Dyeline, Eichner, Extafax, Kalvar, Ozalid, Thermofax, Van Dyke.

Kirk-Othmer's Encyclopedia of Chemical Technology, 3rd ed., Vol. 20, John Wiley & Sons, New York, 1980, 128.

Republic Steel   *See* DR.

RESID-fining   [**Resid**uum re**fin**ing]   A *hydrodesulfurization process adapted for petroleum residues. Developed by Esso Research & Engineering Company and licensed by them and Union Oil Company of California. A proprietary catalyst is used in a fixed bed. As of 1988, eight plants had been designed.

Speight, J. G., *The Desulfurization of Heavy Oils and Residua,* Marcel Dekker, New York, 1981, 190.
Hydrocarbon Process., 1988, **67**(9), 70.
Hydrocarbon Process., 1994, **73**(11), 135.

RESOX A process for converting sulfur dioxide in dilute gas streams to elemental sulfur. The use of coal to remove sulfur dioxide from gas streams was described as early as 1879. Bergbau Forschung developed a process for reversibly adsorbing sulfur dioxide on activated coke in the 1950s, and Foster Wheeler Corporation modified it in the late 1960s in order to produce elemental sulfur. In the first stage, the sulfur dioxide is converted to sulfuric acid in the pores of the coke, and this is reduced back to sulfur dioxide in a second stage. The modified version uses coal instead of coke, and the reduction product is sulfur vapor, which is condensed as a liquid. The first demonstration plant was built by Foster Wheeler in Florida in 1974. *See also* Trail.

British Patent 189 (1879).
Steiner, P., Jüntgen, H., and Knoblauch, K., in *Sulfur Removal and Recovery from Industrial Processes,* Pfeiffer, J. B., Ed., American Chemical Society, Washington, D.C., 1975, 185.

RESS [Rapid Expansion of Supercritical Solutions] A process for depositing a film of solid material on a surface. The substance is dissolved in supercritical carbon dioxide. When the pressure is suddenly reduced, the fluid reverts to the gaseous state and the solute is deposited on the walls of the vessel. Used for size-reduction, coating, and microencapsulation. First described in 1879. Developed in 1983 by R. D. Smith at the Battelle Pacific Northwest Laboratory.

U.S. Patent 4,582,731.
Matson, D. W., Peterson, R. C., and Smith, R. D., *Mat. Lett.,* 1986, **4**, 429
Matson, D. W. and Smith, R. D., *J. Am. Ceram. Soc.,* 1989, **72**, 877.

Resulf A process for removing residual sulfur compounds from refinery tailgases. They are hydrogenated to hydrogen sulfide, which is absorbed in an aqueous solution of an amine such as methyl diethanolamine. Licensed by TPA.

Hydrocarbon Process., 1996, **75**(4), 134.

ReVAP [Reduced volatility alkylation process] A process for improving the safety of *alkylation processes catalyzed by hydrofluoric acid. A proprietary additive curtails the emission of the acid aerosol, which forms in the event of a leak. Developed by Phillips Petroleum Company and Mobil Corporation and first installed at Wood Cross, UT, in 1996. *See also* Alkar.

Chem. Mark. Rep., 1996, **250**(3), 7.

Rexene A process for making polypropylene. Developed by Appryl, a joint venture of BP and Atochem.

Chem. Br., 1996, **32**(8), 7.

Rexforming A petroleum refining process which combines *Platforming with an aromatics extraction process using ethylene glycol. Developed in the 1950s by Universal Oil Products.

Unzelman, G. H. and Wolf, C. J., in *Petroleum Processing Handbook,* Bland, W. F. and Davidson, R. L., Eds., McGraw-Hill, New York, 1967, 3-37.
Asinger, F., *Mono-olefins: Chemistry and Technology,* translated by B. J. Hazzard, Pergamon Press, Oxford, 1968, 391.

Reynolds Metal A process for extracting aluminum from clay by leaching with nitric acid. An essential feature is the pelletizing of the clay by calcination with kaolin in order to provide particles which will not disintegrate during leaching. Aluminum nitrate nonahydrate is crystallized from the leachate and thermally decomposed in several stages designed to conserve the nitric acid and nitrogen oxides. Developed by the Reynolds Metal Company, United States between 1973 and 1988, but not yet commercialized.

U.S. Patents 3,804,598; 4,251,265; 4,256,714.
O'Connor, D. J., *Alumina Extraction from Non-bauxitic Materials,* Aluminium-Verlag, Düsseldorf, 1988, 163.

RH   *See* steelmaking.

Rheniforming   [**Rhenium** re**forming**]   A *catalytic reforming process developed by Chevron Research Company. The catalyst formulation includes rhenium. First announced in 1967 and first commercialized in 1970; by 1988, 73 units had been licensed.

McCoy, C. S. and Munk, P., *Chem. Eng. Prog.,* 1971, **67**(10), 78.
Hughes, T. R., Jacobson, R. L., Gibson, K. R., Schornack, L. G., and McCabe, J. R., *Hydrocarbon Process.,* 1976, **55**(5), 75.
Little, D. M., *Catalytic Reforming,* PennWell Publishing, Tulsa, OK, 1985, 166.
Hydrocarbon Process., 1988, **67**(9), 79.

Rhenipal   A sewage sludge treatment process. One of three proprietary additives is used before dewatering. The resulting filter cake is smaller in volume, has less odor, and its heavy metals are insoluble. Offered by Rhenipal, UK, a joint venture between National Power and Dirk European.

RH-FR   *See* steelmaking.

RH-OB   *See* steelmaking.

Rhodaks   A process for removing hydrogen cyanide from coke-oven gas, developed by Rhodia. *See also* Fumaks-Rhodaks.

Rhône-Poulenc   Also called the RP process. This large French chemical manufacturer is perhaps best known for its process for making oxalic acid by oxidizing propylene with nitric acid. Nitratolactic acid is an intermediate. The process, invented in 1966, is operated on a large scale at Chalampé, France.

U.S. Patent 3,549,696.
French Patent 1,501,725.
British Patents 1,154,061; 1,159,066.

Rhône-Poulenc/Melle Bezons   A process for making acetic acid by oxidizing acetaldehyde with oxygen in air. Removal of the nitrogen would incur a cost penalty.

Weissermel, K. and Arpe, H.-J., *Industrial Organic Chemistry,* 3rd ed., VCH Publishers, Weinheim, Germany, 1997, 172.

R-HYC   A *hydrocracking process.

Marcos, F. and Rosa-Brussin, D., *Catal. Rev. Sci. Eng.,* 1995, **37**(1), 3.

Riedel   A process for making vanillin from guiacol derived from catechol. Invented by J. D. Riedel in 1932.

British Patent 401,562.

Riedel-Pfleiderer   *See* AO.

Riley-Morgan   A coal gasification process, based on a cylindrical gas-producer developed by C. H. Morgan in Worcester, MA, in 1880. By 1964, the Morgan Construction Company had installed more than 9,000 such units in a number of industries worldwide. In 1971 the

Riley Stoker Corporation acquired the manufacturing rights to the technology and made a number of modifications to the design. Air (or oxygen) and steam is passed through a fixed bed of coal, supported on a rotating ash pan. The temperature is kept below the melting point of the ash.

Hebden, D. and Stroud, H. J. F., in *Chemistry of Coal Utilization*, 2nd. Suppl. Vol., Elliott, M. A., Ed., John Wiley & Sons, New York, 1981, 1619.

RIMNAT   A process for making a fertilizer from domestic waste by treatment with the zeolite phillipsite.

Ciambelli, P., Corbo, P., Liberti, L., and Lopez, A., in *Occurrence, Properties, and Utilization of Natural Zeolites; Proceedings of the 2nd. International Conference on Natural Zeolites, Budapest*, Kallo, D. and Sherry, H. S., Eds., 1985, Akad. Kiado, Budapest, 1988, 501.

Rincker-Wolter   A process for making hydrogen by the thermal decomposition of oils and tars over hot coke. Invented by F. G. C. Rincker and L. Wolter in Germany in 1904.

German Patent 174,253.

Ellis, C., *The Chemistry of Petroleum Derivatives*, The Chemical Catalog Co., New York, 1934, 208.

RIP   [Resin in pulp]   A general term for hydrometallurgical processes in which an ion-exchange resin is mixed with a suspension of a ground ore in water. The desired metal is selectively extracted into the resin. *See* CIP (1).

Streat, M. and Naden, D., in *Ion Exchange and Sorption Processes in Hydrometallurgy*, Streat, M. and Naden, D., Eds., John Wiley & Sons, London, 1987, 35.

Rittman   An early process for making aromatic hydrocarbons by thermally cracking petroleum naphtha. *See also* Hall.

British Patents 9,162; 9,163 (1915).

Ellis, C., *The Chemistry of Petroleum Derivatives*, The Chemical Catalog Co., New York, 1934, 165.

Asinger, F., *Mono-olefins: Chemistry and Technology*, translated by B. J., Hazzard, Pergamon Press, Oxford, 1968, 137.

RKN   A process for making hydrogen from hydrocarbon gases (from natural gas to naphtha) by *steam reforming. Developed by Haldor Topsoe in the 1960s; as of 1975, 24 plants were operating.

Hydrocarbon Process., 1975, **54**(4), 132.

RLE   [roasting, leaching, electrowinning]   A process for extracting copper from sulfide ores, using the three named processes. Developed by Hecla Mining Company, AZ, in 1969.

Griffith, W. A., Day, H. E., Jordan, T. S., and Nyman, V. C., *J. Met.*, 1975, **27**(2), 17.

RM   [Ralph M. Parsons]   A process for methanating synthesis gas, i.e. converting a mixture of carbon monoxide and hydrogen to mainly methane and carbon dioxide. Six adiabatic reactors are used in series, and steam is injected at the inlet. Under development by the R. M. Parsons Company in 1975.

Benson, H. E., in *Chemistry of Coal Utilization*, 2nd. Suppl. Vol., Elliott, M. A., Ed., John Wiley & Sons, New York, 1981, 1795.

R-N   [Republic Steel Corp. and National Lead Co.]   An ironmaking process developed by these two American companies in the 1960s. Granulated ore is reduced with coal in a rotating kiln, heated by burning the coal at the lower end. Three plants were operating in 1970.

Robinson-Bindley   *See* Synthetic Oils.

Rodgers   An early process for making potassium cyanide by fusing together potassium ferrocyanide and potassium carbonate.

Roelen   *See* OXO.

Röhm   A process for making sodium cyanide, engineered by Uhde. A plant was commissioned in Kwinana, Western Australia, in 1988.

Roka   A process for making acetone by passing a mixture of ethanol and steam over a catalyst containing iron and calcium:

$$2C_2H_5{\cdot}OH + H_2O = CH_3{\cdot}CO{\cdot}CH_3 + 4H_2 + CO_2$$

Invented by K. Roka at Holzverkohlungs-Industrie in 1924 and operated by British Industrial
Solvents at Hull, UK, in the 1930s.

> German Patent 475,428.
> U.S. Patent 1,663,350.
> Morgan, G. T. and Pratt, D. D., *British Chemical Industry,* Edward Arnold & Co., London, 1938,
> 315.

ROSE (1)   [**R**esiduum **O**il **S**upercritical **E**xtraction]   A process for extracting asphaltenes
and resins from petroleum residues, using supercritical propane or isobutane as the extractant. Developed by Kerr-McGee Corporation in 1979 and sold to the MW Kellog Company
in 1995, at which time 25 units had been licensed.

> Gearhart, J. A. and Garwin, L., *Hydrocarbon Process.,* 1976, **55**(5), 125.
> *Hydrocarbon Process.,* 1978, **57**(9), 200.
> *Chem. Eng. (N.Y.),* 1989, **96**(7), 35.
> *Hydrocarbon Process.,* 1996, **75**(11), 106.

Rose (2)   A process for extracting gold from the residues from zinc production. The
residues are fused with a mixture of borax and silica, and air blown through. The base metals oxidize and pass into the slag.

Rotor   An oxygen steelmaking process, similar to the *Kaldo process but using a furnace
rotating about a horizontal axis. In this method of operation the refractory lining is cooled by
the molten metal and slag and therefore lasts longer. Developed in Oberhausen, Germany, in
the 1950s. *See also* DR.

> British Patent 726,368.
> Osborne, A. K., *An Encyclopedia of the Iron and Steel Industry,* 2nd. ed., The Technical Press,
> London, 1967, 472.

Rozan   A variation of the *Pattinson process for extracting silver from lead, in which steam
is blown through the molten metal. This oxidizes the zinc and antimony, which come to the
surface and are removed.

RPSA   [**R**apid **p**ressure **s**wing **a**dsorption]   A version of the *PSA process which uses fast
pressure-cycles known as parametric pumping. The molecular sieve adsorbent for this duty
has to be of a smaller grain size than that for PSA. Developed by the Union Carbide
Corporation.

> Pritchard, C. L. and Simpson, G. K., *Trans. Inst. Chem. Eng.,* 1986, **64**(6), 467.

R2R   A *catalytic cracking process using an ultrastable zeolite catalyst with two-stage regeneration. Developed by Institut Français du Pétrole and used at Idemitsu Kosan's refineries at Aichi and Hokaido. In 1994, 13 existing plants had been converted to this process.

> *Inf. Chim. Hebdo,* 1994, (1175), 12.
> Chauvel, A., Delmon, B., and Hölderich, W. F., *Appl. Catal. A: Gen.,* 1994, **115**, 173.

RSRP [Richards sulphur recovery process] A proposed modification of the *Claus process in which liquid sulfur is used to cool the catalyst bed. Developed jointly by the Alberta Energy Company and the Hudson's Bay Oil & Gas Company, but not reported to have been commercialized.

Kerr, R. K., Sit, S.-P., Jagodzinski, R. F., and Dillon, J., *Oil & Gas J.*, 1982, **80**(30), 230.

Kohl, A. L. and Riesenfeld, F. C., *Gas Purification*, 4th ed., Gulf Publishing, Houston, TX, 1985, 464.

Rüping Also known as the Empty Cell process. A method for treating timber with a creosote preservative. The wood is first exposed to compressed air and then impregnated at a higher pressure. *See also* Bethell.

Ruhr Chemie Ruhr Chemie was an important German chemical company in the 1930s and 1940s. It was perhaps best known for its process for making acetylene by pyrolyzing hydrocarbons. *See also* Wulff.

Ruhr Chemie-Lurgi A variant of the *Fischer-Tropsch process, developed at Ruhr Chemie and Lurgi Gesellschaft für Warmetechnik in Germany during World War II.

Asinger, F., *Paraffins, Chemistry and Technology*, translated by B. J. Hazzard, Pergamon Press, Oxford, 1968, 168.

Ruhrgas (1) A coal gasification process developed and used in Germany. Pulverized coal is gasified by a blast of preheated air in a vertical shaft, the temperature reaching approximately 1,500°C.

Ruhrgas (2) A process for thermally decomposing oil shale, developed by Lurgi.

Ruhrkohle/VEBA Oel-hydrogenation A coal liquifaction process based on the *IG-Hydrogenation process.

Russell A process for extracting silver from argentite, Ag_2S, using a solution of sodium thiosulfate and cupric sulfate. Invented by E. H. Russell in 1884, following his use of this solution to remove sodium sulfide from soda ash.

U.S. Patent 295,815.

Ruthner A process for recovering hydrochloric acid and iron oxide from steel pickling liquors. Invented in 1968 by A. Hake and P. Borsody at Ruthner Industrieplanungs, Vienna. *See also* Dravo-Ruthner.

U.S. Patent 3,495,945.
Austrian Patent 284,062.

RWD [Reaction with distillation] A general chemical process in which a chemical reaction takes place within a distillation column, of which the packing is also the catalyst. First used in the 1920s. *See* Ethermax.

DeGarmo, J. L., Parulekar, V. N., and Pinjala, V., *Chem. Eng. Prog.*, 1992, **88**(3), 43.

Ryan-Holmes A cryogenic extractive distillation process using liquid carbon dioxide, in which a light hydrocarbon is added in order to suppress the freezing of the carbon dioxide. Licensed by Process Systems International: nine licenses had been granted by 1992.

Chiu, C.-W., *Hydrocarbon Process.*, 1990, **69**(1), 69.
Hydrocarbon Process., 1992, **71**(4), 126.

S

Saarburg-Holter   A *flue-gas desulfurization process.

Sulphur, 1979, (141), 34.

Saarburg-OTTO   A coal gasification process. Powdered coal, together with steam and oxygen, are injected tangentially into a gasifier containing molten slag. Gasification in a bath of molten slag was invented by R. Rummel in the 1950s and developed by Dr. C. Otto & Company in Germany in the early 1960s. In 1976, Saarbergwerke and Otto agreed to a joint development program which culminated in the building of a large demonstration unit at Voelklingen/Fuerstenhausen, Germany, which was operated from 1979 to 1982.

Rummel, R., *Coke Gas,* 1959, **21**(247) 493 (*Chem. Abstr.,* **54,** 11438).
Eur. Chem. News, Petrochem. Suppl., 1981, Dec, 14.
Mueller, R. and Pitz, H., in *Handbook of Synfuels Technology,* Meyers, R. A., Ed., McGraw-Hill, New York, 1984, 3-195.

SABA   [Spherical agglomeration-bacterial adsorption]   A microbiological process for leaching iron pyrites from coal. The bacterium *Thiobacillus ferrooxidans* adsorbs on the surface of the pyrite crystals, oxidizing them with the formation of soluble ferrous sulfate. Developed by the Canadian Center for Mineral and Energy Technology, Ottawa; in 1990 the process had been developed only on the laboratory scale, using coal from eastern Canada.

McCready, G. G. L., in *Bioprocessing and Biotreatment of Coal,* Wise, D. L., Ed., Marcel Dekker, New York, 1990, 685.

SAB   *See* steelmaking.

SABAR   [Strong acid by azeotropic rectification]   A process for making nitric acid by the atmospheric oxidation of ammonia. The nitrous gases from the oxidation are absorbed in azeotropic nitric acid in the presence of oxygen under pressure:

$$2NO_2 + \frac{1}{2}O_2 + H_2O = 2HNO_3$$

Developed by Davy McKee, which built plants from 1974 to 1986. *See also* CONIA.

Hellmer, L., *Chem. Eng. Prog.,* 1972, **68**(4), 67.
Hydrocarbon Process., 1989, **68**(11), 106.
Büchner, W., Schliebs, R., Winter, G., and Büchel, K. H., *Industrial Inorganic Chemistry,* VCH Publishers, Weinheim, Germany, 1989, 63.

Sabatier-Normann   *See* Normann.

saccharification   A general name for processes which convert wood to useful organic chemicals by hydrolysis of the polysaccharides in the wood to monomeric sugars. Exemplified by *Bergius-Rheinau, *Madison, *Scholler-Tornesch.

Riegel's Handbook of Industrial Chemistry, Kent, J.A., Ed. 9th ed., Van Nostrand Reinhold, New York, 1992, 254.

Sachsse   Also called the Flame cracking process, and the Sachsse-Bartholomé process. A process for making acetylene by the partial combustion of methane. The product gases are quenched rapidly and the acetylene is extracted with methyl pyrrolidone. First operated by IG Farbenindustrie at Oppau, Germany, in 1942. Worldwide, 13 plants used the process, of which 7 were still in operation in 1991.

Sachsse, H., *Chem. Ing. Tech.,* 1949, **21,** 129.

Sachsse, H., *Chem. Ing. Tech.,* 1954, **26,** 245.

Miller, S. A., *Acetylene: Its Properties, Manufacture and Uses,* Vol. 1, Ernest Benn, London, 1965, 419.

Tedeschi, R. J., *Acetylene-based Chemicals from Coal and Other Natural Resources,* Marcel Dekker, New York, 1982, 20.

Weissermel, K. and Arpe, H.-J., *Industrial Organic Chemistry,* 3rd ed., VCH Publishers, Weinheim, Germany, 1997, 96.

SAFe   A *BAF process offered by PWT Projects. The biological medium is supported on a bed of expanded shale.

Stephenson, T., Mann, A., and Upton, J., *Chem. Ind. (London),* 1993, (14), 533.

St. Joseph   Also known as St. Joe. A process for extracting zinc from zinc sulfide ore by electrothermic reduction, practised by the St Joseph Lead Company at its Josephtown refinery in the United States, in the 1930s. A mixture of zinc blende with coke was heated by passing electricity through the mixture. The zinc vapor thus produced was condensed in a bath of molten zinc. The name has also been applied to a similar lead extraction process.

Cocks, E. J. and Walters, B., *A History of the Zinc Smelting Industry in Britain,* George G. Harrap, London, 1968, 164.

Morgan, S. W. K., *Zinc and Its Alloys and Compounds,* Ellis Horwood, Chichester, England, 1985, 69.

Salex   A process for purifying sodium chloride. Continuous counter-current extraction with brine removes the impurities without dissolving the salt. There are three variants: SALEX-B, SALEX-C, AND SALEX-M. Developed by Krebs Swiss. First operated in 1982.

Sedivy, V. M., *Ind. Miner. (London),* 1996, (343), 73.

Salsigne   A *cyanide process for extracting gold from ores containing arsenic or antimony. Pre-treatment with a lime slurry reduces cyanide losses.

Yannopoulos, J. C., *The Extractive Metallurgy of Gold,* Van Nostrand Reinhold, New York, 1991, 156.

Salt   An obsolete, two-stage process for obtaining chlorine and sodium nitrate from sodium chloride by the intermediary of nitrosyl chloride. In the first stage, the sodium chloride was reacted with nitric acid, producing nitrosyl chloride and chlorine:

$$3NaCl + 4HNO_3 = NOCl + Cl_2 + 2H_2O + 3NaNO_3$$

In the second, the nitrosyl chloride was either reacted with sodium carbonate:

$$3NOCl + 2Na_2CO_3 = NaNO_3 + 3NaCl + 2CO_2 + 2NO$$

or oxidized with oxygen:

$$2NOCl + O_2 = N_2O_4 + Cl_2$$

The sodium nitrate was used as a fertilizer. The evolution of the process was complex; the book referenced below lists 63 patents relating to it. It was installed by the Solvay Process Company at Hopewell, VA in the 1930s; it was in operation there, subsequently under the management of the Allied Chemical Corporation until the 1950s.

Fogler, M. F., *Chlorine, Its Manufacture, Properties and Uses,* Sconce, J. S., Ed., Reinhold Publishing, New York, 1962, 235.

Salt-cake   One of the two processes comprising the *Leblanc process for making sodium carbonate. Salt-cake was the colloquial name for sodium sulfate.

SAMEX   A process for removing traces of mercury from the waste brine from the *chlor-alkali process.

Nogueira, E. D., Regife, J. M., Melendo, J. F. J., in *Modern Chlor-alkali Technology*, Vol. 1, Coulter, M. O., Ed., Ellis Horwood, Chichester, England, 1980, 85.

Samica   *See* Bardet.

Sandwich desulfurization   A *hydrotreating process for removing sulfur compounds from petroleum streams. The sulfur compounds are first hydrogenated and then absorbed in a train of three catalyst beds—the "sandwich." In the first bed, zinc oxide absorbs hydrogen sulfide and reactive sulfur compounds; in the second, cobalt molybdate on alumina hydrogenates non-reactive thiophenes, forming hydrogen sulfide; in the third, zinc oxide absorbs the hydrogen sulfide from the second bed. Developed and offered by ICI, particularly for use in the *ICI Steam Naphtha Reforming process.

Saniter   A modification of the *Basic Open Hearth process for reducing the sulfur content of the steel product by adding relatively large quantities of limestone and calcium chloride. Invented by E. H. Saniter at the Wigan Coal & Iron Company in England in 1892, and used there and in Germany for approximately 20 years until superseded by the use of calcium fluoride as a flux.

British Patent 8,612 (1892).
Barraclough, K. C., *Steelmaking 1850–1900*, The Institute of Metals, London, 1990, 271.

SAPIC   A process used in metal foundries for curing resin/sand mixtures used in making molds. The resin is usually an unsaturated polyester resin. In the SAPIC process the resin is hardened by means of an organic peroxide, or hydrogen peroxide, which is activated by sulfur dioxide gas when required.

saponification   [Latin, **Sapo,** soap]   A process for making soap by the alkaline hydrolysis of animal or vegetable fats, using aqueous sodium or potassium hydroxide. Glycerol is a co-product. The term is now more generally used for any alkaline ester hydrolysis:

$$R \cdot COOR' + NaOH = R \cdot COONa + R'OH$$

where R is a long-chain alkyl group and R′OH an alcohol or polyol such as glycerol. *See also* Kettle, Twitchell.

Sapoxal   A pulp-bleaching process used in the paper industry. The bleaching agent is oxygen and the process is operated under alkaline conditions.

Sapozhnikov   *See* carbonization.

Sapper   An obsolete process for making phthalic anhydride by oxidizing *o*-xylene, using a mercury sulfate catalyst. Invented by E. Sapper in 1891 in the course of searching for a commercial route to indigo, and used until the catalytic gas-phase oxidation of naphthalene was introduced in 1925.

Franck, H.-G. and Stadelhofer, J. W., *Industrial Aromatic Chemistry*, Springer-Verlag, Berlin, 1988, 266.

SAR   [Sulfuric Acid Recovery]   A process for purifying and concentrating used sulfuric acid for re-use. The acid is heated with oxygen at 1,040°C to convert the acid to sulfur dioxide. This is then oxidized over a vanadium-containing catalyst to sulfur trioxide, which is dissolved in fresh sulfuric acid to give 98 percent acid. Developed by L'Air Liquide and ICI. First demonstrated in 1991 at a methyl methacrylate plant in Taiwan.

Eur. Chem. News, 1991, **57**(1501), 34.
Chem. Br., 1992, **28**(3), 216.

Sarex (1)   [Saccharide extraction]   A version of the *Sorbex process, for separating fructose from mixtures of fructose and glucose. The usual feed is corn syrup. The adsorbent is either a proprietary zeolite or an ion-exchange resin. Unlike all the other Sorbex processes, the solvent is water. The process depends on the tendency of calcium and magnesium ions to complex with fructose. The patents describe several methods for minimizing the dissolution of silica from the zeolite. The process is intended for use with a glucose isomerization unit, so that the sole product from corn syrup is fructose. Invented by UOP in 1976; by 1990, five plants had been licensed.

British Patent 1,574,915.
U.S. Patent 4,248,737.
Ching, C. B. and Ruthven, D. M., *Zeolites,* 1988, **8,** 68.

SAREX (2)   A process for recovering oil from wash liquors. Developed by Separation & Recovery Systems.

Hydrocarbon Process., 1993, **72**(8), 98.

SARP   [Sulphuric acid recovery process]   A method for recovering sulfuric acid which has been used for *alkylation, for re-use. The acid is reacted with propylene, yielding dipropyl sulfate, which is extracted from the acid tar with isobutane. It is not necessary to hydrolyze the sulfate to sulfuric acid because the sulfate itself is an active alkylation catalyst.

Oil & Gas J., 1967, **65**(1), 48.

SAS   [Sasol advanced synthesis]   A process for converting synthesis gas to petroleum and light olefins. Developed and operated by Sasol in South Africa.

Oil & Gas J., 1997, **95**(25), 16.

SASOL   [Suid-Afrikaans Sintetiese Olie]   Not a process but a large coal gasification complex in South Africa, operated by the South African Oil and Gas Corporation. It first operated in 1955 but took several years to be fully commissioned. A Lurgi fixed-bed gasification unit is used for the primary process. Downstream processes include the following ones, described under their respective names: Arge, Fischer-Tropsch, Rectisol, Sulfolin, Synthol.

Report of the Committee on Coal Derivatives, HMSO, CMND 1120, 1960.
Mako, P. F. and Samuel, W. A., in *Handbook of Synfuels Technology,* Meyers, R. A., Ed., McGraw-Hill, New York, 1984, 2-7.
Chem. Eng. (N.Y.), 1995, **102**(12), 70.
Appl. Catal. A: Gen., 1997, **155**(1), N5.

Satco   A process for recovering contaminated sulfuric acid. The acid is cracked at 1,000°C and the resulting sulfur dioxide is cooled, purified, dried, and reconverted to sulfuric acid. Developed by Rhône-Poulenc in the 1990s. Nine plants were operating in 1997.

Chem. Eng. News, 1997, **75**(27), 16.

SBA   [Société Belge de l'Azote]   A process for making acetylene by the partial combustion of methane. It differs from similar processes in using liquid ammonia as a selective solvent for the product. Invented by F. F. A. Braconier and J. J. L. E. Riga at the Société Belge de l'Azote et des Produits Chimiques, Liège, and first operated at Marly, Belgium, in 1958.

U.S. Patent 3,019,271.
Miller, S. A., *Acetylene: Its Properties, Manufacture and Uses,* Vol. 1, Ernest Benn, London, 1965, 465.
Tedeschi, R. J., *Acetylene-based Chemicals from Coal and Other Natural Resources,* Marcel Dekker, New York 1982, 22.

SBA-HT   [Société Belge de l'Azote-Haldor Topsoe]   A process for converting LPG to *syngas rich in hydrogen. Two cracking processes are conducted in two zones of one reactor. In the first zone, the LPG is autothermally cracked with steam and oxygen. In the second, the products from the first are catalytically cracked. It was operated in France and Belgium in the 1960s.

SBA-Kellogg   A pyrolytic process for making ethylene and acetylene from saturated hydrocarbon gases, similar to the *Hoechst HTP process but with the addition of steam which increases the yield of C_2 gases.

>Barry, M. J., Fox, J. M., Grover, S. S., Braconier, F., and Leroux, P., *Chem. Eng. Prog.,* 1960, **56**(1), 39.
>Asinger, F., *Mono-olefins: Chemistry and Technology,* translated by B. J., Hazzard, Pergamon Press, Oxford, 1968, 180.

SBK   [Sinclair-Baker-Kellogg]   A petroleum reforming process which uses a regenerable platinum catalyst.

SCA-Billerud   A variation of the *sulfite process for making paper from wood, in which the waste hydrogen sulfide is burnt to sulfur dioxide and used to make sulfuric acid.

>Rydholm, S. A., *Pulping Processes,* Interscience, New York, 1965, 809.

SCG   *See* Shell Coal Gasification.

Schaffner   Also called Schaffner-Helbig. A process for recovering sulfur from the residues from the Leblanc process. Operated in Aussig, Bohemia, in the 1860s. *See also* Mond.

>*Chem. Ind. (London),* 1985, 3 Jun, 367.

Scheibler   A process for extracting sucrose from molasses residues, based on the precipitation of strontium saccharate. Operated in Germany in the late 19th century.

Scheidemandel   A process for converting bone glue to a bead form. The hot, concentrated glue solution is dropped down a tower filled with an immiscible liquid such as a petroleum fraction or tetrachloroethylene.

Schenk-Wenzel   *See* carbonization.

Schlempe   [German, meaning residues from fermentation processes]   An obsolete process for making sodium cyanide from sugar beet residues.

>*Kirk-Othmer's Encyclopedia of Chemical Technology,* 3rd ed., Vol. 7, John Wiley & Sons, New York, 1979, 324.

Schloesing-Rolland   An early variant of the *ammonia-soda process for making sodium carbonate. Operated near Paris in 1857 and then in Middlesbrough, England, for approximately 20 years until supplanted by the *Solvay process.

>Watts, J. I., *The First Fifty Years of Brunner, Mond & Co.,* Brunner, Mond, Winnington, England, 1923, 18.
>Hardie, D. W. F. and Pratt, J. D., *A History of the Modern British Chemical Industry,* Pergamon Press, Oxford, 1966, 85.

Schmidt   Also called Meissner, and Schmidt-Meissner. The first continuous process for nitrating glycerol to nitroglycerol. Invented by A. Schmidt in Austria in 1927. *See also* Biazzi.

Schmidt-Meissner   *See* Schmidt.

Schoch   A process for making acetylene by subjecting aliphatic hydrocarbons to a silent electric discharge. Developed by E. P. Schoch at the University of Texas.

>Daniels, L. S., *Pet. Refin.,* 1950, **29**(9), 221.

Scholler   A process for making glucose from wood by acid hydrolysis of the cellulose component under pressure. Invented by H. Scholler in Germany in 1929. Operated in Germany in the 1920s and in Russia in the 1970s.

> French Patent 706,678.
> Worthy, W., *Chem. Eng. News*, 1981, **59**(14), 35.

Scholler-Tornesch   A process for making single-cell protein from wood. The wood is *saccharified by heating with dilute sulfuric acid. The resulting sugars are then fermented to ethanol, using the common yeast *Saccharomeces cerevisae*. The process was developed and used in Germany in the 1920s and 1930s. *See also* Heiskenskjold.

> Litchfield, J. H., *CHEMTECH*, 1978, **8**, 218.
> *Riegel's Handbook of Industrial Chemistry*, 9th ed., Kent, J. A., Ed., Van Nostrand Reinhold, New York, 1992, 255.

Schröder-Grillo   Also known as Grillo-Schröder. An early version of the *contact process for making sulfuric acid. The catalyst was magnesium sulfate impregnated with platinum. The process was invented in 1899 by A. Hecksher at the New Jersey Zinc Company and first used at its plant in Mineral Point, WI, in 1901; this was the first use of the contact process in the United States. In the United Kingdom it was first used in Widnes in 1917. *See also* Mannheim (2).

> Miles, F. D., *The Manufacture of Sulfuric Acid (Contact Process)*, Gurney & Jackson, London, 1925, Chap. 8.

Schroeder   An electrolytic process for recovering chlorine from waste hydrochloric acid.

Schulte   A process for removing ammonia from the gases produced in the coking of coal. The ammonia is removed by scrubbing with water, then distilled from the water and incinerated. Invented by E. V. Schulte and commercialized by the Koppers Company.

> Schulte, E. V., *Blast Furn. Coke Oven Raw Mater. Proc.*, 1958, **17**, 237 (*Chem. Abstr.*, **53**, 9944).

Schwarting   Also called Uhde/Schwarting. An anaerobic fermentation process for treating aqueous wastes containing high concentrations of organic materials. Two fermenters are used, operated at different temperatures and acidities. In the first, insoluble materials are brought into solution and most of the organic matter is converted to acids and alcohols. In the second, methane and carbon dioxide are produced. Developed in Germany by Geratebau Schwarting and the Fraunhofer Institute for Boundary Layer Research; engineered and offered by Uhde. Three plants were operating in Germany in 1994 for treating sewage sludges and wastewaters.

Schwenzfeier-Pomelée   A process for purifying beryllium and producing glassy beryllium fluoride. Beryllium hydroxide is dissolved in aqueous ammonium hydrogen fluoride; various metal impurities are removed by successive precipitations, and ammonium fluoroberyllate is crystallized under vacuum. When this is heated, ammonium fluoride vaporizes and molten beryllium fluoride remains.

Sclair   A process for polymerizing ethylene. Depending on the co-monomer used, the product can be linear low-density polyethylene (LLDPE) or high-density polyethylene (HDPE). Developed by DuPont in 1960 and widely licensed. Engineered by Uhde under the name *Sclairtech. Nova Chemicals (Alberta) acquired the technology in 1994.

Sclairtech   An advanced version of the *Sclair ethylene polymerization process, using a *Ziegler-Natta catalyst and multiple reactors. Announced in 1996. The first commercial plant will be built in Alberta by Amoco Canada and Nova, and is scheduled for completion in 2000.

> *Eur. Chem. News*, 1996, **66** (1744), 4.

Scientific Design   A chemical engineering company, founded in New York in 1946 by R. Landau and H. Rehnberg. It developed many processes, of which the first and perhaps the best known was that for oxidizing ethylene to ethylene oxide, using a silver catalyst. Later it merged with Halcon Corporation, to become the Halcon SD Group. *See* Halcon, Oxirane.

> Spitz, P. H., *Petrochemicals, the Rise of an Industry,* John Wiley & Sons, New York, 1988, 319.

Sconox   A catalytic process for oxidizing oxides of nitrogen and carbon monoxide. The catalyst is in the form of a ceramic honeycomb coated with platinum and containing potassium carbonate. The platinum oxidizes the carbon monoxide to carbon dioxide, and the potassium carbonate absorbs the NO_x. Developed in 1995 by Sunlaw Energy Corporation, CA, and Advanced Catalytic Systems, TN.

> *Chem. Eng. (N.Y.),* 1995, **102**(12), 25.

Scot   [**S**hell **C**laus **O**ff-gas **T**reatment]   A variation on the *Claus process for removing hydrogen sulfide from gas streams, in which residual sulfur dioxide in the off-gases is reduced with methane or hydrogen and the resulting hydrogen sulfide is returned to the start of the process. Other features are the catalytic conversion of organic sulfur compounds to hydrogen sulfide, and the use of an alkanolamine for selectively absorbing this. Developed by Shell International Petroleum Maatschappij. In 1996, 130 units were operating. *See also* Beavon.

> Naber, J. E., Wesselingh, J. A., and Groendaal, W., *Chem. Eng. Prog.,* 1973, **69**(12), 29.
> Swaim, C. D., Jr., in *Sulfur Removal and Recovery from Industrial Processes,* Pfeiffer, J. B., Ed., American Chemical Society, Washington, D.C., 1975, 111.
> Kohl, A. L. and Riesenfeld, F. C., *Gas Purification,* 4th ed., Gulf Publishing, Houston, TX, 1985, 749.
> *Hydrocarbon Process.,* 1996, **75**(4), 136.

SCR   [**S**elective **C**atalytic **R**eduction]   A general term for processes which destroy nitrogen oxides in gaseous effluents by reacting them with ammonia in the presence of a catalyst:

$$4NO + 4NH_3 + O_2 = 4N_2 + 6H_2O$$

$$6NO_2 + 8NH_3 = 7N_2 + 12H_2O$$

For clean gaseous effluents, such as those from nitric acid plants, the preferred catalyst is mordenite. For flue-gases containing fly ash, the preferred catalyst is titania-vanadia. The process was developed in Japan in the mid-1970s by a consortium of Hitachi, Babcock-Hitachi, and the Mitsubishi Chemical Company, and by the Sakai Chemical Industry Company. It is widely used in power stations in Japan and Germany. *See also* SNCR.

> Matsuda, S., Takeuchi, M., Hishinuma, T., Nakajima, F., Narita, T., Watanabe, Y., and Imanari, M., *J. Air Pollut. Control Assoc.,* 1978, **28,** 350.
> Offen, G. R., Eskinazi, D., McElroy, M. W., and Maulbetsch, J. S., *J. Air Pollut. Control Assoc.,* 1987, **37,** 864.
> Ritzert, G., *Tech. Mitt.,* 1987, **80,** 602.
> Bosch, H. and Janssen, F., *Catal. Today,* 1988, **2,** 392.
> Mukherjee, A. B., in *Environmental Oxidants,* Nriagu, J. O. and Simmons, M. S., Eds., John Wiley & Sons, New York, 1994, 585.
> *Hydrocarbon Process.,* 1994, **73**(8), 67.
> Radojevic, M., *Chem. Br.,* 1998, **34**(3), 30.

SCWO   [supercritical water oxidation]   A generic name for processes which destroy organic wastes in water by oxidation under supercritical conditions. The first such process was *MODAR, invented in 1980. Since then, several other companies have introduced competing processes.

> *Oil & Gas J.,* 1994, 92(44), 44.
> Luck, F., *Catal. Today,* 1996, **27**(1–2), 195.

Chem. Eng. (N.Y.), 1996, **103**(3), 21.

Ding, Z. Y., Frisch, M. A., Li, L., and Gloyna, E. F., *Ind. Eng. Chem. Res.*, 1996, **35**(10), 3257.

Chem. Eng. (Rugby, England), 1996, (604), 9.

SDA [Spray drier absorber]   A *flue-gas desulfurization process in which an aqueous suspension of lime is injected into a spray drier. Basically similar to *DRYPAC. Developed by Niro Atomiser, Denmark. In 1986 it was in use in 16 plants in Austria, Denmark, Germany, Italy, Sweden, China, and the United States.

SDR   *See* DR.

SDS [Sulzer Dainippon Sumitomo]   A continuous process for polymerizing styrene. The reactants are mixed in a static mixer, which gives a very uniform time/temperature history for the polymer. Developed jointly by Sulzer, Dainippon Ink & Chemicals, and Sumotomo Heavy Industries. Offered for license in 1990.

Eur. Chem. News, 1990, 23 Jul, 20.

SDW [Solvent de-waxing]   A general term for processes which remove linear paraffinic hydrocarbons from petroleum fractions by solvent extraction.

Seabord   A process for removing hydrogen sulfide from coke-oven and oil refinery gases by scrubbing with aqueous sodium carbonate solution. The solution is regenerated by blowing air through it. In its original version the hydrogen sulfide was simply vented to the atmosphere. In later developments, the air containing the hydrogen sulfide was used as the combustion air for boilers or was passed into the coke-oven. Developed by Koppers Company in 1920 and now obsolete.

British Patent 391,833.

Sperr, F. W., Jr., *Proc. Am. Gas. Assoc.*, 1921, Sept. (*Chem. Abstr.*, **16**, 482).

Kohl, A. L. and Riesenfeld, F. C., *Gas Purification*, 4th ed., Gulf Publishing, Houston, TX, 1985, 187.

Seacoke   A process for making tar and coke by carbonizing mixtures of coal and petroleum residuum. The tar would be used in an oil refinery and the coke would be used for generating electricity. The process was sponsored by the U.S. Office of Coal Research 1964–1969; the work was carried out by FMC Corporation, Atlantic Richfield Company, and Blaw-Knox Company. Results from the pilot plant were encouraging but the project was abandoned because the benefits were judged insufficient to justify the complexity.

Aristoff, E., Rieve, R. W., and Shalit, H., in *Chemistry of Coal Utilization*, 2nd Suppl. Vol., Elliott, M. A., Ed., John Wiley & Sons, New York, 1981, 984.

Séailles-Dyckerhoff   A process for extracting aluminum from clays and other aluminous minerals rich in silica. The ore is calcined with limestone in a rotary kiln. The product is leached with aqueous sodium carbonate, yielding sodium aluminate solution, from which alumina is precipitated by carbon dioxide. Invented by J. C. Séailles and W. R. G. Dyckerhoff in 1938; piloted in Tennessee in 1942, South Carolina in 1945, and Germany during World War II. *See also* Pedersen.

U.S. Patent 2,248,826.

O'Connor, D. J., *Alumina Extraction from Non-bauxitic Materials*, Aluminium-Verlag, Düsseldorf, 1988, 233.

Sealosafe   A family of processes for encapsulating inorganic and organic wastes in a cementitious material suitable for landfill. The product, known as Stablex, is made from a cement and an aluminosilicate and may incorporate pulverized fly ash. Developed by C. Chappell in the United Kingdom in the 1970s and now operated in a number of countries. Offered by the Stablex Corporation, Radnor, PA. The environmental acceptability of the product has since been criticized.

Schofield, J. T., in *Toxic and Hazardous Waste Disposal,* Vol. 1, Pojasek, R. J., Ed., Ann Arbor Science, Ann Arbor, MI, 1979, Chap. 15.
Environmental Data Services Report, 1989, (173), 8.
Environmental Data Services Report, 1995, (240), 15.

Sec-Feed   A process for converting used lubricating oils into feeds for catalytic crackers. Essentially it removes water, light hydrocarbons, phosphorus and silicon compounds, and heavy metals. Developed by Chemical Engineering Partners in 1997. Installations in Southern California and North Wales were planned for late 1998.

Oil & Gas J., 1997, **95**(49), 61.

Sedifloc   A water-treatment process based on flocculation. Offered by Lurgi.

Sedimat   A water-treatment process based on sedimentation. Offered by Lurgi.

SEGAS   [Southeastern **Gas**]   Also written Seagas. An oil gasification process in which oil is sprayed into a stream of superheated steam and cracked over a calcia/magnesia catalyst. Developed by the Southeastern Gas Board, United Kingdom.

Stanier, H. and McKean, J. B., *Institution of Gas Engineers,* Publn. No. 457, 1954 (*Chem. Abstr.,* **49**, 3508).
Ward, E. R., *Institution of Gas Engineers,* Publn. No. 515, 1957 (*Chem. Abstr.,* **52**, 3308).
Claxton, G., *Benzoles, Production and Uses,* National Benzole & Allied Products Assoc., London, 1961, 94.
British Petroleum Co., *Gas Making and Natural Gas,* British Petroleum Co., London, 1972, 83.

Selectoforming   A process for increasing the octane rating of a petroleum fraction by selectively cracking the *n*-pentane and *n*-hexane in it. The catalyst is a metal-loaded synthetic zeolite. Developed by Mobil Corporation and first commercialized in the mid-1960s.

Chen, N. Y. and Degnan, T. F., *Chem. Eng. Prog.,* 1988, **84**(2), 32.

Selectopol   A process for converting isobutene into "polymer gasoline," i.e. a mixture of branched-chain $C_6 - C_{12}$ hydrocarbons, using an acid catalyst. Offered for license by the Institut Français du Pétrole.

Hydrocarbon Process., 1980, **59**(9), 219.

Selectox   Also called BSR/Selectox. A process for converting hydrogen sulfide in refinery gases to elemental sulfur. The gases are passed over a fixed bed of a proprietary catalyst (Selectox 33) at 160 to 370°C. Claimed to be better than the *Claus process in several respects. Often used in conjunction with the *Beavon process. Developed by the Union Oil Company of California and the Ralph M. Parsons Company, and first operated in 1978. Thirteen units were operating in 1996.

Beavon, D. K., Hass, R. H., and Muke, B., *Oil & Gas J.,* 1979, **77**(11), 76.
Hass, R. H., Ingalls, M. N., Trinker, T. A., Goar, B. G., and Purgason, R. S., *Hydrocarbon Process.,* 1981, **60**(5), 104.
Kohl, A. L. and Riesenfeld, F. C., *Gas Purification,* 4th ed., Gulf Publishing, Houston, TX, 1985, 743.
Wieckowska, J., *Catal. Today,* 1995, **24**(4), 444.
Hydrocarbon Process., 1996, **75**(4), 136.

Selexol   Also called Selexol Solvent. A process for removing acid gases from hydrocarbon gas streams by selective absorption in polyethylene glycol dimethyl ether (DMPEG). It absorbs carbon dioxide, hydrogen sulfide, carbonyl sulfide, and mercaptans. Absorption takes place in a counter-current extraction column under pressure. The solvent is regenerated by "flashing" (evaporation) or by "stripping" (passing an inert gas through it). It has been used for removing carbon dioxide from *syngas, natural gas, and coal gas. Developed by Allied

Chemical Corporation in the 1960s and now offered for license by Union Carbide Chemicals & Plastics Company. Over 50 units were operating in 1996. *See also* SOLINOX.

Sweny, J. W. and Valentine, J. P., *Chem. Eng. (N.Y.)*, 1970, **77**, 54.

Valentine, J. P., *Oil & Gas J.*, 1974, **72**(46), 60.

Judd, D. K., *Hydrocarbon Process.*, 1978, **57**(4), 122.

Kohl, A. L. and Riesenfeld, F. C., *Gas Purification*, 4th ed., Gulf Publishing, Houston, TX, 1985, 856.

Hydrocarbon Process., 1996, **75**(4), 137.

Selexsorb   A five-stage process for purifying ethylene before converting it to polyethylene. Developed by the Industrial Chemicals Division of the Aluminum Company of America. More than 50 installations were operating in 1996. The name is now used for a family of selective adsorbents based on alumina produced by Alcoa. The range includes Selexsorb CD, CDO, CDX, COS, SPCl, HCl.

Hydrocarbon Process., 1996, **75**(4), 137.

SELOP C4   A process for upgrading the C_4 petroleum fraction by selective catalytic hydrogenation. Different catalysts, containing palladium on alumina, are used for different feedstock compositions. Developed by BASF and used in its Antwerp plant since 1994.

Büchele, W., Roos, H., Wanjek, H., and Müller, H. J., *Catal. Today*, 1996, **30**, 33.

Selox   [Selective oxidation]   A process for selectively oxidizing methane to *syngas using a proprietary heterogeneous catalyst at temperatures up to 1,000°C. Developed on a laboratory scale by TRW, CA, partly financed by the U.S. Department of Energy in 1983.

Chem. Eng. News, 1984, **62**(2), 5.

Chem. Eng. (N.Y.), 1984, **91**(13), 157.

Semet-Solvay   The Semet-Solvay Engineering Corporation, Syracuse, NY, was best known for its coke-oven technology, developed from the end of the 19th century. The eponymous process was a cyclic, non-catalytic process for making fuel gas from oil.

British Petroleum, *Gasmaking* (rev. ed.), British Petroleum, London, 1965, 75.

Sendzimir   *See* metal surface treatment.

SEPACLAUS   [Separation Claus]   An integrated process for removing hydrogen sulfide from coke-oven gases and converting it to elemental sulfur by the *Claus process. It also catalytically decomposes the ammonia present. Offered by Krupp Koppers, Germany.

Sepaflot   A process for removing solids and oil from wastewater by a combination of *Activated Sludge treatment and dissolved air flotation. Offered by Lurgi.

Sepasolv MPE   [Methyl isopropyl ester]   A variation on the *Selexol process, using the methyl isopropyl ethers of polyethylene glycol as the solvent. Developed by BASF. Four commercial plants were operating in 1985, removing hydrogen sulfide from natural gas.

Wölfer, W., *Hydrocarbon Process.*, 1982, **61**(11), 193.

Kohl, A. L. and Riesenfeld, F. C., *Gas Purification*, 4th ed., Gulf Publishing, Houston, TX, 1985, 865.

Serpek   A nitrogen fixation process using aluminum nitride. A mixture of bauxite and coke is heated in nitrogen at 1,800°C to produce aluminum nitride; this yields ammonia on hydrolysis by boiling with aqueous potassium aluminate:

$$Al_2O_3 + 3C + N_2 = 2AlN + 3CO$$

$$2AlN + 3H_2O = 2NH_3 + Al_2O_3$$

Invented by O. Serpek in Germany in 1909.

British Patents 15,996; 15,997 (1909).

SFE [Supercritical Fluid Extraction] *See* ROSE (1).

SFGD [Shell flue gas desulfurization]   A *flue-gas desulfurization process using a fixed bed of copper on alumina. The sulfur dioxide is desorbed with a reducing gas.

Dautzenberg, F. M., Nader, J. E., and van Ginneken, A. J. J., *Chem. Eng. Prog.,* 1991, **67**(8), 86.
Speight, J. G., *Gas Processing,* Butterworth Heinemann, Oxford, 1993, 316.

SGP [Shell gasification process]   A process for converting fuel oils into *syngas; the sulfur is isolated as the element. Developed by Shell, Amsterdam, from the 1950s and now widely used. Licensed by Lurgi Öl Gas Chemie.

shale oil extraction   Those named processes described in this dictionary are: Galoter, HY-TORT, Kiviter, Paraho, Petrosix, PFH, Pumpherston, TOSCO II, TOSCOAL.

Shanks   An obsolete process for extracting sodium nitrate from caliche, a Chilean mineral. The ore was leached with sodium chloride solution at 70°C and the sodium chloride and nitrate were separated by fractional crystallization. *See also* Guggenheim.

Sharple   A process for purifying paraffin wax by crystallization from a petroleum fraction at low temperature.

Sharples   A continuous soapmaking process using centrifuges, invented in 1939 by A. T. Scott at the Sharples Corporation, Philadelphia.

U.S. Patents 2,300,749; 2,300,751; 2,336,893.

Shawinigan   A process for making acetic acid by oxidizing acetaldehyde by atmospheric oxygen in the presence of manganese acetate. Operated by the Shawinigan Chemical Company, at Shawinigan Falls, Quebec, using acetaldehyde made from acetylene.

Cadenhead, A. F. G., *Chem. Metall. Eng.,* 1933, **40**(4), 184.
Dumas, T. and Bulani, W., *Oxidation of Petrochemicals: Chemistry and Technology,* Applied Science Publishers, London, 1974, 26.

Shell-Adip   *See* Adip.

Shell Coal Gasification   Also called SCG and SCGP. A coal gasification process in which powdered coal is reacted with oxygen and steam at high temperature and pressure. The high temperature causes most of the ash to melt and flow down the reactor wall into a water-filled compartment. The product gas is mainly a mixture of carbon monoxide and hydrogen, suitable for the synthesis of ammonia, methanol, and hydrocarbons. Piloted at the Shell laboratory, Amsterdam, and at the Deutsche Shell refinery at Hamburg, since 1978.

Vogt, E. V., Weller, P. J., and Vanderburgt, M. J., in *Handbook of Synfuels Technology,* Meyers, R. A., Ed., McGraw-Hill, New York, 1984, 3-27.
Cornils, B., in *Chemicals from Coal: New Processes,* Payne, K. R., Ed., John Wiley & Sons, Chichester, England, 1987, 19.

Shell Deacon   An improved version of the *Deacon process for oxidizing hydrogen chloride to chlorine, using a catalyst containing the mixed chlorides of copper, potassium, and rare earths. Formerly operated in The Netherlands and still in operation in India.

Tozuka, Y., in *Science and Technology in Catalysis,* Izumi, Y., Aral, Y., and Iwamoto, M., Eds., Elsevier, Amsterdam, 1994, 45.
Wittcoff, H. A. and Reuben, B. G., *Industrial Organic Chemicals,* John Wiley & Sons, New York, 1996, 105.
Weissermel, K. and Arpe, H.-J., *Industrial Organic Chemistry,* 3rd ed. VCH Publishers, Weinheim, Germany, 1997, 219.

Shell Glycerol   The Shell Development Company has developed three routes for making glycerol from propylene. The first begins by chlorinating propylene to make allyl chloride, which is converted to glycerol via epichlorhydrin. The second and third both involve acrolein as an intermediate, the second reacting it with 2-propanol and the third with 2-butanol. The second of these processes became known as the Shell Glycerol process. The successive reactions are:

$$propylene + oxygen \rightarrow acrolein + water$$

$$acrolein + 2\text{-}propanol \rightarrow allyl\ alcohol + acetone$$

$$allyl\ alcohol + hydrogen\ peroxide \rightarrow glycerol$$

The third route was used at the Shell plant at Norco, LA, until it was closed in 1980.

Shell-Koppers   A coal gasification process, using steam and air or oxygen. Operated at the Shell oil refinery in Hamburg.

Eur. Chem. News, Petrochem. Suppl., 1981, 14 Dec.

Shellperm   A process for rendering sandy soils impermeable to water by pumping in a bitumen emulsion with a coagulant which is effective after a predetermined period. Used in the construction of dams.

The Petroleum Handbook, 3rd ed., Shell Petroleum Co., London, 1948, 586.

Sheppard   *See* metal surface treatment.

Sherardizing   [After the inventor, **Sherard** Cowper-Cowles, 1900]   A process for coating iron articles with zinc. The articles are placed in a sealed drum with zinc dust and sand. The drum is rotated and maintained at a temperature below the melting point of zinc. The mechanism is not understood. In 1990 the world consumption of zinc for this process was several thousand tons. *See* metal surface treatment.

British Standard BS 4921.

Sherpol   A process for making polypropylene, developed and licensed by Himont.

Sherritt-Cominco   A process for extracting copper from chalcopyrite, $CuFeS_2$. The ore is reduced with hydrogen, the iron leached out with sulfuric acid, the residual Cu_5FeS_4 dissolved in concentrated sulfuric acid, and the copper isolated by electrowinning or hydrogen reduction. Pilot testing was complete in 1976.

Maschmeyer, D. E. G., Kawulka, P., Milner, E. F. G., and Swinkels, G. M., *J. Met.,* 1978, **27**(7), 27.

Sherritt-Gordon   The Canadian company, Sherritt-Gordon Mines, has developed a number of hydrometallurgical leaching processes known by its company name. The essential feature of these processes is based on the observation of F. A. Forward that sulfide ores will dissolve in aqueous ammonia in the presence of oxygen. One such process is for leaching nickel from sulfide ores, using aqueous ammonia and oxygen under pressure. The hexammino nickel (II) ion is formed:

$$NiS + 2O_2 + 6NH_3 = Ni(NH_3)_6^{2+} + SO_4^{2-}$$

Another Sherritt-Gordon process is for leaching zinc from sulfide ores, again using oxygen under pressure.

Morgan, S. W. K., *Zinc and Its Alloys and Compounds,* Ellis Horwood, Chichester, England, 1985, 124.

Gupta, C. K. and Mukherjee, T. K., *Hydrometallurgy in Extraction Processes,* Vol. 1, CRC Press, Boca Raton, FL, 1990, 26,87.

Shift   *See* reforming.

Shimer   *See* metal surface treatment.

SHOP   [Shell Higher Olefins Process]   A process for producing α-olefins by oligomerizing ethylene, using a proprietary rhodium/phosphine catalyst. The α-olefins can then be isomerized to internal olefins as required. Invented by W. Keim in the Institut für Technische Chemie und Petrolchemie, Aachen, in the 1970s. The first plant was built in Geismar, LA, in 1979; the second in Stanlow, Cheshire, in 1982. Licensed worldwide by a consortium of Union Carbide, Davy-McKee, and Johnson Matthey.

> Freitas, E. R. and Gum, C. R., *Chem. Eng. Prog.*, 1979, **75**(1), 73.
> Sherwood, M., *Chem. Ind. (London)*, 1982, (24), 994.
> Weissermel, K. and Arpe, H.-J. *Industrial Organic Chemistry*, 3rd ed. VCH Publishers, Weinheim, Germany, 1997, 77.

Shoppler   A process for extracting tungsten from scheelite. The ore is fused with sodium carbonate, forming sodium tungstate. This is leached with water, the solution acidified with hydrochloric acid, and hydrated tungsten oxide precipitated by boiling. The metal is produced by reducing the oxide with carbon at a high temperature.

SHP   A process for purifying 1-butene by selective hydrogenation of C_4 streams in petroleum refineries. A hetrogeneous palladium catalyst is used. Developed in Hüls and used in 1989 in Germany, the United States, and Japan. In 1991 the licensing rights were acquired by UOP.

> Derrien, M., Bronner, C., Cosyns, J., and Leger, G., *Hydrocarbon Process.*, 1979, **58**(5), 175.

SHS   [Self-propagating high-temperature synthesis]   A process for manufacturing ceramics and intermetallic compounds by exothermic reactions in which the heat of reaction is large enough to sustain the propagation of a combustion wave through the reactants. The reactants can be mixed powders, or a powder with a gas. It has been used to synthesize TiB_2, ZrB_2, TiC, and AlN from the elements, and to make a number of refractory composites of oxides with metals. Invented in 1967 at the Institute of Chemical Physics, Moscow, by I. P. Borovinskaya, A. G. Merzhanov, and V. M. Shkiro. The Institute of Scientific and Industrial Research at Osaka University has studied the process conducted in various gases under pressure.

> U.S. Patent 3,726,643.
> British Patent 1,321,084.
> Merzhanov, A. G., and Borovinskaya, I. P., *Doklady Akad. Nauk SSSR* (Engl. Transl.), 1972, **204**(2), 429.
> Crider, J. F., *Ceram. Eng. Sci. Proc.*, 1982, **3**, 519.
> Merzhanov, A. G., in *Combustion and Plasma Synthesis of High-temperature Materials*, Munir, Z. A. and Holt, J. B., Eds., VCH Publishers, Weinheim, Germany, 1990, 1.
> *International Journal of Self-propagating High-temperature Synthesis*, Allerton Press, New York, from 1992.
> Avakayan, P. B., Nereseyan, M. D., and Merzhanov, A. G., *Amer. Ceram. Soc. Bull.*, 1996, **75**(2), 50.
> Parkin, I. P., *Chem. Ind. (London)*, 1997, (18), 725.
> He, C. and Stangle, G. C., *J. Mater. Res.*, 1998, **13**(1), 135.

SHU   [Saarberg-Holter-Lurgi]   A *flue-gas desulfurization process using wet limestone as the scrubbing medium, assisted by the addition of dilute formic acid. Developed by the companies named, and used in 11 power stations in Germany and Turkey in 1987.

SIAPE   [Société Industrielle d'Acide Phosphorique et d'Engrais]   A *Wet process for making phosphoric acid, based on a pair of coupled, stirred reactors. Operated by the eponymous company in Tunisia.

French Patent 1,592,005.

Becker, P., *Phosphates and Phosphoric Acid,* 2nd. ed., Marcel Dekker, New York, 1989, 349.

Siemens   A method for making ultra-pure silicon for semiconductors by thermally decomposing trichlorosilane. Invented in 1954 by F. Bischof at Siemens-Halska. In 1993 it was the major process used worldwide.

German Patent 1,102,117.

Yaws, C. L. and Hopper, J. R., in *Chemical Processing Handbook,* Marcel Dekker, New York, 1993, 939.

Siemens Open Hearth   *See* Open Hearth.

Siemens-Martin   A predecessor of the Siemens Open Hearth process. *See* Open Hearth.

Sieurin   An early process for making sponge iron. Layers of iron ore concentrate, lime, and coal were heated in covered crucibles.

SIIL   A direct reduction ironmaking process, using coal as the reductant. In 1997, five plants were operating in India and one in Peru. *See* DR.

Silamit P3   A cyclic *catalytic reforming process for making town gas from oil, similar to the *UGI Process. Developed by Gaz de France and built by Silamit Indugas, Düsseldorf, Germany.

Gas Making and Natural Gas, British Petroleum Co., London, 1972, 94.

Silver II   An electrolytic oxidation process for destroying traces of organic substances in water. The oxidizing agent is the silver ion in a nitric acid environment. Developed by AEA Technology, Oxford, and used for destroying war gases.

Chem. Eng. (N.Y.), 1996, **103**(5), 41.

Eur. Chem. News, 1996, **66**(1726), 23.

Eur. Chem. News, CHEMSCOPE, 1997, Jul, 18.

Simons   An electrochemical method for fluorinating organic compounds. First developed by J. H. Simons at Pennsylvania State College in 1941 but not announced until 1948 for reasons of national security. A direct current is passed through a solution of an organic compound in anhydrous hydrofluoric acid; hydrogen is evolved at the cathode and the organic material is fluorinated at the anode.

U.S. Patents 2,519,983; 2,594,272; 2,616,927.

Simons, J. H., *J. Electrochem. Soc.,* 1949, **95,** 47.

Simons, J. H., in *Fluorine Chemistry,* Vol. 1, Simons, J. H., Ed., Academic Press, New York, 1950, 414.

Burdon, J. and Tatlow, J. C., *Adv. Fluorine Chem.,* 1960, **1,** 129.

Pletcher, D. and Walsh, F. C., *Industrial Electrochemistry,* 2nd. ed., Chapman & Hall, London, 1960, 319.

Simplex   A process for reducing the carbon content of ferrochrome, an alloy of iron and chromium. Some of the alloy is oxidized by heating in air, and this is mixed in appropriate proportions with the remainder; on heating the mixture in a vacuum furnace the carbon volatilizes as carbon monoxide.

Sinclair-Baker   A naphtha reforming process, catalyzed by platinum on alumina. *See also* reforming.

SINI   Also known as the Double Steeping process. A variation of the *viscose process for making regenerated cellulose fibers, in which the treatment with sodium hydroxide is done in two stages, at different concentrations. Invented by H. Sihtola, around 1976.

Sihtola, H., *Pap. Puu.,* 1976, **58**(9), 534 (*Chem. Abstr.,* **86,** 56986).

SINOx [Siemens NO_x system]   A process for removing nitrogen oxides and dioxins from the exhausts of stationary diesel engines and truck engines. The catalyst is based on titania and is in the form of a honeycomb. The reducing agent is ammonia, generated from an aqueous solution of urea.

> Hums, E., Joisten, M., Müller, R., Sigling, R., and Spielmann, H., *Catal. Today,* 1996, **27**(1–2), 29.

Sinterna A process for stabilizing the calcium sulfate/sulfite waste produced by *FGD so that it may be used for landfill. The waste is mixed with ash from a coal-fired power station, pelletized, and sintered. Developed in the 1970s by Battelle Columbus Laboratories, OH, under contract with Industrial Resources. *See also* Fersona.

> U.S. Patent 3,962,080.
> Dulin, J. M., in *Toxic and Hazardous Waste Disposal,* Vol. 1, Pojasek, R. J., Ed., Ann Arbor Science, Ann Arbor, MI, 1979, Chap. 18.

Siroc [**Si**licate **roc**k]   Also known as the one-shot system. A chemical grouting system for hardening ground formations. Aqueous solutions of sodium silicate and formamide are mixed and injected into the ground; the formamide slowly reacts with the silicate, precipitating hydrated silica, which binds the soil particles together. Invented in 1961 by the Diamond Alkali Company. *See also* Joosten.

> U.S. Patent 2,968,572.

Sirofloc [**CSIRO floc**culation]   A process for purifying municipal water supplies by flocculation, developed by the Commonwealth Scientific & Industrial Research Organization (CSIRO), Australia. Powdered magnetite, which has been given a special surface treatment, is added to the water. Particulate impurities such as clays adhere to these particles by colloidal forces. The magnetite particles are removed by a magnetic separator, chemically cleaned, and re-used. The effluent from this process occupies a much smaller volume than the traditional alum sludge. The first installation outside Australia was at the Redmires waterworks, Yorkshire, UK, in 1988. The process should also be applicable to the effluents from sewage works, and pilot trials of such a process were being held in Melbourne and Sydney in 1992. In 1995 it was in use by South West Water, UK.

> *New Sci.,* 1990, **127**(1725), 44.
> *Environ. Data Serv. Report,* 1995, (240), 23.

Sirola A variation of the *Sulfite process for making paper from wood, in which two "cooking" processes are used; the first is alkaline and the second acid. *See also* Kramfors, Stora.

Sirosmelt [**CSIRO smelt**ing]   A copper smelting process developed by the Commonwealth Scientific & Industrial Research Organization (CSIRO), Australia. Used in Miami, AZ. *See* Isasmelt.

> *Australian Bus. Rev. Weekly,* 1991, 29 March.

Sirotherm An ion-exchange process for desalinating brackish waters, in which regeneration is accomplished by heating the resin to approximately 80°C, rather than by reverse ion-exchange. Developed in the 1960s by D. E. Weiss and others at the Chemical Technology Division of the Commonwealth Scientific and Industrial Research Organization, Melbourne.

> Calomon, C., in *Ion Exchange for Pollution Control,* Calomon, C. and Gold, H., Eds., CRC Press, Boca Raton, FL, 1979, 96.

Siurin *See* Hoganas.

Skarstrom *See* PSA.

SKF A *DR process for making iron. Powdered iron ore and coal are injected through a plasma arc heater into a vertical shaft furnace. *See* DR.

SKIP [Skeletal isomerization process] A process for converting linear butenes into isobutene. Developed by Texas Olefins in the 1990s and operated by that company in Houston, TX.

> Morgan, M. L., *Chem. Ind. (London)*, 1998, (3), 90.

Skygas A process for gasifying carbon-containing wastes. The wastes are passed through a shaft furnace, heated by an arc burning between carbon electrodes. The wastes are gasified with the addition of water and the off-gases are further processed in a secondary coke-filled electrically-heated reactor. The product is a medium-BTU gas which can be used for raising steam, generating electricity, or making ammonia or methanol. Under development jointly by Xytel-Techtel and Montana Precision Mining.

> *Processing (Sutton, England)*, 1989, Aug/Sept, 25.

Slow-cooled matte A process for extracting platinum metals from copper-nickel matte. The molten matte is cooled slowly, over several days. This causes the platinum metals to enter a nickel-iron phase which can be separated magnetically from the other components. Operated by Rustenberg Platinum Mines in South Africa, and INCO in Canada.

SL/RN A *DR ironmaking process, using coal as the reductant in a rotary kiln. First operated in New Zealand in 1970 and now in operation in several other countries. Six plants were in operation as of 1997. The Codir process is similar to this process. *See* DR.

SM *See* SMART SM.

SMART *See* SMART SM.

SMART H₂ [Steam Methane Advanced Reformer Technology] A process for making hydrogen by the steam reforming of methane. It differs from similar systems in housing the catalyst within a proprietary heat exchanger. Developed by Mannesmann KTI in 1996; it was planned for installation in Salisbury, MD, in late 1997.

SMART SM [Styrene monomer advanced reheat technology] A process for making styrene by dehydrogenating ethylbenzene. It uses "oxidative reheating" – *in-situ* reheating of process gas between endothermic stages of dehydrogenation, which uses less energy than previous processes. Developed and licensed jointly by UOP and ABB Lummus Crest. It had not been commercialized by 1995. *See also* Styro-Plus.

> Ward, D. J., Black, S. M., Imai, T., Sato, Y., Nakayama, N., Tokano, H., and Egawa, K., *Hydrocarbon Process.*, 1987, **66**(3), 47.
> *Hydrocarbon Process.*, 1987, **66**(11), 87.
> *Eur. Chem. News*, 1990, 1/8 Jan, 23.
> Cavani, F. and Trifiro, F., *Appl. Catal. A: Gen.*, 1995, **133**, 237.

SMDS [Shell middle distillate synthesis] A three-stage process for converting natural gas to liquid fuels. The first stage uses *reforming to convert the natural gas to *syngas. The syngas is converted to heavy paraffins by the *Fischer-Tropsch process; this mixture is converted to hydrocarbons of the required boiling range by hydroconversion. The process was commercialized in Sarawak in the early 1990s by a consortium of Shell Gas, Mitsubishi Corporation, Petronas, and the Sarawak State Government, but suffered an explosion in 1997.

> van der Burgt, M., van Klinken, J., and Sie, T., *Pet. Rev.*, 1990, **44**(516), 204.
> *Oil & Gas J.*, 1990, **88**(40), Suppl., 15.
> Sie, S. T., Senden, M. M. G., and van Wechem, H. M. H., *Catal. Today*, 1991, **8**, 371.
> Chauvel, A., Delmon, B., and Hölderich, W. F., *Appl. Catal. A: Gen.*, 1994, **115**, 186.

SMPO [styrene **m**onomer **p**ropylene **o**xide] A process for making propylene oxide by the catalytic epoxidation of propylene. The catalyst contains a compound of vanadium, tungsten, molybdenum, or titanium on a silica support. Developed by Shell and operated in The Netherlands since 1978.

> Weissermel, K. and Arpe, H.-J., *Industrial Organic Chemistry*, 3rd ed., VCH Publishers, Weinheim, Germany, 1997, 270.

SNAM (1) [Societa **N**azionale Metandotti] A urea synthesis process, developed in the late 1960s and now widely used worldwide.

> Slack, A. V. and Blouin, B. M., *CHEMTECH*, 1971, **1**, 32.
> *Hydrocarbon Process.*, 1979, **58**(11), 248.

SNAM (2) An *ammoxidation process for converting propylene to acrylonitrile. The catalyst is based on molybdenum/vanadium or bismuth, operated in a fluidized bed. Operated in Europe in 1968.

SNCR [**S**elective **n**on-**c**atalytic **r**eduction] A generic term for processes which remove oxides of nitrogen from flue-gases by non-catalytic chemical reactions. These include the reaction with ammonia at high temperature (1,300 to 1,900°C), and the reaction with urea. *See* NOxOut, SCR.

> Radojevic, M., *Chem. Br.*, 1998, **34**(3), 30.

SNOX A combined *flue-gas desulfurization and denitrification process. The NO_x is first removed by the *SCR process, and then the SO_2 is catalytically oxidized to SO_3 and converted to sulfuric acid by the *WSA process. Developed by Haldor Topsoe and first operated at a power station in Denmark in the 1990s.

> *Chem. Eng. (Rugby, England)*, 1989, (462), 29.
> Rostrup-Nielsen, J. R., *Catal. Today*, 1993, **18**, 141.
> *Hydrocarbon Process.*, 1993, **72**(8), 80.
> *Appl. Cat.*, 1994, **3**(4), N28.
> *Environ. Sci. Technol.*, 1994, **28**(2), 88.

SNPA *See* Sulfreen.

SNPA-DEA A process for removing hydrogen sulfide and carbon dioxide from gas streams by absorption in aqueous diethanolamine. Developed by the Société Nationale des Pétroles d'Aquitaine at the gasfield in Lacq, France, and still used there and in Canada.

> Canadian Patent 651, 379.
> Wendt, C. J., Jr. and Dailey, L. W., *Hydrocarbon Process.*, 1967, **46**(10), 155.
> *Hydrocarbon Process.*, 1975, **54**(4), 95.
> Kohl, A. L. and Riesenfeld, F. C., *Gas Purification*, 4th ed., Gulf Publishing, Houston, TX, 1985, 39.

Socony Vacuum This United States oil company, now the Mobil Oil Corporation, invented many processes, but the one bearing the company name was that for making thiophene from butane and elemental sulfur at 560°C. It was operated by the Pennwalt Company in the 1950s and 1960s but then abandoned.

Sobolevsky A process for converting native platinum to malleable platinum by pressing and heating. Developed by P. G. Sobolevsky in Russia in the 1820s.

> McDonald, D., *A History of Platinum*, Johnson Matthey Co., London, 1960, 159.

SOFT [**S**plit-**o**lefin **f**eed **t**echnology] An improved version of Phillips' HF *alkylation process.

> Rhodes, A. K., *Oil & Gas J.*, 1994, **92**(34), 56.

SOHIO [Standard **Ohio**] The Standard Oil Company of Ohio (later BP Chemicals America) has developed many processes, but its *ammoxidation process, for converting propylene to acrylonitrile, is the one mostly associated with its name. First operated in the United States in 1960, it is the predominant process for making acrylonitrile used in the world today.

> Jacobs, M., *Ind. Eng. Chem.*, 1996, **74**(41), 40.
> Weissermel, K. and Arpe, H.-J., *Industrial Organic Chemistry*, 3rd ed., VCH Publishers, Weinheim, Germany, 1997, 305.

Soldacs *See* PuraSiv HR.

Solexol A process for refining fats and oils by solvent extraction into propane. Developed by MW Kellogg Company. Not to be confused with Selexol.

> Passino, H. J., *Ind. Eng. Chem.*, 1949, **41**, 280.

Sol-Gel A family of processes for making oxide ceramics and glasses from colloidal hydrated oxides or hydroxides. The colloids are converted to gels by several methods and the gels are then dehydrated and calcined. First used in the late 1950s in the nuclear energy industry at Oak Ridge National Laboratory, TN. It was then used for making uranium dioxide in the form of ceramic spheres for fuel elements. Later, many other metal oxides were made in this way. The process has also been used for making metal oxides in the form of thin films, fibers, and special shapes for electronics. *See also* Stoeber.

> Klein, L. C., Ed., *Sol-Gel Technology*, Noyes Publications, Park Ridge, NJ, 1988.
> Segal, D., *Chemical Synthesis of Advanced Ceramic Materials*, Cambridge University Press, Cambridge, 1989, Chaps. 4, 5.

Solid Lime *See* Imatra.

SOLINOX [SO_x Linde NO_x] A process for removing both NO_x and SO_x from fluegases. The SO_x is removed by scrubbing with tetra-ethylene glycol dimethyl ether, circulated in a packed tower (the *Selexol process). The NO_x is destroyed by Selective Catalytic Reduction (*SCR). The sorbent is regenerated with steam; the SO_x is recovered for conversion to sulfuric acid. Developed by Linde in 1985 and used in a lead smelter in Austria and several power stations in Germany. In 1990 it was announced that it would be used at the titanium pigment plant in The Netherlands operated by Sachtleben.

> Merrick, D. and Vernon, J., *Chem. Ind. (London)*, 1989, (3), 56.
> *Hydrocarbon Process.*, 1990, **69**(4), 92.

Soliroc A process for solidifying aqueous wastes, converting them to a solid form suitable for landfill. The waste is initially acidic. Sodium silicate, cement, and lime are added, converting the liquid to a gel which hardens in several days. Used in France, Norway, Belgium, and Canada.

Solutizer Also spelled Solutiser. An early process for removing sulfur compounds from fuel oil by solvent extraction with alkaline sodium isobutyrate. Developed by Shell Development Company.

> *Hydrocarbon Process.*, 1964, **43**(9), 213.
> Unzelman, G. H. and Wolf, C. J., in *Petroleum Processing Handbook*, Bland, W. F. and Davidson, R. L., Eds., McGraw-Hill, New York, 1967, 3-116.

SOLVAHL A petroleum de-asphalting process.

> Chauvel, A., Delmon, B., and Hölderich, W. F., *Appl. Catal. A: Gen.*, 1994, **115**, 186.

Solvay (1) An important process for making sodium carbonate, invented by E. Solvay in Belgium in 1861 and still used worldwide. *See* Ammonia-soda.

> British Patent 3,131 (1863).

Solvay (2) A pulp-bleaching process using chlorine dioxide, generated by reducing sodium chlorate with methanol. In 1962, 20 paper mills in the United States were using this process. *See also* Mathieson (1).

Sheltmire, W. H., in *Chlorine, Its Manufacture, Properties and Uses,* Sconce, J. S., Ed., Reinhold Publishing, New York, 1962, 304,538.

Solvex [**Solv**ent **ex**traction] A process for extracting platinum metals from their ores by solvent extraction. Invented by Jonson Mathey in 1980 and piloted by that company, together with Rustenburg Platinum Holdings, since 1983. Commercialized in South Africa in the late 1980s.

Bruce, J. T., *Platinum Met. Rev.,* 1996, **40**(1), 4.

Solv-X A pulp-bleaching process which prevents the transition-metal catalyzed side reactions of hydrogen peroxide which would otherwise occur.

Pulp Pap. Eur., 1997, **5**(2), 32.

SONOX A process for simultaneously removing SO_2 and NO_x from flue-gas. A slurry of lime or limestone, with a proprietary nitrogen-based additive, is injected into the furnace. Developed by Ontario Hydro (the largest electricity supplier in North America) and marketed by Research-Cottrell.

Eur. Chem. News, 1990, **55**(1444), 27.
Eur. Chem. News, 1991, **56**(1471), 33.

Soot A process for making a very pure "preform" of silica glass for drawing into optical fiber. Invented in 1970 by D. B. Keck and P. C. Schultz at Corning Glass Works, New York. The silica made in this way was so much purer than the silicate glasses used previously that it revolutionized the fiber optics communications business, permitting the progressive replacement of copper wire by optical fiber for telephone links.

U.S. Patents 3,711,262; 3,737,292.
Schultz, P. C., *Am. Ceram. Soc. Bull.,* 1973, **52**, 383.
Kirk-Othmer's Encyclopedia of Chemical Technology, 4th ed., Vol. 12, John Wiley & Sons, New York, 1991–1998, 615.

SOR [**S**elective **o**lefin **r**ecovery] A process for recovering or purifying olefins by selective absorption in silver nitrate solution. The silver solution is passed through hollow-fiber microporous membranes and the olefin-containing gases are passed along the outside of the fibers. Developed by BP Chemicals and engineered by Stone & Webster; piloted in Grangemouth, Scotland, and Toledo, OH.

Eur. Chem. News, 1996, **65**(1709), 59.
Hydrocarbon Process., 1996, **75**(5), 29.

Sorbex A family of separation processes, based on continuous liquid chromatography. The adsorbent, chosen for its selectivity and usually a zeolite, is contained in a chromatographic column with a number of ports spaced along its length. These ports, operated by a proprietary rotary valve, function alternately as inlets and outlets and thus permit simulated countercurrent liquid–solid adsorption to be operated in a continuous mode. Variations of this process for specific mixtures, each with its preferred adsorbent and desorbent, are described under their own names, viz. Cymex, Ebex, Molex, Olex, Parex, Sarex, Sorbutene. Invented in 1957 by C. G. Gerhold, D. B. Broughton, and their colleagues at Universal Oil Products Company (now UOP).

U.S. Patent 2,985,589.
Broughton, D. B., *Chem. Eng. Prog.,* 1968, **64**(8), 60.
Spitz, P. H., *Petrochemicals, the Rise of an Industry,* John Wiley & Sons, New York, 1988, 191.

Johnson, J. A. and Oroskar, A. R., in *Zeolites as Catalysts, Sorbents and Detergent Builders,* Karge, H. G. and Weitkamp, J., Eds., Elsevier, Amsterdam, 1989.

Johnson, J. A. and Kabza, R. G., in *Advances in Separation Processes,* American Institute of Chemical Engineers, New York, 1990, 35.

Jeanneret, J. J. and Mowry, J. R., in *Handbook of Petroleum Refining Processes,* 2nd. ed., Meyers, R. A., Ed., McGraw-Hill, New York, 1997, 10.45.

Sorbutene   A version of the *Sorbex process, for extracting 1-butene from mixed C_4 hydrocarbons. Offered for license by UOP.

Friedlander, R. H., in *Handbook of Petroleum Refining Processes,* Meyers, R. A., Ed., McGraw-Hill, New York, 1986, 8-101.

Sovafining   [**So**cony **Va**cuum re**fining**]   A hydrofining process developed by the Socony-Vacuum Company, now Mobil Corporation. *See* Hydrofining.

Sovaforming   [**So**cony **Va**cuum re**forming**]   A *catalytic reforming process, using a platinum catalyst in a fixed bed. Developed by the Socony-Vacuum Oil Company in 1954. Subsequently renamed Platinum Reforming, or PR.

Pet. Process., 1955, **10**, 1174.

Unzelman, G. H. and Wolf, C. J., in *Petroleum Processing Handbook,* Bland, W. F. and Davidson, R. L., Eds., McGraw-Hill, New York, 1967, 3-30.

SOXAL   A *flue-gas desulfurization process, basically similar to the *Wellman-Lord process but regenerating the bisulfite solution in a membrane electrochemical cell. Developed by Allied Signal Group and piloted at the Florida Power & Light Company, Miami, FL, in 1991. A larger demonstration plant was planned for Dunkirk, New York.

Chem. Week, 1986, **139**(22), 106.

Chem. Mark. Rep., 1991, **239**(5), 5.

Chem. Mark. Rep., 1991, **240**(22), 9.

SPD   [**S**lurry **p**hase **d**istillate]   A process for making diesel fuel, kerosene, and naphtha from natural gas. Developed by Sasol and first commercialized in South Africa in 1993. A joint venture with Haldor Topsoe for the further development and commercialization of the process was announced in 1996. Commercialization in Nigeria was announced in 1998.

Hydrocarbon Process., 1996, **75**(4), 46.

SPGK   [**S**hell **P**oly**G**asoline and **K**erosene]   A process for oligomerizing $C_2 - C_5$ olefins to liquid transport fuels. The reaction takes place at 200 to 280°C and is catalyzed by a zeolite. Developed by the Shell Petroleum Company, Amsterdam, in 1990.

Eur. Chem. News, 1991, **56**(1455), 23.

Chem. Eng. (Rugby, England), 1991, (489), 12.

SPHER   [**S**hell **P**ellet **H**eat **E**xchange **R**etorting]   A process for extracting oil from shale. The process is conducted in a fluidized bed in which heat is transferred by inert pellets of two sizes.

Chem. Eng. News, 1980, **58**(37), 42.

Spherilene   [**Spher**ical polyethy**lene**]   A process for making polyethylene in the form of granules. The catalyst combines a *Ziegler-Natta catalyst with a silane. Developed by Montecatini, Italy. The first plant was due for startup by Himont, in Lake Charles, LA, in 1993.

Eur. Chem. News, 1991, **56**(1476), 27.

Eur. Chem. News, 1992, **58**(1530), 18.

Spheripol A process for making polypropylene and propylene co-polymers. Homopolymerization is conducted in the liquid phase in a loop tubular reactor; co-polymerization is conducted in the gas phase in a fluidized-bed reactor. The catalyst is treated with a special silane. The product is in the form of beads of up to 5 mm in diameter. Developed by Montecatini, Italy, and first licensed by Himont, United States, and Mitsui Petrochemical Industries, Japan. In 1989, 29 licenses had been granted worldwide. Now offered for license by Montell, a joint venture between Montedison and Shell. *See also* Addipol.

> *Chem. Week,* 1987, **141**(21), 99.
> *Hydrocarbon Process.,* 1989, **68**(11), 109.
> *Hydrocarbon Process.,* 1991, **70**(3), 173.

SPIREX A *DR process for making iron powder or hot briquetted iron from iron ore fines. Three stages are used. The first is a circulating fluidized bed preheater whose turbulent conditions reduce the particle size of the ore. The second and third stages achieve the reduction in fluidized beds, fed by reducing gases from a *MIDREX reformer. Developed by Midrex Direct Reduction Corporation and Kobe Steel. A demonstration plant was scheduled to be built at the Kobe Steel plant in Venezuela in 1997.

SPM *See* DR.

SPOR [Sulfur removal by partial oxidation and reduction] A process for converting H_2S and SO_2 to elemental sulfur by the use of stoichiometric amounts of oxygen and sulfur.

> Chung, J. S., Paik, S. C., Kim, H. S., Lee, D. S., and Nam, I. S., *Catal. Today,* 1997, **35**(1–2), 37.

SSPD [Sasol slurry phase distillate] A process for converting natural gas to diesel fuel, kerosene, and naphtha. Operated by Sasol in South Africa since 1993. Three stages are involved. In the first, natural gas is converted to synthesis gas by *reforming. In the second, the synthesis gas is converted to waxy hydrocarbons in a slurry-phase reactor. In the third, the waxes are upgraded to middle distillates. *See also* Arge.

> *Oil & Gas J.,* 1997, **95**(25), 16.

Spühlgas [German — rinsing gas] Also known as **Lurgi Spühlgas.** A low-temperature process for carbonizing lump or briquetted materials such as coal, lignite, peat, wood, and oil shale. The heat is provided by burning the gas which is also generated. Developed by Lurgi in Germany in 1925, originally to provide liquid hydrocarbons and later to provide industrial coke too; now offered for license. As of 1981, 270 plants had been built in a number of countries.

> Rammler, R. W. and Weiss, H.-J., in *Handbook of Synfuels Technology,* Meyers, R. A., Ed., McGraw-Hill, New York, 1984, 4-12.

Squire and Messel An early process for making oleum from sulfuric acid produced by the *Chamber process. The acid was decomposed at red heat to sulfur dioxide, oxygen, and steam; the steam was condensed out, and the remaining gases passed over platinized pumice to form sulfur trioxide, which was absorbed in more chamber acid. Invented by W. S. Squire and R. Messel in 1875 in London and first commercialized there. Messel was one of the founders of the Society of Chemical Industry and is still commemorated in that society by the biennial award of a medal.

> British Patent 3,278 (1875).
> Morgan, G. T. and Pratt, D. D., *British Chemical Industry,* Edward Arnold & Co., London, 1938, 39.

SRB [Sulfate reducing bacteria] A process for removing heavy metals and sulfates from contaminated water by the use of special bacteria. Ethanol is provided as a cheap source of energy for the bacteria. The sulfur is precipitated as heavy metal sulfides. Developed by Shell Research, Sittingbourne, United Kingdom, and first operated in The Netherlands in 1991.

Barnes, L. J., Janssen, F. J., Sherrin, J., Versteegh, J. H., Koch, R. O., and Scheerin, P. J. H., *Trans. Inst. Chem. Eng., Chem. Eng. Res. Dev.,* 1991, **69A**, 184.
Chem. Br., 1991, **27**, 884.

SRC [Solvent refined coal] Also known as P&M and PAMCO. A coal liquification process developed by the Spencer Chemical Company and the Pittsburgh & Midway Coal Mining Company, which later became part of the Gulf Oil Corporation, based on the *Pott-Broche process. Two versions of the process became known as Gulf SRC I and Gulf SRC II. The purpose is to produce a fuel of low sulfur content. Crushed coal is suspended in a mixture of cresol and tetrahydronaphthalene and hydrogenated under pressure with no added catalyst. The process is actually catalyzed by the finely divided iron pyrites naturally present in the coal. The undissolved residue is filtered off under pressure. Piloted in the 1970s at Tacoma and Fort Lewis, WA, and at Wilsonville, AL. SRC II is a variant, using more severe cracking conditions, which produces heavy heating oil.

Brant, V. L. and Schmid, B. K., *Chem. Eng. Prog.,* 1969, **65**(12), 55.
Pastor, G. R., Keetley, D. J., and Naylor, J. D., *Chem. Eng. Prog.* 1976, **72**(8), 67.
Alpert, S. B. and Wolk, R. H., in *Chemistry of Coal Utilization,* 2nd Suppl. Vol., Elliott, M. A., Ed., John Wiley & Sons, New York, 1981, 1926.
Phillips, M. T., Bronfenbrenner, J. C., Kuhns, A. R., O'Leary, J. R., Snyder, G. D., and Znaimer, S., in *Handbook of Synfuels Technology,* Meyers, R. A., Ed., McGraw-Hill, New York, 1984, 1-3.

SREX [Strontium extraction] A process for removing strontium-90 from aqueous wastes from nuclear fuel processing, by solvent extraction into a solution of 18-crown-6 in octanol. Developed by E. P. Horwitz at the Argonne National Laboratory, Chicago, IL, in 1990.

SRP [Sulfonation-Reduction-Peroxide] A pulp-bleaching process which combines reduction with sodium borohydride and oxidation with hydrogen peroxide. Developed by Atochem in 1991 and licensed to Morton International, Chicago.

Eur. Chem. News, 1991, **57**(1480), 38.

SRU A variation of the *Claus process, for use in the presence of ammonia. It uses a special reactor, designed to avoid plugging by compounds of ammonia with sulfur trioxide. Designed by JGC Corporation. Thirty one units were in operation as of 1992.

Hydrocarbon Process., 1996, **75**(4), 138.

SSC [Statione Sperimentale per i Combustibili] A cyclic catalytic reforming process for making gas from oil. Developed by SSC, Milan.

Gas Making and Natural Gas, British Petroleum Co., London, 1972, 96.

SSF [Simultaneous saccharification and fermentation] *See* Emert.

S-T *See* VAW-Sulfite.

Staatsmijnen-Otto Also known as the Pieters process, after the inventor, H. A. J. Pieters. A process for removing hydrogen sulfide from coal gas by scrubbing with an aqueous solution containing a suspension of iron cyanide complexes known as iron blue. The product is elemental sulfur, which separates as a froth and is purified by heating with water in an autoclave. Staatsminjen is a Dutch producer of smokeless fuels which has also given its name to a briquetting process.

U.S. Patent 2,169,282.
Kohl, A. L. and Riesenfeld, F. C., *Gas Purification,* 4th ed., Gulf Publishing, Houston, TX, 1985, 512.

Stack   *See* Dutch.

Stackpol 150   Also known as **IFP Stackpol 150.** A *flue-gas desulfurization process. The sulfur dioxide is removed by scrubbing with aqueous ammonia, and the sulfur is then recovered by a four-stage process. Developed by the Institut Français du Pétrole.

> *Sulphur,* 1976, (125), 43.
> *Env. Sci. Technol.,* 1977, **11**(1), 22.
> *Hydrocarbon Process.,* 1986, **65**(4), 97.

Stamicarbon   [**Staat**s**mij**nen **carbon**]   Stamicarbon bv is the licensing subsidiary of the Dutch chemical company DSM. It offers a number of processes, including *HPO and *HSO. Historically, the process for which the company was best known was a coal carbonization process; today, a urea-manufacturing process is probably its most important one.

Standard Oil   A process for polymerizing ethylene and other linear olefins and di-olefins to make linear polymers. This is a liquid-phase process, operated in a hydrocarbon solvent at an intermediate pressure, using a heterogeneous catalyst such as nickel oxide on carbon, or vanadia or molybdena on alumina. Licensed to Furukawa Chemical Industry Company at Kawasaki, Japan.

> Raff, R. A. V. and Allison, J. B., *Polyethylene,* Interscience, New York, 1956, 68.
> Peters, E. F., Zletz, A., and Evering, B. L., *Ind. Eng. Chem.,* 1957, **49,** 1879.

STAR   [**St**eam **A**ctive **Re**-forming]   A *catalytic reforming process for converting aliphatic hydrocarbons to olefins or aromatic hydrocarbons. Hydrocarbons containing five or fewer carbon atoms are converted to olefins. Those containing six or more are dehydrocyclized to aromatic hydrocarbons. The reactions take place in the vapor phase, in a fixed catalyst bed containing a noble metal catalyst, in the presence of steam. Demonstrated on a semi-commercial scale and offered for license by Phillips Petroleum Company. The first commercial plant was built for Coastal Chemicals in Cheyenne, WY, in 1992; another for Polibutenos Argentinos in 1996.

> Brinkmeyer, F. M., Rohr, D. F., Olbrich, M. E., and Drehman, L. E., *Oil & Gas J.,* 1983, **81**(13), 75.
> Little, D. M., *Catalytic Reforming,* PennWell Publishing, Tulsa, OK, 1985, 171.
> Hutson, T., Jr. and McCarthy, W. C., in *Handbook of Petroleum Refining Processes,* Meyers, R. A., Ed., McGraw-Hill, New York, 1986, 4-29.
> Hu, Y. C., in *Chemical Processing Handbook,* Marcel Dekker, New York, 1993, 807.

STDP   [**S**elective **t**oluene **d**isproportionation **p**rocess]   A process for converting touene to mixed xylenes, predominately *p*-xylene. It takes place in the presence of hydrogen over a ZSM-5–type catalyst. Developed by Mobil in the 1980s and first operated by Enichem.

> Weissermel, K. and Arpe, H.-J., *Industrial Organic Chemistry,* 3rd ed. VCH Publishers, Weinheim, Germany, 1997, 333.

STEAG   [**St**einkohlen **E**lektrizitat **A.G.**]   A process for producing both electric power and gas from coal, developed by the German company named. Installed at Leuna, Germany.

> Dainton, A. D., in *Coal and Modern Coal Processing,* Pitt, G. J. and Millward, G. R., Eds., Academic Press, London, 1979, 135.

steam cracking   *See* reforming.

steam reforming   *See* reforming. Those steam reforming processes with special names which are described in this dictionary are: Fauser-Montecatini, ICI Low Pressure Methanol, ICI Steam Naphtha Reforming, Kureha/Union Carbide, MRG, MS, Onia-Gegi, Pyrotol, Remet.

steelmaking   Because of its size and importance, the iron and steel industry has attracted many inventors and developers and many of their processes are known by special names. The more important of these, listed as follows, are mostly described in individual entries: Acid Bessemer, Acid open hearth, AOD, Basic Bessemer, Basic open hearth, Bertrand Thiel, Bessemer, Chenot, Corex, Crucible, DH, D-LM, FIOR, HIB, Hoesch, Huntsman, LBE, L-D, Open Hearth, Perrin, Pig and ore, Pig and scrap, Puddling, Saniter, Siemens Martin, SL-RN, Talbot, Thomas, Uchatius, Wiberg-Sodefors.

The steelmaking processes listed as follows are more specialized or of lesser importance and are not described elsewhere in this work: **ACAR** [Allis-Chalmers Agglomeration-Reduction], **ASEA-SKF** [named after two Swedish companies], **CAB** [Capped oxygen bubbling], **CAS** [Composition adjustment by sealed oxygen bubbling], **Hornsey**, **KIP** [Kimetsu injection process], **Leckie**, **LF** [Ladle-furnace], **Moffat**, **Monell**, **Morse**, **Nesbitt**, **Nippon Steel**, **OBM**, **QQ-BOP**, **RH** [Ruhrstahl-Heraeus], **RH-FR** [Ruhrstahl-Heraeus], **RH-OB** [Ruhrstahl-Heraeus, oxygen blowing], **SAB** [Sealed argon bubbling], **TN** [Thiessen Niederrhein], **Tysland-Hole**, **VAD** [Vacuum arc degassing], **VAR** [Vacuum arc remelting], **VOD** [Vacuum oxygen decarburization].

Osborne, A. K., *An Encyclopedia of the Iron & Steel Industry*, 2nd. ed., Technical Press, London, 1967.
Kirk-Othmer's Encyclopedia of Chemical Technology, 4th ed., Vol. 22, John Wiley & Sons, New York, 1991–1998, 765.

Steffens   A process for separating sugar from beet sugar molasses by adding calcium hydroxide to precipitate calcium saccharate. Treatment of the liquor with carbon dioxide precipitates calcium carbonate and regenerates the sucrose. Invented in Vienna in 1883 by C. Steffens. *See also* Boivan-Louiseau, Scheibler.

British Patents 967 (1883); 2,416 (1883).

Stelling   *See* DR.

Stengel   A process for making ammonium nitrate by reacting ammonia vapor with 60 percent nitric acid in a packed tower reactor. Air blown through the reactor reduces the moisture content to the desired value, and the product flows to the bottom of the reactor where it is discharged on to a moving, water-cooled belt. Invented by A. Stengel and first operated by Commercial Solvents Corporation in Sterlington, LA, in the 1950s.

U.S. Patent 2,568,901.
Dorsey, J. J., Jr., *Ind. Eng. Chem.*, 1955, **47**, 11.
Hester, A. S., Dorsey, J. J., Jr., and Kaufman, J. T., in *Modern Chemical Processes*, Vol. 4, Reinhold Publishing, New York, 1956, 36.

STEREAU   A *BAF process.

Stephenson, T., Mann, A., and Upton, J., *Chem. Ind. (London)*, 1993, (14), 533.

Stevens   *See* carbonization.

STEX   [Styrene extraction]   A process for extracting styrene from pyrolysis gasoline. Developed by Toray.

Weissermel, K. and Arpe, H.-J., *Industrial Organic Chemistry*, 3rd ed. VCH Publishers, Weinheim, Germany, 1997, 341.

Still   A method for increasing the yield of light oil formed in the carbonization of coal. Some of the gas produced is passed through the partially carbonized coal in a cooler part of the bed. Developed by C. Still and used in Recklinghausen, Germany in the 1930s for producing motor fuel. *See also* Carl Still.

U.S. Patents 1,810,629; 1,937,853.
Dean, H., *Fuel,* 1934, **13,** 112.

Stockbarger   *See* Bridgman.

Stöber   A process for making metal oxides in the form of small spheres of uniform diameter by the controlled hydrolysis of metal alkoxides. First used in 1968 to make silica spheres from alkyl silicates. The products can be used to make high quality oxide ceramics. *See also* Sol-Gel.

Stöber, W., Fink, A., and Bohn, E., *J. Colloid Interface Sci.,* 1968, **26,** 62.

Stoic   Also called **Foster Wheeler-Stoic** and **FW-Stoic.** A two-stage, nonslagging coal gasification process, operated under atmospheric pressure and using air as the oxidant. Initially developed by Stoic Combustion Limited Pty, South Africa; licensed and further developed by Foster Wheeler Corporation, United States. First used in South Africa in 1950; now widely used in Europe, the United States, and South Africa.

Brand, R. G. and Bress, D. F., in *Handbook of Synthetic Fuels Technology,* Meyers, R. A., Ed., McGraw-Hill, New York, 1984, 3-8.

Stone & Webster/Ionics   A *flue-gas desulfurization process in which the sulfur dioxide is absorbed in aqueous sodium hydroxide, forming sodium sulfite and bisulfite, the sulfur dioxide is liberated by the addition of sulfuric acid, and the reagents are regenerated electrolytically. Designed by Stone & Webster Engineering Corporation and Ionics Incorporated and operated in a demonstration plant in Milwaukee in 1974 but not yet commercialized.

Humphries, J. J., Jr. and McRae, W. A., *Proc. Am. Power Conf.,* 1970 **32,** 663 (*Chem. Abstr.,* **74,** 34348).
Kohl, A. L. and Riesenfeld, F. C., *Gas Purification,* 4th ed., Gulf Publishing, Houston, TX, 1985, 357.

Stora   A two-stage variation on the *Sulfite papermaking process, in which the acidity of the second stage is increased by adding sulfur dioxide. *See also* Kramfors, Sirola.

Higham, R. R. A., *A Handbook of Papermaking,* 2nd. ed., Business Books, London, 1968, 261.

Stratco   A process for making a high-octane gasoline component by alkylation of $C_3 - C_5$ hydrocarbons with isobutane, catalyzed by sulfuric acid. The product is known as an alkylate. Operated in several oil refineries in the United States.

Weitkamp, J. and Maixner, S., *Erdoel Kohle,* 1983, **36,** 523.

Strategic-Udy   *See* DR.

Stretford   A process for removing hydrogen sulfide and organic sulfur compounds from coal gas and general refinery streams by air oxidation to elementary sulfur, using a cyclic process involving an aqueous solution of a vanadium catalyst and anthraquinone disulfonic acid. Developed in the late 1950s by the North West Gas Board (later British Gas) and the Clayton Aniline Company, in Stretford, near Manchester. It is the principle process used today, with over 150 plants licensed in Western countries and at least 100 in China.

U.S. Patent 2,997,439.
Nicklin, T. and Brunner, E., *Hydrocarbon Process.,* 1961, **40**(12), 141.
Kohl, A. L. and Riesenfeld, F. C., *Gas Purification,* 4th ed., Gulf Publishing, Houston, TX, 1985, 521.
Wilson, B. M. and Newell, R. D., in *Acid and Sour Gas Treating Processes,* Newman, S. A., Ed., Gulf Publishing, Houston, TX, 1985, 342.
Dalrymple, D. A., Trofe, T. W., and Evans, J. M., *Chem. Eng. Prog.,* 1989, **85**(3), 43.
Hydrocarbon Process., 1996, **75**(4), 138.

Sturzelberg   *See* DR.

Styro-Plus   An early version of the *SMART SM process.

Sucro-Blanc   A process for decolorizing sugar solution by the addition of calcium hypochlorite.

> Spencer, G. L. and Meade, G. P., *Cane Sugar Handbook,* 8th ed., John Wiley & Sons, New York, 1945, 331.

Suida   An extractive distillation process for concentrating the dilute acetic acid obtained from the manufacture of cellulose acetate. It was originally used for separating the products of wood pyrolysis. Invented in 1926 by H. Suida in Vienna and operated in the 1930s.

> U.S. Patents 1,624,812; 1,697,738; 1,703,020.
> Faith, W. L., Keyes, D. B., and Clark, R. L., *Industrial Chemicals,* 1st ed., John Wiley & Sons, New York, 1950, 11.

Sulfa-Check   A process for removing acid gases from hydrocarbon gas streams. Developed by NL Industries. Eighty units were in operation in the United States in 1986.

> *Chem. Eng. (N.Y.),* 1987, **94**(2), 159.

SULFACID   A process for removing sulfur dioxide from effluent gases using active carbon. The gas is first contacted with dilute sulfuric acid from the adsorption step. It then passes through a fixed bed of active carbon on which water is being sprayed. The resulting dilute sulfuric acid is recovered. Suitable for effluents from the *Contact process and flue-gases. Developed by Lurgi, Bergbau-Forschung, and Babcock & Wilcox.

> *Oil Week,* 1971, 5 July, 9.
> Kohl, A. L. and Riesenfeld, F. C., *Gas Purification,* 4th ed., Gulf Publishing, Houston, TX, 1985, 405.

Sulfa Guard   Similar to *Sulfa-scrub, but reportedly more cost-effective.

Sulfa-scrub   A process for removing low concentrations of hydrogen sulfide from gas streams by reaction with hexahydrotriazine. The product is water-soluble, non-corosive, and non-hazardous. Developed by the Quaker Petroleum Chemical Company in 1991.

> U.S. Patent 4,978,512.
> *Chem. Eng. (N.Y.),* 1991, **98**(9), 43.
> Dillon, E. T., *Hydrocarbon Process.,* 1991, **70**(12), 65.

Sulfate (1)   A process for making titanium dioxide pigment from ilmenite. The ilmenite is digested with sulfuric acid, yielding a solution of titanyl and ferrous sulfates:

$$FeTiO_3 + 2H_2SO_4 = TiO{\cdot}SO_4 + FeSO_4 + 2H_2O$$

Much of the ferrous sulfate is crystallized out and discarded. The titanium is hydrolyzed by boiling, yielding hydrated titanium dioxide, which is then calcined:

$$TiO{\cdot}SO_4 + (x+1)H_2O = TiO_2{\cdot}xH_2O + H_2SO_4$$

$$TiO_2{\cdot}xH_2O = TiO_2 + xH_2O$$

The process can be adjusted to yield the product in either the anatase or rutile crystal modification, by use of proprietary nuclei at the precipitation stage and by adding small quantities of other materials at the calcination stage. The whole process is much more complex than has been indicated, involving at least 20 processing steps, every one of which is critical for the development of optimum pigment properties. Disposal of the waste sulfuric acid and ferrous sulfate has been a major problem for the industry for many years. In the 1980s, international legislation compelled the manufacturers to recycle or neutralize the acid.

Barksdale, J., *Titanium: Its Occurrence, Chemistry, and Technology,* 2nd. ed., Ronald Press, New York, 1966, Chaps. 13–16.
Egerton, T. A. and Tetlow, A., in *Industrial Inorganic Chemicals: Production and Use,* Thompson, R., Ed., Royal Society of Chemistry, Cambridge, 1995, 360.

Sulfate (2)   An acid papermaking process, also known as the *Kraft process.

SulfaTreat   A process for removing hydrogen sulfide and mercaptans from natural gas or carbon dioxide streams, using a proprietary solid absorbent, which is subsequently dumped. Over 600 plants were in operation or planned in 1996. Licensed by Gas Sweetner Associates.

Samuels, A., *Oil & Gas J.,* 1990, **88**(6), 44.
Hydrocarbon Process., 1996, **75**(4), 140.

SulFerox   A process for removing hydrogen sulfide and organic sulfur compounds from hydrocarbons, similar to the *Stretford process but using an aqueous solution containing chelated iron and proprietary additives. The product is elemental sulfur. The basic reactions are:

$$2Fe^{3+} + H_2S = 2Fe^{2+} + 2H^+ + S$$

$$2Fe^{2+} + 2H^+ + \tfrac{1}{2}O_2 = 2Fe^{3+} + H_2O$$

Developed and jointly licensed by Shell Oil Company and Dow Chemical Company. Introduced in 1987; by 1996, more than 20 units were operating and 10 were in design or under construction. The first application in coke making, in a plant near Pittsburgh, PA, was announced in 1996.

Fong, H. L., Kushner, D. S., and Scott, R. T., *Oil & Gas J.,* 1987, **85**(21), 54.
Dalrymple, D. A., Trofe, T. W., and Evans, J. M., *Chem. Eng. Prog.,* 1989, **85**(3), 43.
Chem. Eng. (N.Y.), 1991, **98**(9), 44.
Hydrocarbon Process., 1996, **75**(4), 140.
Eur. Chem. News, 1996, **65**(1720), 24.

Sulfex   [Sulfide extraction]   A process for removing heavy metals from waste streams by adding ferrous sulfide to precipitate them as their sulfides. Developed by the Permutit Company and now owned by U.S. Filter/Warrendale. Not to be confused with SULF-X or Sulph-X.

Chem. Eng. (N.Y.), 1983, **90**(10), 23.

Sulfiban   A process for removing hydrogen sulfide from coke-oven gases by scrubbing with monoethanolamine. Developed by the Bethlehem Steel Corporation and B. S. & B. Process Systems and tested in a demonstration plant of the former company in the mid 1970s. Not commercialized.

Massey, M. J. and Dunlap, R. W., *J. Air Pollut. Control Assoc.,* 1975, **25**(10), 1019.
Kohl, A. L. and Riesenfeld, F. C., *Gas Purification,* 4th ed., Gulf Publishing, Houston, TX, 1985, 104.

SULFICAT   A method for presufiding *HDS catalysts. Developed by Eurocat. Piloted in 1982 and commercialized in France in 1986.

Wilson, J. H. and Berrebi, G., in *Catalysis 1987,* Ward, J. W., Ed., Elsevier, Amsterdam, 1988, 393.
Chauvel, A., Delmon, B., and Hölderich, W. F., *Appl. Catal. A: Gen.,* 1994, **115**, 184.

Sulfidine   A process for removing sulfur dioxide from smelter gases by reaction with a suspension of xylidene in water. Developed by the Gesellschaft für Chemische Industrie, Basel and Metallgesellschaft, Frankfurt, and used in Germany in the 1930s; now probably obsolete.

Katz, M. and Cole, R. J., *Ind. Eng. Chem.,* 1950, **42**, 2263.

Kohl, A. L. and Riesenfeld, F. C., *Gas Purification*, 4th ed., Gulf Publishing, Houston, TX, 1985, 380.

Sulfining   A process for removing sulfur compounds from petroleum distillates by treatment with sulfuric acid, followed by electrostatic phase separation.

Unzelman, G. H. and Wolf, C. J., in *Petroleum Processing Handbook*, Bland, W. F. and Davidson, R. L., Eds., McGraw-Hill, New York, 1967, 3-137.

Sulfinol   A process for removing hydrogen sulfide, carbon dioxide, carbonyl sulfide, and organic sulfur compounds from natural gas by scrubbing with di-isopropanolamine dissolved in a mixture of sulfolane and water. Developed in the 1960s by Shell International Research Mij N.V., The Netherlands and Shell Development Company, Houston. In 1996, over 180 commercial units were operating or under construction.

Deal, G. H., Jr., Evans, H. D., Oliver, E. D., and Papadopoulos, M. N., *Pet. Refin.*, 1959, **38**(9), 185.
Dunn, C. L., Freitas, E. R., Goodenbour, J. W., Henderson, H. T., and Papadopoulos, M. N., *Hydrocarbon Process.*, 1964, **43**(3), 150.
Kohl, A. L. and Riesenfeld, F. C., *Gas Purification*, 4th ed., Gulf Publishing, Houston, TX, 1985, 867.
Hydrocarbon Process., 1996, **75**(4), 142.

Sulfint   A process for removing hydrogen sulfide from industrial gases by scrubbing with an aqueous solution of the iron-EDTA complex. The solution is regenerated by air oxidation, liberating sulfur dioxide, and reverse osmosis which separates the dissolved salts from the iron complex. Developed by Integral Engineering, Vienna, and licensed by Le Gaz Integral Enterprise. Several units were in operation as of 1992.

Mackinger, H., Rossati, F., and Schmidt, G., *Hydrocarbon Process.*, 1982, **61**(3), 169.
Kohl, A. L. and Riesenfeld, F. C., *Gas Purification*, 4th ed., Gulf Publishing, Houston, TX, 1985, 516.
Hydrocarbon Process., 1992, **71**(4), 136.

Sulfite   An acid papermaking process. Wood chips are digested in a hot sulfite solution, made by dissolving sulfur dioxide in a suspension of calcium or magnesium carbonate or hydroxide. The lignin in the wood is thereby converted to soluble calcium or magnesium lignosulfonate. Invented by B. C. Tilghman in Philadelphia in the 1860s; commercialized by C. D. Ekman in Sweden in the 1870s and widely used thereafter.

U.S. Patent 70,485.
Grant, J., *Cellulose Pulp and Allied Products*, Leonard Hill, London, 1958, Chaps. 2, 7.
Wenzel, H. F. J., *Sulfite Pulping Technology*, Lockwood Trade Journal Co., New York, 1965.

Sulfite/bisulfite   *See* Burkheiser.

Sulfolane   A process for removing aromatic hydrocarbons from petroleum fractions by liquid–liquid extraction using sulfolane (tetramethylene sulfone; tetrahydrothiophene-1,1-dioxide) at approximately 190°C. Developed by Shell Development Company in 1959 and first commercialized in 1962; now licensed through UOP. It replaced the *Udex process. Sulfolane is used for another purpose in the *Sulfinol process.

Deal, C. H., Evans, H. D., Oliver, E. D., and Papadopulos, M. N., *Pet. Refin.*, 1959, **38**(9), 185.
Beardmore, F. S. and Kosters, W. C. G., *J. Inst. Pet.*, 1963, **49**(469), 1.
Unzelman, G. H. and Wolf, C. J., in *Petroleum Processing Handbook*, Bland, W. F. and Davidson, R. L., Eds., McGraw-Hill, New York, 1967, 3-108.
Wheeler, T., in *Handbook of Petroleum Refining Processes*, Meyers, R. A., Ed., McGraw-Hill, New York, 1986, 8-53.

Sulfolin   A process for removing sulfur compounds from hydrocarbons, similar to the *Stretford process, but including vanadium and an organic nitrogen compound in the catalytic solution. Developed jointly by SASOL and Linde. First commercialized in 1978: six plants were operating in 1992, including one for treating the *Rectisol off-gas in the *SASOL plant in South Africa, and one in the Dakota Gasification Plant.

> Dalrymple, D. A., Trofe, T. W., and Evans, J. M., *Chem. Eng. Prog.*, 1989, **85**(3), 43.
> *Hydrocarbon Process.*, 1992, **71**(4), 137.

Sulfosorbon   A process for removing hydrogen sulfide and carbon disulfide from the gaseous effluent from the *Viscose process. Offered by Lurgi.

SULFOX   A UOP process for the catalytic oxidation of aqueous sulfides. Not commercialized.

> *Sulphur,* 1974, (117), 40.
> *Hydrocarbon Process.*, 1975, **54**(4), 106.

Sulfreen   Also known as the SNPA process. A variation of the *Claus process for removing hydrogen sulfide from gas streams by reaction with sulfur dioxide to produce elemental sulfur. It differs from the Claus process in using a lower temperature, causing the sulfur to be retained on the catalyst. The catalyst was originally carbon but alumina is used now. Developed by Lurgi Gesellschaft für Warme und Chemotechnik and the Société National des Pétroles d'Aquitane. As of 1996 nearly 50 units were in operation or under construction. *See also* Hydrosulfreen, Oxysulfreen.

> Krill, H. and Storp, K., *Chem. Eng. (N.Y.)*, 1973, **80**(17), 84.
> Davis, G. W., *Oil & Gas J.*, 1985, **83**(8), 108.
> Kohl, A. L. and Riesenfeld, F. C., *Gas Purification,* 4th ed., Gulf Publishing, Houston, TX, 1985, 446.
> *Hydrocarbon Process.*, 1996, **75**(4), 142.

SULFREX   A catalytic process for removing sulfur compounds from LPG, gasoline, and kerosene. Developed by Total/IFP in the 1980s and operated in Tenguiz, CIS, since 1991.

> Chauvel, A., Delmon, B., and Hölderich, W. F., *Appl. Catal. A: Gen.*, 1994, **115**, 1783.

Sulften   A process for removing hydrogen sulfide from the tail gases from the *Claus process. The solvent (Ucarsol HS 103) was developed by the Union Carbide Corporation; the process was developed by Ford Bacon and Davis, Dallas, TX.

> *Chem. Eng. (N.Y.)*, 1984, **91**(13), 150.

SULF-X   [**Sulf**ur extraction]   A regenerable *flue-gas desulfurization process in which the sulfur dioxide is absorbed by aqueous sodium sulfide in a bed packed with pyrite. Ferrous sulfate is produced; this is removed by centrifugation and calcined with coke and fresh pyrite. Sulfur vapor is evolved and condensed, and the residue is re-used in the scrubber. Piloted in the mid-1980s. Not to be confused with Sulfex or Sulph-X.

Sulpel   [**Sul**phur **pel**letization]   A process for making sulfur pellets. Molten sulfur is injected through nozzles into water containing a trace of a proprietary additive which gives the resulting pellets a smooth, waterproof surface. Developed and offered by Humphreys and Glasgow, United Kingdom; nine plants had been engineered as of 1992.

Sulph-X   A process for trapping sulfur in coal combustion. A proprietary mixture of inorganic salts, including sodium chloride, is mixed with the coal and combines with the sulfur dioxide so that it remains fixed in the ash instead of evolving with the combustion gases. Invented in China and developed in Australia in the 1990s by Coal Corporation Pty. Not to be confused with Sulfex or SULF-X.

Sultrol A *flue-gas desulfurization process, similar to the *Flakt-Boliden process, but using potassium citrate instead of sodium citrate. Developed by Pfizer and announced in 1985, but not known to have been commercialized.

Sulzer A family of processes for purifying organic chemicals by melt-crystallization without using solvents. Two systems are in use: static crystallization, and falling-film crystallization. The latter is proprietary to Sulzer Chemtec, Switzerland.

> Wynn, N. P., *Chem. Eng. Prog.,* 1992, **88**(3), 52.

Sulzer-MWB [Metalwerk A. G. Buchs] An obsolete process for recovering naphthalene from a coal tar fraction by multi-stage fractional crystallization from the melt. Formerly operated by Rutgerswerke at Castrop-Rauxel, Germany.

> U.S. Patent 3,621,664.
> Franck, H.-G. and Stadelhofer, J. W., *Industrial Aromatic Chemistry,* Springer-Verlag, Berlin, 1988, 303.

Sumitomo-BF A *PSA hydrogen purification process using a carbon molecular sieve as the selective adsorbent. Developed by Sumitomo, Japan.

> Suzuki, M., in *Adsorption and Ion Exchange: Fundamentals and Applications,* LeVan, M. D., Ed., American Institute of Chemical Engineers, New York, 1998, 122.

Superclaus [**super**ior **Claus**] A superior version of the *Claus process. Hydrogen sulfide is catalytically oxidized to elemental sulfur using air and water. The first generation of catalysts used iron and chromium oxides on α-alumina. The second generation used iron oxide on silica. The latest version, Superclaus-99, uses a different catalyst which produces less sulfur dioxide in the oxidation stage. Developed by in The Netherlands by Comprimo, V.E.G. Gasinstitut, and the University of Utrecht, and operated in Germany since 1988. Sixty units were licensed and/or built between 1991 and 1997.

> Lagas, J. A., Borsboom, J., and Heijkoop, G., *Hydrocarbon Process.,* 1989, **68**(4), 40.
> Goar, B. G., Lagas, J. A., Borsboom, J., and Heijkoop, G., *Sulphur,* 1992, (220), 44.
> Wieckowska, J., *Catal. Today,* 1995, **24**(4), 442.
> *Hydrocarbon Process.,* 1996, **75**(4), 144.

Supercondensed Mode A method of operating gas-phase olefin polymerization plants. *See* High Productivity.

Super Drizo *See* Drizo.

Superflex A process for converting mixed hydrocarbon streams to olefins. Developed by Arco Chemical.

> *Chem. Br.,* 1993, 110.

SUPER-SCOT An improved version of the *SCOT process which emits smaller quantities of sulfur compounds. Engineered by Stork Engineers and Contractors.

Supersorbon A process for removing organic vapors from gas streams by adsorption on activated carbon. The organic materials are subsequently recovered by steam treatment. Developed by Lurgi in 1923 and still used today.

SuRe [**Su**lphur **re**covery] A version of the *Claus process in which the capacity of the plant is increased by using air enriched in oxygen in the production of the sulfur dioxide. There are two versions: **SURE SSB** [Side Stream Burner], and **SURE DC** [Double Combustion]. In the first, a small portion of the feed stream containing hydrogen sulfide is burnt sub-stoichiometrically in a second burner; in the second, the hydrogen sulfide is oxidized in two stages, with cooling and sulfur separation between them. Both of these modifications are to provide the hydrogen sulfide/sulfur dioxide mixture at the optimum ratio and temperature for the Claus reaction. Developed by the Ralph M. Parsons Company and BOC Group. One plant was operating in Japan in 1992. A large development plant was installed at a carbon disulfide plant at Stretford, United Kingdom in 1996.

Chem. Eng. (Rugby, England), 1991, (494), 13.
Eur. Chem. News, 1997, **67**(1746), 19.
Hydrocarbon Process., 1996, **75**(4), 144.

SURECAT A method for prereducing and passivating nickel catalysts. Developed in 1990 by Eurocat for ATOCHEM.

Rommelaere, F., Rondi, J. C., Dufresne, P., Rabehasaina, H., Boitiaux, J. P., and Sarrazin, P., *Bull. Soc. Chim. Belg.,* 1991, **100,** 897.

Suspensoid An early *catalytic cracking process in which the silica-alumina catalyst was suspended in the petroleum. First operated in Ontario in 1940.

Asinger, F., *Mono-olefins: Chemistry and Technology,* translated by B. J., Hazzard, Pergamon Press, Oxford, 1968, 22.
Unzelman, G. H. and Wolf, C. J., in *Petroleum Processing Handbook,* Bland, W. F. and Davidson, R. L., Eds., McGraw-Hill, New York, 1967, 3-11.

SVP A process for making chlorine dioxide by reacting sodium chlorate with hydrochloric acid. Invented in 1971 by the Hooker Chemical Corporation, Niagara Falls.

U.S. Patent 3,816,077.

Swift A process for making monoammonium phosphate from liquid ammonia and phosphoric acid. The reactants are mixed with a special nozzle. The slurry product from the neutralization is injected into the top of a heated tower; water flashes off as the product falls and the powdered product collects at the base of the column. *See also* Gardinier.

SWITGTHERM A catalytic process for oxidizing volatile organic compounds (VOCs). It involves regenerative heat exchange, which permits autothermal operation at VOC concentrations in the range 250 to 650 ppm. Developed in Poland and now used in over 100 installations there.

Haber, J. and Borowiak, M., *Appl. Catal. A: Gen.,* 1997, **155**(2), 293.

SX A common abbreviation for solvent extraction, as used in hydrometallurgy.

SYDEC [Selective yield delayed coking] A process which converts petroleum residues to petroleum coke and lighter hydrocarbons. Developed by Foster Wheeler USA Corporation.

Elliott, J. D., *Oil & Gas J.,* 1991, **89**(5), 41.
Hydrocarbon Process., 1994, **73**(11), 96.

Sydox [Sydney oxidation] A process for destroying polychlorinated biphenyls by oxidation. A catalyst containing ruthenium is used, and the temperature is kept below 100°C to prevent the formation of dioxins. Developed by J. Beattie at the University of Sydney in the 1980s; by 1991 it had not been piloted.

Beder, S., *New Sci.,* 1991, **130**(1772), 36.

syngas Also called synthesis gas. Not a process, but a general name for mixtures of carbon monoxide and hydrogen made from petroleum fractions or coal. Widely used as a feed for synthesizing organic chemicals by catalytic processes.

Cornils, B., in *Chemicals from Coal: New Processes,* Payne, K. R., Ed., John Wiley & Sons, Chichester, England, 1987, 1, 45, 93.
Rostrup-Nielsen, J. R., *Catal. Today,* 1993, **18**(4), 305.
Eur. Chem. News CHEMSCOPE, 1997, Sept., 24.

Synol A version of the *Fischer-Tropsch process developed in Germany during World War II. It used a different catalyst and produced a larger fraction of alcohols and olefins.

Storch, H. H., Golumbic, N., and Anderson, R. B., *The Fischer-Tropsch and Related Syntheses,* John Wiley & Sons, New York, 1951, 559.

Synroc [**Syn**thetic **roc**k] A process for immobilizing radioactive wastes by incorporating them in a synthetic rock. Invented in 1978 by A. E. Ringwood in the Australian National University, Canberra, and subsequently developed further in many other laboratories in several countries. The "rock" has four main components, all containing titanium: perovskite, zirconolite, hollandite, and rutile. A non-radioactive pilot plant was designed and built at the Australian Nuclear Science and Technology Organization (ANSTO), Sydney, and operated from 1988 to 1991. The project is continuing at ANSTO with emphasis on Synroc processing science and active plant design. In 1990 the Australian company Nuclear Waste Management Pty, which has nonexclusive rights, proposed to build a pilot plant in The USSR, but this plan was probably abandoned.

Ringwood, A. E., Kesson, S. E., Reeve, K. D., Levins, D. M., and Ramm, E. J., in *Radioactive Waste Forms for the Future,* Lutze, W. and Ewing, R. C., Eds., North-Holland, Amsterdam, 1988, 233.

SynSat [**Syn**ergetic **Sat**uration] A process for removing aromatic hydrocarbons and sulfur compounds from diesel fuel. Developed by ABB Lummus Crest and Criterion Catalyst Company. Six units were operating in 1996.

Suchanek, A. J., *Oil & Gas J.,* 1990, 7 May, 109.
Stanislaus, A. and Cooper, B. H., *Catal. Revs., Sci. Eng.,* 1994, **36**(1), 113.
Hydrocarbon Process., 1996, **75**(11), 130.

Synthane A coal gasification process using steam and oxygen in a fluidized bed. An unusual feature is the large volume of hot gas recycled. Developed by the U.S. Bureau of Mines from 1961. A pilot plant, designed by the C. E. Lummus Company, was built at Bruceton, PA, in 1976.

Hebden, D. and Stroud, H. J. F., in *Chemistry of Coal Utilization,* 2nd Suppl. Vol., Elliott, M. A., Ed., John Wiley & Sons, New York, 1981, 1692.

synthesis gas *See* syngas.

Synthetic Oils Also known as the Robinson-Bindley process. A variation on the *Fischer-Tropsch process which uses a different catalyst, a different H_2/CO ratio, and yields a higher proportion of olefins in the product. Piloted by Synthetic Oils, United Kingdom, in the 1930s, but not commercialized.

Mydleton, W. W., *J. Inst. Fuel,* 1938, **11**, 477.

Synthine [**Synth**etic **benzin**] An early version of the *Fischer-Tropsch process in which a mixture of carbon monoxide and hydrogen was passed over an iron catalyst and thereby converted to a complex mixture of oxygenates.

Fischer, F. and Tropsch, H., *Ber. Dtsch. Chem. Ges.,* 1923, **56**, 2428.
Lane, J. C., *Pet. Refin.,* 1946, **25**(8), 87; **25**(9), 423; **25**(10), 493; **25**(11), 587 (*Chem. Abstr.,* **42**, 9118).

Synthoil A coal liquifaction process in which coal, suspended in oil from the process, is hydrogenated over a cobalt/molybdenum catalyst on alumina. The process was piloted by the Pittsburgh Energy Research Center at Bruceton, PA in the 1970s using several types of coal, but it was abandoned in 1978. *See also* CSF, H-Coal.

Yavorsky, P. M., Akhtar, S., and Freidman, S., *Chem. Eng. Prog.,* 1973, **69**(3), 51.
Yavorsky, P. M., Akhtar, S., Lacey, J. J., Weintraub, M., and Reznik, A. A., *Chem. Eng. Prog.,* 1975, **71**(4), 79.

Synthol   A version of the *Fischer-Tropsch process, for making liquid fuels and organic chemicals from *syngas. Developed by Pullman Kellogg between 1940 and 1960. First operated at the SASOL plant in South Africa in 1955. The name was used also for the product from the original Fischer-Tropsch process, developed in the 1920s. *See also* Synol.

> *Hydrocarbon Process.,* 1963, **42**(11), 225.
> Garrett, L. W., Jr., *Chem. Eng. Prog.,* 1960, **56**(4), 39.
> Asinger, F., *Paraffins, Chemistry and Technology,* translated by B. J. Hazzard, Pergamon Press, Oxford, 1968, 96.
> Mako, P. F. and Samuel, W. A., in *Handbook of Synfuels Technology,* Meyers, R. A., Ed., McGraw-Hill, New York, 1984, 2-13.

Synthracite   *See* carbonization.

Syntroleum   A *GTL process developed by Syntroleum Corporation, specially suitable for small plants in remote locations. Licensed to Texaco and Marathon Oil. One such plant is scheduled to be built by Brown and Root for Texaco in 1999.

> *Appl. Catal. A: Gen.,* 1997, **155**(1), N5.
> *Hydrocarbon Process.,* 1998, **77**(1), 41.

T

Taciuk   *See* carbonization.

Tainton   A metallurgical process in which sulfides are converted to sulfates by heating in a controlled quantity of air, and the sulfates so produced are dissolved out in water. Used for removing zinc from silver and lead ores.

Takahax   A variation of the *Stretford process for removing hydrogen sulfide from gas streams, in which naphthaquinone sulfonic acid is used in place of anthraquinone disulfonic acid. Four variants have been devised: types A and B use ammonia as the alkali, types C and D use sodium hydroxide or carbonate. Developed by the Tokyo Gas Company and licensed in the United States by Ford Baken and Davis, Dallas, TX. Many plants are operating in Japan.

> Kohl, A. L. and Riesenfeld, F. C., *Gas Purification,* 4th ed., Gulf Publishing, Houston, TX, 1985, 537.
> *Hydrocarbon Process.,* 1975, **54**(4), 105.

Takasago   A catalytic process for the enantioselective isomerization of allylic amines. The catalyst is a chiral rhodium complex. Used in the manufacture of (-)menthol. Named after Takasago International Corporation, the Japanese company which commercialized the process in 1983.

> *Catalytic Asymmetric Synthesis,* Ojima, I., Ed., VCH Publishers, New York, 1993, 42.

Talafloc   A process for decolorizing cane sugar syrup by precipitating the coloring impurities with a long-chain quaternary ammonium salt. Invented in 1967 by M. C. Bennett at Tate and Lyle, United Kingdom.

> British Patent 1,224,990.
> U.S. Patent 3,698,951.

Talalay   A process for making foam rubber. The foaming gas is either carbon dioxide or oxygen generated from hydrogen peroxide and an enzyme. After foaming, the latex particles are coagulated by freezing and introducing carbon dioxide. Invented by T. Talalay at the B. F. Goodrich Company in 1959. *See also* Dunlop.

> U.S. Patent 2,984,631.
> Madge, E. W., *Latex Foam Rubber,* John Wiley & Sons, New York, 1962.

Talbot   A semi-continuous steelmaking process which combines the Bessemer and *Open Hearth processes. Molten pig iron from a Bessemer converter is poured into an Open Hearth furnace containing fresh ore and lime. Impurities in the pig iron oxidize and enter the slag. The process improves the yield of steel and the throughput of the plant. Introduced by B. Talbot at Pencoed, PA, in 1900 and subsequently adopted in Europe.

> Barraclough, K. C., *Steelmaking 1850–1900,* The Institute of Metals, London, 1990, 283.

Talodura   A process for clarifying cane sugar syrup. Calcium phosphate is precipitated in it, and flocculated with a polyacrylamide. Invented in 1973 by J. T. Rundell and P. R. Pottage at Tate & Lyle, United Kingdom.

> British Patent 1,428,790.

TAREX   A process for destroying hydrocarbons, organic wastes, and hydrogen chloride by controlled combustion in a special combustion chamber which can withstand sudden pressure surges. Engineered by KEU Energie-&-Umwelttechnik, Germany.

TAS   [Autothermal aerobic stabilization]   An autothermal, aerobic process for stabilizing activated sewage sludge. Offered by Linde, Munich.

Tatoray   [Transalkylation aromatics **Toray**]   A process for transalkylating toluene, and/or trimethylbenzenes, into a mixture of benzene and xylenes. Operated in the vapor phase, with hydrogen, in a fixed bed containing a zeolite catalyst. Developed jointly by Toray Industries and UOP and now licensed by UOP. First operated commercially in Japan in 1969; as of 1992, 23 units were operating and 6 more were in design and construction.

> Otani, S., *Chem. Eng. (N.Y.),* 1970, **77**(16), 118.
> Iwamura, T., Otani, S., and Sato, M., *Bull. Jpn. Pet. Inst.,* 1971, **13,** 116 (*Chem. Abstr.,* **75,** 131292).
> *Hydrocarbon Process.,* 1975, **54**(11), 115.
> Jenneret, J. J., in *Handbook of Petroleum Refining Processes,* Meyers, R. A., Ed., McGraw-Hill, New York, 1997, 2.55.

T2BX   [Toluene "to" benzene and xylenes]   A process for converting toluene to a mixture of benzene and xylenes. Toluene vapor, mixed with hydrogen, is passed over a zeolite catalyst at 430 to 370°C; the hydrogen is separated and the products fractionated. The production of benzene is its main purpose. Developed by Cosden Technology, United States.

> *Chem. Eng. (N.Y.),* 1987, **94**(2), 159.

TCC   [Thermofor Catalytic Cracking]   *See* Thermofor.

TCF   [Totally Chlorine-Free]   A generic term for pulp-bleaching processes which do not use chlorine in any form. Oxidants and enzymes are used. *See also* ECF.

> *Chem. Eng. (N.Y.),* 1997, **104**(4), 33.
> Nelson, P. J., in *Environmentally Friendly Technologies for the Pulp and Paper Industries,* Young, R. A. and Akhar, M., Eds., John Wiley & Sons, New York, 1998, 215.

TDP   [Toluene disproportionation process]   A general name for catalytic processes for converting toluene to a mixture of xylene isomers and benzene. One proprietary version is *MTDP.

TEES [Thermochemical environmental energy systems] A catalytic process for destroying organic wastes in aqueous systems by thermochemical gasification. High temperatures and pressures are used. The catalyst is nickel metal supported on sodium carbonate; the products are mostly methane, carbon dioxide, and hydrogen. Developed by Battelle Pacific Northwest Laboratory, Richland, WA, in the late 1980s and now licensed by *Onsite *Ofsite, Inc.*

U.S. Patent 5,019,135.
Process Eng. (London), 1989, Oct, 25.

Teijin A process for oxidizing *p*-xylene to terephthalic acid. The catalyst used is a soluble cobalt compound, used at a high concentration. Developed by Teijin, Tokyo.

Yoshimura, T., *Chem. Eng. (N.Y.),* 1969, **76**(10), 78.
Raghavendrachar, P. and Ramachandran, S., *Ind. Eng. Chem. Res.,* 1992, **31**, 453.

Tekkosha An electrolytic process for obtaining sodium from the sodium amalgam formed in the *chlor-alkali process. The electrolyte is a fused mixture of sodium hydroxide, sodium iodide, and sodium cyanide. The sodium deposits at the iron cathode. Developed by Tekkosha Company, Japan, in the 1960s and commercialized in 1971.

Yamaguchi, T., *Chem. Econ. Eng. Rev.,* 1972, **4**(1), 24.
Nakamura, T. and Fukuchi, Y., *J. Met.,* 1972, **24**(8), 25.

Tenex-Plus A process for increasing the octane number of gasoline by a combination of hydrogenation and isomerization. *See also* Ben-Sat.

Weissermel, K. and Arpe, H.-J., *Industrial Organic Chemistry,* 3rd ed, VCH Publishers, Weinheim, Germany, 1997, 347.

Tennessee Eastman *See* Eastman.

Tenteleff Also spelled Tentelew. An early version of the *Contact process for making sulfuric acid. The catalyst was platinum supported on asbestos. Invented in 1907 and operated by the Gesellschaft der Tentelewschen Chemischen Fabrik, St. Petersberg.

British Patents 12,213 (1907); 14,670 (1911).
Miles, F. D., *The Manufacture of Sulfuric Acid (Contact Process),* Gurney & Jackson, London, 1925, Chap. 9.

Terra-Crete A process for stabilizing the calcium sulfate/sulfite waste from *flue-gas desulfurization, so that it may be used for landfill. Calcination converts the calcium sulfite to cementitious material to which proprietary additions are made. Developed by SFT Corporation, York, PA. *See also* Terra-Tite.

Valiga, R., in *Toxic and Hazardous Waste Disposal,* Vol. 1, Pojasek, R. J., Ed., Ann Arbor Science, Ann Arbor, MI, 1979, Chap. 10.

Terra-Tite A process for stabilizing the calcium sulfate/sulfite waste from *flue-gas desulfurization, so that it may be used for landfill. Proprietary cementitious additives are used. Developed by the Stabatrol Corporation, Norristown, PA. *See also* Terra-Crete.

Smith, R. H., in *Toxic and Hazardous Waste Disposal,* Vol. 1, Pojasek, R. J., Ed., Ann Arbor Science, Ann Arbor, MI, 1979, Chap. 8.

TeRRox A process for decontaminating soil which has been polluted by hydrocarbons by treating it with hydrogen peroxide. Developed by DeGussa and operated at its plant in Knapsack, Germany, from 1996.

Chem. Process. SA, 1996, **3**(4), 8.

TERVAHL A *visbreaking process, developed by the Institut Français du Pétrole for ASVAHL and commercialized in a joint venture between Société Nationale Elf Aquitane, Institut Français du Pétrole, and Total Oil Company. One plant was operating in France in 1988. *See also* HYVAHL.

> *Hydrocarbon Process.,* 1994, **73**(11), 148.
> Chauvel, A., Delmon, B., and Hölderich, W. F., *Appl. Catal. A: Gen.,* 1994, **115,** 186.

Testrup *See* Normann.

Tetra [**Tetra**ethylene glycol] Also called Tetra-extraction. A process for removing aromatic hydrocarbons from petroleum fractions by liquid–liquid extraction using tetraethylene glycol. Developed by Union Carbide Corporation as an improvement on the *Udex process. In 1981, 30 Udex-type units had been converted to this process. Now licensed by UOP.

> *Hydrocarbon Process.,* 1980, **59**(9), 204.
> Symoniac, M. F., Ganju, Y. N., and Vidueira, J. A., *Hydrocarbon Process.,* 1981, **60**(9), 139.
> Vidueira, J. A., in *Handbook of Solvent Extraction,* Lo, C. C., Baird, M. H. I., and Hanson, C., Eds., John Wiley & Sons, Chichester, England, 1983, 18.2.2.

TETRA HDS [High density solids] A process for aiding the removal of heavy metals from wastewaters. It is a physical process which controls the characteristics of heavy metal hydroxide precipitates so that they settle quicker. The precipitates have a hydrophobic surface, so they are easy to de-water. Developed and licensed by Tetra Technologies, Houston, TX. Widely used by the iron and steel industry in the United States. Not to be confused with *hydrodesulfurization, often abbreviated to HDS.

Tetronics A process for treating dusts from electrical arc furnaces for making steel and nonferrous metals. Volatile metals (zinc, lead, cadmium) are recovered, and residual slag is nontoxic and suitable for landfill. The dusts, mixed with coal dust and a flux, are fed to a furnace heated by a plasma gun. The metal oxides present are selectively reduced and the vapors of zinc, lead, and cadmium are condensed in a modified *Imperial Smelting furnace. Developed by Tetronics Research & Development Company, United Kingdom, and first commercialized for steel dusts at Florida Steel, Jackson, TN, in 1989. Seven plants were operating in several countries in 1992.

> Chapman, C. D. and Cowx, P. M., *Steel Times,* 1991, Jun, 301.

Texaco The Texaco Oil Company has developed many processes, but the one mostly associated with its name is a coal gasification system. Powdered coal, in the form of a water slurry, together with oxygen, is fed to the gasifier. The water moderates the temperature of the reaction. Development began at Texaco's laboratory in Los Angeles in 1948 and has continued ever since. The first demonstration unit was built in Morgantown, WV, in 1957. The process is now operated in the United States, Germany, and Japan, and is licensed in China.

> Schlinger, W. D., in *Handbook of Synfuels Technology,* Meyers, R. A., Ed., McGraw-Hill, New York, 1984, 3-5.
> *Chem. Week,* 1987, **141**(3), 36.

Texaco Selective Finishing *See* TSF.

Thann A process for making crystalline calcium hypochlorite by passing chlorine into an aqueous slurry of calcium hydroxide. There are several such processes; in this one, some of the filtrate is recycled in order to produce larger crystals. Invented by J. Ourisson in France in 1936.

> French Patent 825,903.
> British Patent 487,009.

THD [Toluene **h**y**d**rodealkylation] A process for converting toluene to benzene, developed by the Gulf Oil Corporation.

THDA [Thermal **h**y**d**rodealkylation] A process for dealkylating alkyl benzenes to produce benzene. The by-product is mainly methane. Developed by UOP and licensed by that company.

> Mowry, J. R., in *Handbook of Petroleum Refining Processes,* Meyers, R. A., Ed., McGraw-Hill, New York, 1986, 2-3.

Thénard A process for making white lead pigment (basic lead carbonate) by boiling litharge (lead monoxide) with lead acetate solution and passing carbon dioxide gas into the suspension.

Thermal Black One of the processes used to make carbon black. The feedstock is usually natural gas. The gas is pyrolyzed in one of a pair of refractory reactors which has been preheated by burning part of the feed and hydrogen from the process. When the temperature has fallen, the functions of the reactors are interchanged.

> Kühner, G. and Voll, M., in *Carbon Black Science and Technology,* Donnet, J.-B., Bansai, R. C., and Wang, M.-J., Eds., Marcel Dekker, New York, 1993, 59.

thermal cracking The pyrolysis of petroleum fractions to produce lower molecular weight materials. Developed by J. A. Dubbs and C. P. Dubbs in 1909 and demonstrated on a larger scale in Kansas in 1919. Such processes with special names which are described in this dictionary are: ACR, Burton, Carburol, Cross, Dubbs, Dubrovai, Fleming, FLEXICOKING, Flexicracking, FLUID COKING, Gyro, Hall, Hoechst HTP, HSC, Jenkins, Knox, MHDV, TERVAHL, TPC, Tube and Tank, TVP, Visbreaking. *See also* catalytic cracking.

Thermatomic An early process for making carbon black by the incomplete combustion of natural gas. Operated in the 1920s and 1930s by the Thermatomic Carbon Company, Pittsburgh.

> Ellis, C., *The Chemistry of Petroleum Derivatives,* The Chemical Catalog Co., New York, 1934, 210.

Thermatrix A process for destroying organic vapors in waste streams by oxidation in a packed-bed reactor containing a porous inert ceramic matrix. The operating temperature is 870 to 980°C, at which temperature hydrocarbons are oxidized to carbon dioxide and water, and chlorinated, fluorinated, or sulfonated hydrocarbons are converted to HCl, HF, or SO_2, respectively. Quantities of carbon monoxide and NO_x produced are minimal. Developed by Thermatrix, CA, in the 1990s. By 1996 it had been installed in over 50 plants in a variety of industries. Although the capital cost is greater than that of a similar catalytic oxidation unit, the running cost is claimed to be less.

> Hohl, H. M., *Oil & Gas J.,* 1996, **94**(45), 77.

Thermit Also spelled Thermite, and also called the Goldschmidt process. The reaction of metallic aluminum with a metal oxide is very exothermic and can be used to liberate other metals from their oxides, or simply as a source of heat. In the latter case, iron oxide is used:

$$2Al + Fe_2O_3 = Al_2O_3 + 2Fe$$

Refractory metals such as chromium, manganese, and cobalt are made in this way. The process was invented by H. Goldschmidt at the German company Th. Goldschmidt, Essen, in 1898. *See also* Goldschmidt.

THERMOCAT A petroleum cracking process which combines fixed-bed catalytic cracking with steam cracking. Developed by Veba Oel and Linde from 1994. *See* PYROCAT.

Thermofax   An early thermographic copying process using paper impregnated with a ferric salt, a heavy metal sulfide, and a phenolic compound.

U.S. Patents 2,663,654; 2,663,655; 2,663,656; 2,663,657.
Chem. Eng. News, 1964, 13 Jul, 115; 20 Jul, 85.

Thermofor   This name was first used in the 1930s for the equipment and process for burning off the carbon which deposits on the clays used for purifying mineral oils. The lumps of clay were regenerated by passing through a hot reactor. The Socony Vacuum Oil Company subsequently used this technology as the basis for its range of processes for regenerating cracking catalysts. These included: Thermofor Catalytic Cracking (*TCC), Thermofor Catalytic Reforming (*TCR), Thermofor Pyrolytic Cracking (*TPC), *Airlift Thermofor Catalytic Cracking. The first Thermofor cracking process was commercialized in 1943. Socony operated a Thermofor Catalytic Reformer from 1955; the catalyst is chromia on alumina.

Enos, J. L., *Petroleum Progress and Profits,* MIT Press, Cambridge, MA, 1962, 165.
Unzelman, G. H. and Wolf, C. J., in *Petroleum Processing Handbook,* Bland, W. F. and Davidson, R. L., Eds., McGraw-Hill, New York, 1967, 3-7, 3-34.
The Petroleum Handbook, 6th ed., Elsevier, Amsterdam, 1983, 286.

Thermosoft   A water-softening process for treating waters that are high in dissolved solids and alkalinity, intended for oilfield steam injection. Naturally occurring bicarbonate is added and the temperature is raised to 180 to 200°, causing the hydroxides of calcium and magnesium to precipitate. Invented by T. Bertness and licensed to U.S. Filter Corporation in 1995.

Chem. Eng. (N.Y.), 1995, **102**(11), 23.

Thermosorption   A process for recovering hydrocarbons from wet natural gas by adsorption on activated carbon. Offered by Lurgi. *See also* Supersorbon.

THGP   [Texaco **h**ydrogen **g**eneration **p**rocess]   A process for making pure, high-pressure hydrogen from various gaseous and light hydrocarbons. Partial combustion of the hydrocarbons yields *syngas. The carbon monoxide in this is converted to carbon dioxide by the *shift reaction, and removed by *PSA. Seen as an alternative to steam reforming. Offered by Texaco Development Corporation.

Hydrocarbon Process., 1996 **75**(4), 149.

Thiofex   An early process for refining benzole by treatment with sulfuric acid at a rising temperature, followed by sodium carbonate. Invented in 1947 by T. Scott at Refiners Limited, and used in the United Kingdom.

British Patent 642,772.
Claxton, G., *Benzoles, Production and Uses,* National Benzole & Allied Products Assoc., London, 1961, 434.

Thiolex   A process for removing hydrogen sulfide from a light hydrocarbon liquid by extraction with aqueous sodium hydroxide passed through a bundle of hollow fibers immersed in it. Developed by the Merichem Company, Houston, TX. In 1991, 52 units were operating. Variations are known as Thiolex/Regen and Thiolex/Regen/Mericat. *See also* Mericat.

Hydrocarbon Process., 1984, **63**(4), 87.
Hydrocarbon Process., 1996, **75**(4), 126.

Thionate   *See* Feld.

Thiopaq   A process for treating the spent alkaline solutions used in oil refineries for scrubbing hydrogen sulfide from gas streams. Developed by UOP.

Oil & Gas J., 1997, **95**(29), 54.

Thoma   A process for alkylating aniline with methanol or ethanol, to produce mixtures of mono- and di-alkylanilines. Operated in hot, concentrated phosphoric acid in a vertical tubular reactor. The proportions of secondary and tertiary amines can be partly controlled by controlling the ratios of the reactants; the products are separated by fractional distillation. Invented in 1954 by M. Thoma in Germany.

U.S. Patent 2,991,311.

Thomas   Also called the Basic Bessemer process. A variation of the *Bessemer process for steelmaking, for use with ores rich in phosphorus, in which the converter is lined with calcined dolomite (magnesium and calcium oxides), and limestone is added to the charge; the phosphorus remains with the slag. Invented by the cousins Sidney Gilchrist Thomas and Percy Carlisle Gilchrist (hence the alternative name for the process: Thomas and Gilchrist) in 1877. Developed at Blaenavon steelworks in South Wales and first commercialized at the works of Bolckow, Vaughan & Company in Middlesbrough, England. Widely used for the treatment of iron ores rich in phosphorus. *See also* Bessemer.

British Patent 4,422 (1877).
Barraclough, K. C., *Steelmaking 1850–1900*, The Institute of Metals, London, 1990, 207, 222.

Thomas and Gilchrist   *See* Thomas.

Thompson-Stewart   A process for making basic lead carbonate ("white lead," $2PbCO_3 \cdot Pb(OH)_2$), by reacting lead monoxide ("litharge") with acetic acid and then with carbon dioxide. Basic lead acetate is an intermediate. *See also* Dutch, Carter.

Thorex   [Thorium extraction]   A process for separating the products from the nuclear breeder reaction in which uranium-233 is produced by the neutron bombardment of thorium-232. It uses solvent extraction into tri-*n*-butyl phosphate. Developed at the Oak Ridge National Laboratory, TN, in the early 1960s. *See also* Butex, Purex, Redox.

Thoroughbred   A family of *flue-gas desulfurization processes. Commercialized versions have been called Thoroughbred-101, Thoroughbred-102, and Thoroughbred-121. Another name used for the last is CT-121. They are integrated *flue-gas desulfurization processes which achieve limestone neutralization, oxidation, and conversion to gypsum, in a complex jet bubbling reactor. Developed by the Chiyoda Chemical Engineering and Construction Company, Japan, and widely used in Japan and the United States. As of 1986, CT-121 had been installed in eight plants in the United States and Japan.

U.S. Patents 4,099,925; 4,156,712 (reactor); 4,178,348; 4,203,954 (process).
Idemura, H., *Chem. Econ. Eng. Rev.*, 1974, **6**(8), 22.
Tamaki, A., *Chem. Eng. Prog.*, 1975, **71**(5), 55.
Kohl, A. L. and Riesenfeld, F. C., *Gas Purification*, 4th ed., Gulf Publishing, Houston, TX, 1985, 333, 374.

Thum   A variation of the *Balbech process for separating silver from gold in which the electrodes are held vertically. The anodes are contained in cloth bags to retain the slimes; silver deposits at the cathodes and is periodically scraped off.

Thylox   A process for removing hydrogen sulfide from refinery and coke-oven gases by absorption in a solution of sodium ammonium thioarsenate. The solution is regenerated by blowing air through it, precipitating elemental sulfur, which is filtered off:

$$(NH_4)_3AsO_2S_2 + H_2S = (NH_4)_3AsOS_3 + H_2O$$

$$2(NH_4)_3AsOS_3 + O_2 = 2(NH_4)_3AsO_2S_2 + 2S$$

Invented in 1926 by H. A. Gollmar and D. L. Jacobson at the Koppers Company. Although the process had shortcomings – it removed only 90 to 96 percent of the hydrogen sulfide and the sulfur produced was quite impure – it had been used in ten installations in the United States by 1950.

> U.S. Patents 1,719,177; 1,719,180; 1,719,762.
> Kohl, A. L. and Riesenfeld, F. C., *Chem. Eng. (N.Y.)*, 1959, **66**, 152.
> Kohl, A. L. and Riesenfeld, F. C., *Gas Purification*, 4th ed., Gulf Publishing, Houston, TX, 1985, 497.

Thyssen-Galoczy   A slagging coal gasification process.

TIGAS   [Topsoe integrated gasoline synthesis]   A multi-stage process for converting natural gas to gasoline. Developed by Haldor Topsoe and piloted in Houston from 1984 to 1987. Not commercialized, but used in 1995 as the basis for a process for making dimethyl ether for use as a diesel fuel.

> Topp-Joergensen, J., *Stud. Surf. Sci. Catal.*, 1988, **36**, 473.
> Rouhi, A. M., *Chem. Eng. News*, 1995, **73**(22), 37.

Tin Sol   A process for making a stannic oxide sol by electrodyalysis. Invented in 1973 by H. P. Wilson of the Vulcan Materials Company, Birmingham, AL.

> U.S. Patent 3,723,273.

TIP   [Total isomerization process]   *See* Total Isomerization.

Tisco   A direct reduction ironmaking process, using coal as the reductant. Operated on a small scale in India since 1986.

Titanizing   [From the Greek, Titan, meaning a person of superhuman strength; often incorrectly assumed to be derived from the name of the element titanium]   A process for hardening the surfaces of glass vessels by coating them with a layer of titanium dioxide or tin dioxide. The oxides are deposited from the vapors of the respective tetrachlorides or tetraalkoxides by Chemical Vapor Deposition (see CVD). Invented by S. M. Budd at United Glass, United Kingdom, and widely used for strengthening glass bottles.

> British Patents 1,115,342; 1,187,784.

Titanox FR   A process for making cellulose textiles flame-resistant by treating them with titanyl acetate chloride and antimony oxychloride. Invented in 1951 by W. F. Sullivan and I. M. Panik at the National Lead Company, New York. *See also* Erifon.

> U.S. Patent 2,658,000.

TN   *See* steelmaking.

Topnir   Not a chemical process but an instrumental process for on-line monitoring of hydrocarbon process streams by infrared spectroscopy. Developed by BP and offered for license in 1997.

> *Eur. Chem. News*, 1997, **67**(1753), 23.

Toray (1)   A large Japanese chemicals manufacturer, perhaps best known for its process for synthesizing *l*-lysine for use as a dietary supplement. The starting material is cyclohexene which is converted in five steps to racemic lysine. An enzymic process isolates the desired optical isomer; the other is recycled.

Toray (2)   process for making terephthalic acid by oxidizing *p*-xylene, using a cobalt catalyst promoted by paraldehyde.

Raghavendrachar, P. and Ramachandran, S., *Ind. Eng. Chem. Res.,* 1992, **31**, 453.

Toray Aromax   *See* Aromax.

TOSCO II   A process for extracting oil from shale, by contacting it with hot ceramic or steel balls in a rotating drum. Based on an invention made in 1922 by F. Puening in Pittsburgh, who used hot iron balls to provide heat for the destructive distillation of lignite, shale, peat, and bituminous coal. Further developed by the Oil Shale Corporation in the 1950s. The Tosco Corporation, in association with Exxon Corporation, continued the work in Colorado from the 1960s until the U.S. oil shale projects were discontinued in 1982.

U.S. Patents 1,698,345–1,698,349, inclusive.
Klass, D. L., *CHEMTECH,* 1975, **5**, 499.
Waitman, C. S., Braddock, R. L., and Siebert, T. E., in *Handbook of Synfuels Technology,* Meyers, R. A., Ed., McGraw-Hill, New York, 1984, 4-45.

TOSCOAL   [The Oil Shale Corporation]   A low-temperature carbonization process for producing liquid fuels from oil shales. Developed by the Oil Shale Corporation in the 1960s. *See* TOSCO II.

Carlson, F. B., Yardumian, L. H., and Atwood, M. T., *Chem. Eng. Prog.,* 1973, **69**(3), 50.
Probstein, R. F. and Hicks, R. E., *Synthetic Fuels,* McGraw-Hill, New York, 1982, 259.

Total Isomerization   Also called TIP. An integrated process which combines light paraffin isomerization, using a zeolite catalyst, with the *IsoSiv process, which separates the unconverted normal paraffins so that they can be returned to the reactor. Developed by Union Carbide Corporation and now licensed by UOP. The first plant was operated in Japan in 1975; by 1992, more than 25 units had been licensed.

Hydrocarbon Process., 1980, **59**(5), 110.
Hydrocarbon Process., 1988, **67**(9), 82.
Cusher, N. A., in *Handbook of Petroleum Refining Processes,* Meyers, R. A., Ed., McGraw-Hill, New York, 1997, 9.29.

Toth   A method proposed for making aluminum metal from clay. The dried clay, mixed with coke, is chlorinated to yield aluminum trichloride and silicon tetrachloride. The volatile chlorides are separated by distillation and the aluminum chloride then reduced with manganese metal:

$$Al_xSi_yO_z + (3x + 4y)/2\ Cl_2 + z/2\ C = xAlCl_3 + ySiCl_4 + z/2\ CO_2$$

$$2AlCl_3 + 3Mn = 2Al + 3MnCl_2$$

The manganese chloride is then reduced with hydrogen:

$$MnCl_2 + H_2 = Mn + 2HCl$$

and the manganese metal re-used. In another version, the manganese chloride is oxidized in oxygen to manganese oxide, which is then reduced to the metal using carbon. Invented in 1969 by C. Toth and piloted by the Applied Aluminium Research Corporation in Baton Rouge, LA, in the mid-1970s but not commercialized. The practicality and economics of this complex cyclic process remain controversial. Meanwhile, a plant for making aluminum trichloride by this process, intended for use as a catalyst, is planned for completion in Louisiana in 1999.

U.S. Patent 3,918,960.
Chem. Eng. News, 1973, **51**(9), 11.
Grjotheim, K., Krohn, C., Malinovsky, M., Matiaskovsky, K., and Thonstad, J., *Aluminium Electrolysis—Fundamentals of the Hall-Hérault Process,* CRC Press, Boca Raton, FL, 1982, 13.

Eur. Chem. News, 1984, **42**(1120), 10.
Chem. Mark. Rep., 1997, **251**(13), 5.

Tower Biology A biological waste-treatment process, developed from the *Activated Sludge process. The sludge is contained in a tall tower, at the base of which oxygen is injected as small bubbles. The bubbles are almost completely absorbed by the time they reach the surface of the liquid. The system uses less energy than does surface aeration. Developed by Bayer in 1980 for its plant at Leverkusen, Germany; subsequently adopted in India and then elsewhere.

Chem. Eng. (N.Y.), 1992, **99**(12), 101.

Townsend A process for removing hydrogen sulfide from natural gas by absorption in triethylene glycol containing sulfur dioxide. Part of the sulfur produced is burnt to sulfur dioxide in order to provide this solution. The hydrogen sulfide and sulfur dioxide react in the presence of water to generate elemental sulfur. Invented in 1959 by F. M. Townsend.

U.S. Patent 3,170,766.
Townsend, F. M. and Reid, L. S., *Pet. Refin.,* 1958, **37,** 263.
Kohl, A. L. and Riesenfeld, F. C., *Gas Purification,* 4th ed., Gulf Publishing, Houston, TX, 1985, 539.

TPC [Thermofor pyrolytic cracking] A continuous process for thermally cracking petroleum fractions on a moving bed of hot pebbles. Developed by the Socony Vacuum Oil Company (now a part of Mobil Corporation). *See also* Thermofor.

Eastwood, S. C. and Potas, A. E., *Pet. Eng.,* 1948, **19**(12), 43.
Asinger, F., *Mono-olefins: Chemistry and Technology,* translated by B. J. Hazzard, Pergamon Press, Oxford, 1968, 163.

Trail A process for recovering elemental sulfur from sulfur dioxide by reduction with carbon:

$$SO_2 + C = CO_2 + S$$
$$CO_2 + C = 2CO$$

Carbonyl sulfide is an intermediate in this reaction. A mixture of sulfur dioxide and oxygen was blown into the bottom of coke-fired reduction furnace and sulfur vapor condensed from the off-gases. Trail is the location of a large mine and smelter in British Columbia. The process was originally used in the 1930s for abating air pollution from the smelter, but when the demand for sulfuric acid for fertilizer production increased in 1943 it became obsolete. *See also* Boliden (1), RESOX.

Katz, M. and Cole, R. J., *Ind. Eng. Chem.,* 1950, **42,** 2264.

Tramex [Transuranic metal (or amine) extraction] A process for separating transuranic elements from fission products by solvent extraction from chloride solutions into a tertiary amine solution. Developed at Oak Ridge National Laboratory, TN, for processing irradiated plutonium.

Leuze, R. E. and Lloyd, M. H., *Prog. Nucl. Energy,* 1970, (Ser. III), **4,** 596.

transalkylation In organic chemistry, any reaction in which an alkyl group is transferred from one molecule to another. In process chemistry the word has a more limited meaning, generally restricted to aromatic rearrangements such as the conversion of toluene to benzene and C_8 aromatic hydrocarbons. Examples of such processes are *LTD, *Tatoray.

Transcat An *oxychlorination process for making vinyl chloride from ethane and chlorine:

$$2C_2H_6 + Cl_2 + \tfrac{1}{2}O_2 = 2CH_2{=}CHCl + 3H_2O$$

The catalyst, and the source of the oxygen, is cupric oxide dissolved in a molten mixture of cupric chloride and potassium chloride. Developed by Lummus Corporation.

Weissermel, K. and Arpe, H.-J., *Industrial Organic Chemistry,* 3rd ed., VCH Publishers, Weinheim, Germany, 1997, 221.

Transplus   A transalkylation process for making mixed xylenes from heavy aromatics and toluene. Developed by Mobil Technology and Chinese Petroleum Corporation.

Eur. Chem. News, 1997, **68**(1783), 33.

TRC   [Thermal Regenerative Cracking]   Also known as Quick Contact and QC. A process for making olefins from petroleum fractions by rapid thermal cracking of petroleum residues. The feed is cracked by passing through a hot fluidized bed of micron-sized refractory particles, with a contact time of a quarter of a second. Developed by Gulf Chemical, Gulf Canada, and Stone & Webster, and piloted at Cedar Bayou, TX, from 1981.

Hu, Y. C., in *Chemical Processing Handbook,* Marcel Dekker, New York, 1993, 782.
Eur. Chem. News, 1996, **65** (1710), 24.
Picciotti, M., *Oil & Gas J.,* 1997, **95**(25), 55.

Treadwell   A process for extracting copper from chalcopyrite by leaching with the stoichiometric quantity of sulfuric acid:

$$CuFeS_2 + 4H_2SO_4 = CuSO_4 + FeSO_4 + 2SO_2 + 2S + 4H_2O$$

Developed by the Anaconda company in 1968 but not commercialized.

Prater, J. D., Queneau, P. B., and Hudson, T. J., *J. Met.,* 1970, **22**(12), 23.
Gupta, C. K. and Mukherjee, T. K., *Hydrometallurgy in Extraction Processes,* Vol. 1, CRC Press, Boca Raton, FL, 1990, 78.

treating   A general name for processes which remove S-, N-, and O-compounds from petroleum streams.

Trencor   Also called Trentham Trencor. A wet-scrubbing process for removing residual sulfur dioxide and hydrogen sulfide from the tail gas from the *Claus process.

Speight, J. G., *Gas Processing,* Butterworth Heinemann, Oxford, 1993, 327.

Trentham Trencor   *See* previous entry.

Trickle Hydrodesulfurization   A process for removing sulfur-, nitrogen-, and heavy-metal-compounds from petroleum distillates before *catalytic cracking. The preheated feed is hydrogenated, without a catalyst, in an adiabatic reactor at 315 to 430°C. Developed by Shell Development Company. As of 1978, 91 units had been installed.

Hoog, H., Klinkert, H. G., and Schaafsma, A., *Pet. Refin.,* 1953, **32**(5), 137.
Hydrocarbon Process., 1964, **43**(9), 194.

Tri-NOx   A process for removing NO_x and nitric acid mists from the waste gases from the manufacture of electronic devices. Developed by Wacker Siltronic Corporation, based on a scrubber engineered and manufactured by Tri-Mer Corporation.

Chem. Eng. (N.Y.), 1996, **103**(12), 123.

Triolefin   A process for disproportionating propylene into a mixture of ethylene and 2-butene. The reaction takes place at 160°C over a cobalt/molybdenum catalyst on an alumina base. Developed by the Phillips Petroleum Company from 1963. A commercial plant was built by Gulf Oil Canada in 1966 and operated by Shawinigan between 1966 and 1972 before closing for economic reasons.

U.S. Patent 3,236,912.
Logan, R. S. and Banks, R. L., *Hydrocarbon Process.,* 1968, **47**(6), 135.
Hydrocarbon Process., 1971, **50**(11), 140.

Hydrocarbon Process., 1991, **70**(3), 144.

Weissermel, K. and Arpe, H.-J., *Industrial Organic Chemistry*, 3rd ed., VCH Publishers, Weinheim, Germany, 1997, 67.

Truex [Transuranium extraction]   A process for removing transuranic elements during the processing of nuclear fuel by solvent extraction. Developed by E. P. Howitz at the Argonne National Laboratory, Chicago, IL. *See also* SREX.

TRW Gravichem   A modification of the TRW Meyers process (see next entry) in which the coal fraction which contains less pyrites is first removed by sedimentation in the ferric sulfate solution.

IEA Coal Research, *The Problems of Sulphur*, Butterworths, London, 1989, 30.

TRW Meyers [Named after the three CalTech professors who founded the company: Thompson, Ramo, and Wooldridge]   A chemical method for desulfurizing coal. The iron pyrites is leached out with a hot aqueous solution of ferric sulfate, liberating elemental sulfur. The resulting ferrous sulfate solution is re-oxidized with air or oxygen:

$$Fe_2(SO_4)_3 + FeS_2 = 3FeSO_4 + 2S$$

$$7Fe_2(SO_4)_3 + 8H_2O + FeS_2 = 15FeSO_4 + 8H_2SO_4$$

$$4FeSO_4 + 2H_2SO_4 + O_2 = 2Fe_2(SO_4)_3 + 2H_2O$$

Meyers, R. A., *Hydrocarbon Process.*, 1975, **54**(6), 93.

IEA Coal Research, *The Problems of Sulphur*, Butterworths, London, 1989, 28.

TSA [Thermal (or Temperature) swing adsorption]   A method for separating gases by cyclic adsorption and desorption from a selective adsorbent, at alternating temperatures. Less commonly used than *PSA.

Yang, R. T., *Gas Separation by Adsorption Processes*, Butterworths, Guildford, England, 1987, 204.

Sherman, J. D. and Yon, C. M., in *Kirk-Othmer's Encyclopedia of Chemical Technology*, 4th ed., Vol. 1, John Wiley & Sons, New York, 1991–1998, 546.

TSF [Texaco Selective Finishing]   A process for separating linear from branched-chain aliphatic hydrocarbons by *PSA, using zeolite 5A as the adsorbent. The desorbent is a hydrocarbon having two to four carbon atoms less than the feed. Developed by Texaco in the late 1950s. Believed to be still in operation in its Trinidad refinery as of 1990.

Franz, W. F., Christensen, E. R., May, J. E., and Hess, H. V., *Oil & Gas J.*, 1959, **57**(15), 112.

Cooper, D. E., Griswold, H. E., Lewis, R. M., and Stokeld, R. W., *Eur. Chem. News*, 1966, **10**(254), Suppl., 23.

TSR (1) [Trace sulphur removal]   A process for removing sulfur compounds from naphtha so that they will not poison catalytic reformers. A proprietary solid absorbent is used. Developed by the Union Oil Company of California and first commercialized in 1983 at the Unocal oil refinery in San Francisco.

TSR (2) [Tiomin synthetic rutile]   A process for removing much of the iron from ilmenite in order to make a feedstock for titanium pigment manufacture. Developed by Tiomin Resources, Canada, in the 1990s.

T-Star   A *hydrotreating process introduced by Texaco in 1993.

Hydrocarbon Process., 1997, **76**(2), 45.

TTH [Tieftemperaturhydrierung; German, meaning low-temperature hydrogenation]   A petroleum refining process for converting tars and middle distillates into lower-boiling fractions. The catalyst is a combination of tungsten and nickel sulfides. Developed in Germany by IG Farbenindustrie. *See also* MTH.

> Weisser, O. and Landa, S., *Sulphide Catalysts, Their Properties and Applications,* Pergamon Press, Oxford, 1973, 333.

Tube and Tank   A continuous process for thermally cracking petroleum, developed by Standard Oil of Indiana in the 1920s.

> Ellis, C., *The Chemistry of Petroleum Derivatives,* The Chemical Catalog Co., New York, 1934, 109.
> Enos, J. L., *Petroleum Progress and Profits,* MIT Press, Cambridge, MA, 1962, Chap. 3.

TVP   [True vapor-phase]   A *thermal cracking process in which vaporized petroleum oil is contacted with a hotter gas such that the temperature of the gas mixture is approximately 500°C. Used in the 1930s, but supplanted by various *catalytic cracking processes.

> *The Petroleum Handbook,* 3rd ed., Shell Petroleum Co., London, 1948, 172.

Twitchell   An early process for the acid-catalyzed hydrolysis of animal and vegetable fats for the production of glycerol and soap. The catalyst was a mixture of sulfonated oleic and naphthenic acids and sulfuric acid, known as "Twitchell saponifier." Invented in 1897 by E. Twitchell and commercialized by Joslin, Schmidt & Company, Cincinnati, OH. The British soapmakers at that time, Joseph Crosfield & Sons, did not use it because the products were considered to be too dark in color.

> U.S. Patents 601,603; 628,503; 1,170,468.
> Twitchell, E., *J. Am. Chem. Soc.,* 1900, **22,** 22.
> Twitchell, E., *J. Am. Chem. Soc.,* 1906, **28,** 196.
> Mills, V. and McClain, H. K., *Ind. Eng. Chem.,* 1949, **41,** 1982.
> Musson, A. E., *Enterprise in Soap and Chemicals,* Manchester University Press, Manchester, 1965, 174.

Tyrer   A process for making phenol by first sulfonating benzene. Benzene vapor was passed through hot sulfuric acid; the excess of benzene served to remove the water formed in the reaction. The benzene sulfonic acid was then hydrolyzed by fusion with sodium hydroxide. Invented by D. Tyrer in 1916. *See also* Dennis-Bull.

> U.S. Patents 1,210,725; 1,210,726.
> Kenyon, R. L. and Boehmer, N., in *Modern Chemical Processes,* Vol. 2, Reinhold Publishing, New York, 1952, 35.

Tysland-Hole   *See* steelmaking.

U

UCAP A process for selectively removing residual sulfur dioxide from the tail gas from the *Claus process. It had not been commercialized by 1983.

Yon, C. M., Atwood, G. R., and Swaim, C. D., Jr., *Hydrocarbon Process.*, 1979, **58**(7), 197.

UCAR [Union Carbide Carbon dioxide] A process for removing carbon dioxide from gas streams by scrubbing with methyl diethanolamine. Use of a proprietary corrosion inhibitor permits higher concentrations of the amine to be used than in similar processes.

Hawkes, E. N. and Mago, B. F., *Hydrocarbon Process.*, 1971, **50**(8), 109.

UCARSOL [Union Carbide solvent] *See* HS.

UCB-MCI [Union Chimique—Chemische Bedrijven and Ministry of Chemical Industry for the USSR] An *EHD process for making adiponitrile, differing from the *Monsanto process in using an emulsion of acrylonitrile and in not using a membrane.

Uchatius A modification of the *Huntsman process for making steel. Cast iron was first granulated by pouring the molten metal into water. The granules, mixed with fresh iron ore containing manganese, and fireclay, were then heated as in the Huntsman process. The molten product was poured into molds. Invented in Austria by F. Uchatius, a captain in the Austrian Army, and operated at the Newburn steelworks on Tyneside, England, from 1856 to 1876. In Viksmanshyttan, Sweden, it was operated from 1859 to 1929.

British Patent 2,189 (1855).
Barraclough, K. C., *Steelmaking Before Bessemer, Vol. 2, Crucible Steel,* The Metals Society, London, 1984, 71.
Barraclough, K. C., *Steelmaking 1850–1900,* The Institute of Metals, London, 1990, 36.

U-COAL A coal gasification process developed by UBE Industries, Japan.

Udex [Universal Dow extraction] A process for removing aromatic hydrocarbons from petroleum fractions by liquid–liquid extraction with glycols, followed by extractive distillation. The glycol used originally was diethylene glycol, but later this was replaced by triethylene glycol and tetraethylene glycol. Extraction is conducted at 140 to 150°C under a pressure of 10 atm. Developed jointly by Dow Chemical Company and UOP in the 1940s and used on a large scale since 1952. Largely replaced by the *Sulfolane process in the 1960s. *See also* Tetra.

Grote, H. W., *Chem. Eng. Prog.*, 1958, **54**(8), 43.
Beardmore, F. S. and Kosters, W. C. G., *J. Inst. Pet.*, 1963, **49**(469), 1.
Somekh, G. S. and Friedlander, B. I., *Hydrocarbon Process.*, 1969, **48**(12), 127.
Hoover, T. S., *Hydrocarbon Process.*, 1969, **48**(12), 131.
Achilladelis, B., *Chem. Ind. (London)*, 1975, (8), 343.
Franck, H.-G. and Stadelhofer, J. W., *Industrial Aromatic Chemistry,* Springer-Verlag, Berlin, 1988, 107.

Ufer A proces for refining the light oil produced in coal carbonization. The oil is washed with sulfuric acid and a controlled amount of water is then added to the mixture. The complex mixture of reaction products ("resins") enters the oil phase; the dilute sulfuric acid can be used directly for making ammonium sulfate. Invented in Germany in 1924 by A. Ufer; operated in Germany and Canada in the 1920s and 1930s.

German Patent 489,753.
British Patent 251,117.
Claxton, G., *Benzoles, Production and Uses,* National Benzole & Allied Products Assoc., London, 1961, 433.

U-GAS [Utility **gas**]   A process for gasifying carbonaceous products such as coal, oil, forest wastes, and municipal solid wastes, by reacting them with steam and oxygen (or air) at 950 to 1,100°C, under 3 to 35 atm. The product gases are hydrogen, carbon monoxide, carbon dioxide, and some methane. The ash is agglomerated into rough spheres, hence the name of the equipment – Ash Agglomerating Gasifier. Developed by the Institute of Gas Technology, Chicago, from 1974 and tested there on a variety of coals. First licensed in 1989 to Tampella, a Finnish corporation involved in forest products, which built a plant at Messukyla, Finland, in 1990.

Dainton, A. D., in *Coal and Modern Coal Processing,* Pitt, G. J. and Millward, G. R., Eds., Academic Press, London, 1979.
Patel, J. G., *Int. J. Energy Res.,* 1980, **4,** 149.
Chem. Eng. (N.Y.), 1996, **103**(3), 41.

UGI [United **G**as **I**mprovement Company]   Also called Ugite. A regenerative catalytic reforming process for making town gas and liquid hydrocarbons from oil. The catalyst was a fixed bed of hot, refractory pebbles. Developed by UGI Company, Philadelphia, PA, in the early 1940s.

Asinger, F., *Mono-olefins: Chemistry and Technology,* translated by B. J. Hazzard, Pergamon Press, Oxford, 1968, 168.
Gas Making and Natural Gas, British Petroleum Co., London, 1972, 93.

UGINE   *See* DR.

Ugite   *See* UGI.

Uhde-Hibernia   A process for making a mixed ammonium nitrate – ammonium sulfate fertilizer (ASN) – which is less liable to explode than ammonium nitrate. Sulfuric acid is added to aqueous ammonium nitrate and ammonia gas passed in. The double salt crystallizes out. Additives are used to improve the handling characteristics of the product. Developed by Hibernia and licensed to Friedrich Uhde. *See also* Victor.

Nitrogen, 1968, (53), 27.

Uhde/Schwarting   *See* Schwarting.

Ultracat   A version of the *FCC process, developed by Standard Oil of Indiana in the 1970s.

Ultrafining   Two *hydrodesulfurization processes developed by Standard Oil of Indiana, one for petroleum residua and one for vacuum gas oil.

Unzelman, G. H. and Wolf, C. J., in *Petroleum Processing Handbook,* Bland, W. F. and Davidson, R. L., Eds., McGraw-Hill, New York, 1967, 3-40.

Ultraforming   A *catalytic reforming process developed by Standard Oil of Indiana and licensed by Amoco Oil Company. The catalyst contains platinum and rhenium, contained in a "swing" reactor – one that can be isolated from the rest of the equipment so that the catalyst can be regenerated while the unit is operating. The first unit was commissioned in 1954.

Unzelman, G. H. and Wolf, C. J., in *Petroleum Processing Handbook,* Bland, W. F. and Davidson, R. L., Eds., McGraw-Hill, New York, 1967, 3-32.
Little, D. M., *Catalytic Reforming,* PennWell Publishing, Tulsa, OK, 1985, 169.

Ultra-Orthoflow An *FCC process which converts petroleum distillates and heavier fractions to products of lower molecular weight. Developed by MW Kellogg Company. Over 100 units were operating in 1988.

Hydrocarbon Process., 1988, **67**(9), 67.

ULTROX A process for removing traces of organic compounds from groundwater or wastewater by oxidizing them with ozone and/or hydrogen peroxide, under the influence of ultraviolet radiation. Invented and developed by Ultrox International in the 1970s and first demonstrated in 1989 with contaminated groundwater from a former drum recycling plant in San Jose, CA. Further developed with the General Electric Company as its *GEODE process. Ultrox International was bought by Zimpro Environmental (now U.S. Filter) in 1993.

U.S. Patent 4,792,407.
Chem. Eng. (N.Y.), 1989, **96**(4), 19.
Hughes, S., *Water Waste Treat.*, 1992, **35**(7), 26.
Masten, S. J. and Davies, S. H. R., in *Environmental Oxidants*, Nriagu, J. O. and Simmons, M. S., Eds., John Wiley & Sons, New York, 1994, 533.

UMATAC A process for extracting hydrocarbons from tar sands. The sand is heated in a rotating kiln in which the tar is thermally cracked. Developed in Calgary, Canada, in the 1970s by UMATAC Industrial Processes. It had not been commercialized by1984.

Bowman, C. W., Phillips, R. S., and Turner, L. R., in *Handbook of Synfuels Technology*, Meyers, R. A., Ed., McGraw-Hill, New York, 1984, 5-42.

Unibon A family of related processes offered by UOP for inter-converting hydrocarbons using combinations of *hydrotreating and *hydrocracking. *See* AH Unibon, HB Unibon, HC Unibon, LPG Unibon, LT Unibon, RCD Unibon.

UNICARB [Union Carbide carbon dioxide] A process for spray painting objects using supercritical carbon dioxide as the solvent. Developed by Union Carbide Company and workers at Johns Hopkins University.

Ind. Health Hazards Update, 1995, Apr.

Unicoil An early thermal process for cracking petroleum.

Asinger, F., *Mono-olefins: Chemistry and Technology*, translated by B. J. Hazzard, Pergamon Press, Oxford, 1968, 339.

Unicracking A *hydrocracking process for simultaneously hydrogenating and cracking various liquid petroleum fractions to form hydrocarbon mixtures of lower molecular weight. The catalyst is either zeolite-containing or an amorphous aluminosilicate. Jointly developed and marketed by UOP and Union Oil Company of California. In 1990, 84 units were operating. The technology was acquired by UOP in 1995.

Adams, N. R., Watkins, C. H., and Stine, L. O., *Chem. Eng. Prog.*, 1961, **57**(12), 55.
Speight, J. G., *The Desulfurization of Heavy Oils and Residua*, Marcel Dekker, New York, 1981, 181.
Hydrocarbon Process., 1994, **73**(11), 128.
Reno, M., in *Handbook of Petroleum Refining Processes*, Meyers, R. A., Ed., McGraw-Hill, New York, 1997, 7.41.

Unicracking/DW [Dewaxing] A version of the *Unicracking process developed for upgrading waxy petroleum fractions.

UNIDAK A process for extracting naphthalene from reformer residues in petroleum refining. It includes a dealkylation stage to convert the naphthalene homologues to naphthalene.

The process temperature is approximately 600°C; the catalyst is based on cobalt/molybdenum. Developed by the Union Oil Company of California.

Hydrocarbon Process., 1963, **42**(11), 232.

Unifining A *hydrodesulfurization process developed jointly by UOP and Union Oil Company of California. It is now incorporated in the UOP *hydrotreating and UOP *Unibon processes.

Claxton, G., *Benzoles, Production and Uses*, National Benzole & Allied Products Assoc., London, 1961, 452.
Unzelman, G. H. and Wolf, C. J., in *Petroleum Processing Handbook*, Bland, W. F. and Davidson, R. L., Eds., McGraw-Hill, New York, 1967, 3-39.

Unionfining A group of petroleum *hydrodesulfurization and *hydrodenitrogenation processes developed by the Union Oil Company of California, primarily for making premium quality diesel fuel. In 1991, 90 such units were operating. One variant is for purifying naphthalene by selective hydrogenation. The naphthalene vapor is hydrogenated at 400°C over a cobalt/molybdenum catalyst, thereby converting the sulfur in thionaphthalene to hydrogen sulfide. The technology was acquired by UOP in 1995.

Hydrocarbon Process., 1988, **67**(9), 79.
Eur. Chem. News, 1995, **63**(1653), 24.
Kennedy, J. E., in *Handbook of Petroleum Refining Processes*, Meyers, R. A., Ed., McGraw-Hill, New York, 1997, 8.29.

UNIPOL [**Uni**on Carbide **Pol**ymerization] A process for polymerizing ethylene to polyethylene, and propylene to polypropylene. It is a low-pressure, gas-phase, fluidized-bed process, in contrast to the *Ziegler-Natta process, which is conducted in the liquid phase. The catalyst powder is continuously added to the bed and the granular product is continuously withdrawn. A co-monomer such as 1-butene is normally used. The polyethylene process was developed by F. J. Karol and his colleagues at Union Carbide Corporation; the polypropylene process was developed jointly with the Shell Chemical Company. The development of the ethylene process started in the mid 1960s, the propylene process was first commercialized in 1983. It is currently used under license by 75 producers in 26 countries, in a total of 96 reactors with a combined capacity of over 12 million tonnes/y. It is now available through Univation, the joint licensing subsidiary of Union Carbide and Exxon Chemical. A supported metallocene catalyst is used today.

U.S. Patents 4,003,712; 4,011,382.
Chem. Eng. Int. Ed., 1979, **86**(26), 80.
Karol, F. J., *CHEMTECH*, 1983, **13**, 222.
Karol, F. J. and Jacobsen, F. I., in *Catalytic Polymerization of Olefins*, Keii, T. and Soga, K., Eds., Elsevier, Amsterdam, 1986, 323.
Hydrocarbon Process, 1991, **70**(3), 173.
Burdett, I. D., *CHEMTECH*, 1992, **22**(10), 616.

Unisar [**Uni**on **s**aturation of **ar**omatics] A process for hydrogenating aromatic hydrocarbons in petroleum fractions, using a noble metal heterogeneous catalyst. Developed by the Union Oil Company of California. The first commercial unit opened in Beaumont, TX, in 1969; eight commercial plants were in operation in 1991.

Hydrocarbon Process., 1970, **49**(9), 231.
Hydrocarbon Process., *Refinery Process Handbook*, 1982, 137.
Gowdy, H. W., in *Handbook of Petroleum Refining Processes*, Meyers, R. A., Ed., McGraw-Hill, New York, 1997, 8.55.

Unisol A process for extracting organic sulfur and nitrogen compounds from petroleum fractions by solvent extraction with aqueous sodium or potassium hydroxide containing methanol. First operated in Montana in 1942.

> Unzelman, G. H. and Wolf, C. J., in *Petroleum Processing Handbook,* Bland, W. F. and Davidson, R. L., Eds., McGraw-Hill, New York, 1967, 3-118.

Unisulf [**Un**ocal **sulf**ur removal] A process for removing sulfur compounds from petroleum fractions, similar to the *Stretford process, but including in the catalytic solution: vanadium, a thiocyanate, a carboxylate (usually citrate), and an aromatic sulfonate complexing agent. Developed by the Union Oil Company of California in 1979, commercialized in 1985, and operated in three commercial plants in 1989.

> U.S. Patent 4,283,379.
> Dalrymple, D. A., Trofe, T. W., and Evans, J. M., *Chem. Eng. Prog.,* 1989, **85**(3), 43.
> Kohl, A. L. and Riesenfeld, F. C., *Gas Purification,* 4th ed., Gulf Publishing, Houston, TX, 1985, 536.

UNOX A modification of the *Activated Sludge sewage treatment process for treating domestic effluents, based on the use of oxygen instead of air, in closed reaction tanks. The preferred source of oxygen depends on the size of the plant; small plants use liquid oxygen; medium-sized plants use the *PSA process; and large plants have cryogenic generators. Developed by the Union Carbide Corporation in the late 1960s and now licensed to a number of other companies.

> Lewandowski, T. P., *Water Pollut. Control,* 1974, **73**(6), 647.
> Gray, N. F., *Activated Sludge: Theory and Practice,* Oxford University Press, Oxford, 1990, 119.

Urbain A process for activating charcoal by heating it with phosphoric acid and then washing with hydrochloric acid. Invented by E. Urbain in 1923; the product was used in the 1920s and 1930s for recovering benzole vapor.

> British Patent 218,242.
> French Patent 579,596.
> Claxton, G., *Benzoles, Production and Uses,* National Benzole & Allied Products Assoc., London, 1961, 417.

Urea 2000plus A process for making urea from ammonium carbamate, using a novel pool reactor. Developed by DSM and offered for license in 1996.

> *Eur. Chem. News,* 1996, **65**(1716), 22.

USC [**U**ltra **S**elective **C**onversion] A front-end process for improving the operation of catalytic crackers for making ethylene. Developed and offered by Stone & Webster Engineering Corporation.

USCO A direct reduction ironmaking process, using coal gas as the reductant. Operated in South Africa for several years since 1985, but now abandoned. *See* DR.

USS Phosam *See* Phosam.

UTI A process for making urea from ammonia and carbon dioxide, using heat-recycle. Invented in 1970 by I. Mavrovic in New York.

> U.S. Patents 3,759,992; 3,952,055.
> Frayer, J. Y., *Chem. Eng. (Rugby, England),* 1973, **80**(12), 72.

UVOX [**U**ltra**v**iolet **ox**idation] A process for purifying seawater before clarification, using chlorine and exposure to sunlight. Developed in South Africa.

> Hebden, D. and Botha, G. R., *Desalination,* 1980, **32**, 115.

V

VACASULF   [**Vac**uum de**sulf**urization]   A process for removing hydrogen sulfide from coke-oven gas by scrubbing with aqueous potassium carbonate:

$$K_2CO_3 + H_2S = KHS + KHCO_3$$

Hydrogen sulfide is recovered from the scrubbing solution under vacuum, hence the name. It is then either oxidized with air and the sulfur dioxide used for making sulfuric acid, or converted to elemental sulfur by the *Claus process. The process is suitable only for gases not containing ammonia. Developed by Krupp Koppers, Germany. Three units were being built in 1993.

Vacuum carbonate   An improved version of the *Seabord process for removing hydrogen sulfide from refinery gases, in which the hydrogen sulfide is stripped from the sodium carbonate solution by steam instead of by air. Developed by the Koppers Company, Pittsburgh, in 1939; two plants were using this process in the United States in 1950.

> U.S. Patents 2,379,076; 2,464,805; 2,242,323.
> Reed, R. M. and Updegraff, N. C., *Ind. Eng. Chem.*, 1950, **42**, 2271.
> Claxton, G., *Benzoles, Production and Uses,* National Benzole & Allied Products Assoc., London, 1961, 212.

VAD   *See* steelmaking.

Valorga   A process for treating household waste by anaerobic digestion with the production of methane. Developed in the 1990s at the Languedoc University of Science & Technology.

> *Waste Manag. Environ.*, 1996, **6**(3), 40.

van Arkel and de Boer   Also called the Iodide process. A process for producing or purifying a metal by thermal decomposition of its iodide on a hot tungsten filament. Modern high-intensity halide lamps are based on this reaction. Originally used for making small quantities of tungsten, titanium, and zirconium, but today probably used only for the preparation of ultrapure chromium. First used in 1925 by A. E. van Arkel at the Philips Gloelampenfabrik in The Netherlands for making tungsten; subsequently used by him and J. H. de Boer for making other metals. An extension of the process has been to make metal nitrides by passing a mixture of the metal halide with nitrogen and hydrogen over a hot tungsten wire.

> U.S. Patent 1,671,213.
> van Arkel, A. E., *Physica*, 1923, **3**, 76.
> van Arkel, A. E. and de Boer, J. H., *Z. Anorg. Allg. Chem.*, 1925, **148**, 345.
> Rolsten, R. F., *Iodide Metals and Metal Iodides,* John Wiley & Sons, New York, 1961.

Van Dyke   A reprographic process, based on the photoreduction of ferric ammonium oxalate to ferrous amonium oxalate, which in turn reduces silver nitrate to silver metal.

> Kosar, J., *Light Sensitive Systems,* John Wiley & Sons, New York, 1965.

Van Ruymbeke (1)   A process for recovering glycerol from the residual liquor from the kettle soapmaking process. After separating the solid soap, the liquor is heated with basic ferric sulfate, thereby precipitating the residual carboxylic acids as their insoluble ferric salts. After removing these by filtration, the liquor is concentrated by vacuum evaporation and the glycerol distilled out under vacuum. *See also* Garrigue.

> Martin, G. and Cooke, E. I., in *Industrial and Manufacturing Chemistry,* Cooke, E. I., Ed., Technical Press, Kingston Hill, Surrey, 1952, 122.

Van Ruymbeke (2) A process for dehydrating 95 percent aqueous ethanol by counter-current extraction of the vapor with glycerol.

British Patents 184,036; 184,129.
U.S. Patent 1,459,699.

VaporSep A family of separation processes, based on membranes which are selectively permeable to organic vapors. Developed by Membrane Technology & Research, CA, in the 1990s and used by DSM in its polypropylene plant to separate propylene from nitrogen. The membrane is a three-layer sandwich, packaged in spiral modules.

Eur. Chem. News, 1996, **65**(1700), 24.
Eur. Chem. News, 1997, **67**(1771), 16.

VAR *See* steelmaking.

Varga A complex process for hydrogenating brown coal and high-molecular-weight asphaltenes. The process uses hydrogen at a high pressure, in the presence of an iron oxide catalyst. Invented by J. Varga in Budapest and operated in Germany.

Varga, J., *Brennstoff-Chem.,* 1928, **9**, 277.
Varga, J., Rabo, G., and Zalai, A., *Brennstoff-Chem.,* 1956, **37**, 244.
Weisser, O. and Landa, S., *Sulphide Catalysts, Their Properties and Applications,* Pergamon Press, Oxford, 1973, 296.

VAROX An air separation process, allowing for variable oxygen demand. Developed by Linde from 1984.

VAW Lurgi [Vereinigte Aluminiumwerke] An energy-efficient process for extracting aluminum from bauxite. Extraction is done in a pipe reactor, and the aluminum hydroxide is converted to the oxide in a fluidized bed.

VAW Sulfite [Vereinigte Aluminiumwerke] Also known as the S-T process. A process for extracting aluminum from clay by sulfurous acid. Basic aluminum sulfite, $Al_2O_3 \cdot 2SO_2 \cdot 5H_2O$, is crystallized from the purified leachate and thermally decomposed. The process was operated on a commercial scale by Vereinigte Aluminiumwerke at Lauterwerk, Germany, during World War II.

O'Connor, D. J., *Alumina Extraction from Non-bauxitic Materials,* Aluminium-Verlag, Düsseldorf, 1988, 188.

VCC *See* Veba-Combi Cracking.

Veba A process for *hydrocracking bitumen, developed by Veba Oel, Germany.

Graeser, U. and Niemann, K., *Oil & Gas J.,* 1982, **82**(12), 121.

Veba-Combi Cracking Also called VCC. A *Bergius-Pier high-pressure thermal hydrotreating process. The catalyst is usually a promoted iron oxide, operated in a slurry, but an added catalyst may not be necessary. Used in Germany during World War II. A version developed by Veba Oel Entwicklungsgesellschaft mbH was operated in Bottrop, Germany, from 1988 to 1994, using vacuum residues from crude oil distillation and visbreaking operations, with plastics from municipal wastes, and with chlorinated organic compounds.

U.S. Patent 4,851,107.
Bowman, C. W., Phillips, R. S., and Turner, L. B., in *Handbook of Synfuels Technology,* Meyers, R. A., Ed., McGraw-Hill, New York, 1984, 5-74.
Chauvel, A., Delmon, B., and Hölderich, W. F., *Appl. Catal. A: Gen.,* 1994, **115**, 173.

Ventron   A process for removing mercury from aqueous wastes containing organic mercury compounds. Chlorine is passed in, converting organic mercury compounds to inorganic compounds, and the mercury is then reduced to the metallic state with sodium borohydride.

   Rosenzweig, M. D., *Chem. Eng., (N.Y.),* 1971, **78**(5), 70.

VERA   [Verglasungsanlage für Radioactive Abfälle]   A continuous process for immobilizing nuclear waste by incorporating it in a borosilicate glass made by spray calcination. Developed at the Kernforschungszentrum Karlsruhe, Germany, from the mid-1960s. The process was abandoned in the mid-1970s, but some of the technology was later used in *PAMELA.

   Lutze, W., in *Radioactive Waste Forms for the Future,* Lutze, W. and Ewing, R. C., Eds., North-Holland, Amsterdam, 1988, 7.

Verneuil   A process for growing single crystals of refractory compounds. The powdered material is dropped through an oxy-hydrogen flame and the product, consisting of microscopic molten droplets, is collected on a seed crystal. As the liquid mass reaches a cooler zone it crystallizes in the form of a single crystal known as a boule. Invented by A. V. L. Verneuil at the Museum of Natural History, Paris, who made synthetic rubies and sapphires in this way.

   Verneuil, A. V. L., *Ann. Chim. Phys.,* 1904, **3**, 20.
   Merker, L., *Min. Eng. (N.Y.),* 1955, **7**, 645.
   Elwell, D., *Man-made Gemstones,* Ellis Horwood, Chichester, England, 1979, 34.
   Vere, A. W., *Crystal Growth: Principles and Progress,* Plenum Press, New York, 1987, 67.

VerTech   A sewage-treatment process in which the sludge is digested under aerobic conditions at high temperature and pressure in a vertical shaft approximately one mile deep. Developed in the 1980s by VerTech Treatment Systems, The Netherlands. Piloted in 1985 at Longmont, CO, and installed at Apeldoorn, The Netherlands, in 1991. *See also* Deep Shaft.

   Water Bull., 1994, 31 Mar, 6.
   Downie, N. A., *Industrial Gases,* Blackie Academic & Professional, London, 1997, 438.

VGO Isomax   [Vacuum gas oil]   A *hydrodesulfurization process adapted for treating vacuum gas oil, a petroleum fraction. Developed by Chevron Research Company in the early 1970s. In 1972, five plants were in operation and six were under construction. *See also* RDS Isomax and VRDS Isomax.

   Hydrocarbon Process., 1972, **51**(9), 184.

Viad   An early, two-stage coal gasification process.

Victor   A process for making a mixed ammonium nitrate – ammonium sulfate fertilizer (ASN) which is less liable to explode than ammonium nitrate. Ammonium sulfate is mixed with nitric acid and the mixture ammoniated in an evaporator. Developed by Gewerkschaft Victor – Chemische Werke and used by that company at Castrop Rauxel, Germany. *See also* Uhde – Hibernia.

   Nitrogen, 1968, (53) 27.

Visbreaking   A thermal cracking process which reduces the viscosity of the residues from petroleum distillation, so that they may be handled at lower temperatures. It is essentially a high-temperature, noncatalytic pyrolytic process conducted in the presence of steam. *See also* HSC.

   Ballard, W. P., Cottingham, G. I., and Cooper, T. A., in *Encyclopedia of Chemical Processing and Design,* McKetta, J. J. and Cunningham, W. A., Eds., Marcel Dekker, New York, 1981, **13**, 172.
   The Petroleum Handbook, 6th ed., Elsevier, Amsterdam, 1983, 280.

Viscose   Also known as the Cross-Bevan-Beadle process. A process for making regenerated cellulose fibers. The product has been known by the generic name "rayon" since 1924. Cellulose, from cotton or wood, is first reacted with sodium hydroxide (*mercerization), yielding alkali cellulose. This is dissolved in carbon disulfide, yielding cellulose xanthate, which is dissolved in sodium hydroxide solution. Injection of this solution (known as viscose because of its high viscosity) into a bath of acid regenerates the cellulose. The process was invented by C. F. Cross, E. J. Bevan, and C. Beadle in London in 1882, further developed in the United States in the 1890s, and then widely adopted worldwide. *See also* Sini.

Cross, C. F., Bevan, E. J., and Beadle, C., *J. Soc. Chem. Ind.*, 1892, **12**, 516.
Moncrieff, R. W., *Man-made Fibres*, 6th ed., Butterworth Scientific, London, 1975, 162.

VITOX   A process for providing oxygen to a microbiological process such as sewage treatment. The heart of the process is a sub-surface mixer-oxygenator, developed by the British Oxygen Company for uprating the oxygenation capacity of overloaded sewage plants. It is now used as an integral design feature in new plants. The oxygen is introduced at the neck of a venturi and forms extremely fine bubbles. The process operates in open tanks, unlike the *Unox process, which uses closed tanks. The equipment has been used also for dissolving carbon dioxide in water, for hardening it with lime. In 1991 the process was used in over 100 sewage plants in the United Kingdom and another 200 in the rest of the world.

Gould, F. J. and Stringer, P. R., in *Effluent Treatment and Disposal,* Institution of Chemical Engineers, Rugby, England, 1986, 33.
Gray, N. F., *Activated Sludge: Theory and Practice,* Oxford University Press, Oxford, 1990, 120.

Vitrifix   A vitrification process for converting asbestos to a harmless glassy substance, suitable for use as a construction material. Developed in the United Kingdom.

Vniios   A process for catalytically pyrolyzing hydrocarbons to low molecular weight alkenes, similar to catalytic cracking but more efficient. The catalyst is either potassium vanadate on corundum or indium oxide on pumice. Developed by the All-Union Research Institute for Organic Synthesis, Moscow.

Oil & Gas J., 1997, **95**(25), 54.

VOD   *See* steelmaking.

Voest   *See* DR.

Volto   *See* Elektrion.

von Heyden   One of several processes for oxidizing naphthalene to phthalic anhydride. It operates with a fixed bed of vanadium/molybdenum oxide catalyst. Another version of the von Heyden process has been developed by Wacker-Chemie for oxidizing *o*-xylene to phthalic anhydride, and is licensed by that company. In 1989, 65 plants had been built or were under construction.

Hydrocarbon Process., 1989, **68**(11), 107.

VPSA   [Vacuum pressure swing adsorption]   Also known as **VSA.** A version of *PSA in which the adsorbed gas fraction is desorbed by reducing the pressure, rather than by displacement. Proprietary versions developed by the Linde Division of Union Carbide Corporation are known as *OxyGEN and *NitroGEN. Invented by L'Air Liquide, France, in 1957.

French Patent 1,223,261.
U.S. Patent 3,155,468.
Young, R. T., *Gas Separation by Adsorption Processes,* Butterworths, Guildford, England, 1987.
Chem. Eng. (N.Y.), 1989, **96**(10), 17.

VRC [Valorisation des résidues chlorés] A process for incinerating chlorinated organic residues. The hydrochloric acid produced is condensed and sold (hence the name). Developed by Atochem, France in 1975, operated at Saint-Auban, and planned for installation in Poland in 1992.

Robin, A., in *Chemical Waste Handling and Treatment*, Muller, K. R., Ed., Springer-Verlag, Berlin, 1985, 268.
Eur. Chem. News, 1990, **55**(1450), 30.

VRDS Isomax [Vacuum residua desulphurization] A *hydrodesulfurization process adapted for processing the residues from the vacuum distillation of the least volatile fraction of petroleum. An extension of the *RDS Isomax process, developed and piloted by Chevron Research Company in the early 1970s. In 1988, one unit was under construction and one was being engineered.

Speight, J. G., *The Desulfurization of Heavy Oils and Residua*, Marcel Dekker, New York, 1981, 194.

VSA [Vacuum swing adsorption] *See* VPSA.

Vulcanization The treatment of natural rubber with sulfur to reduce its tackiness and improve its strength and elasticity. Invented independently by C. Goodyear and N. Hayward in the United States in 1839, and by T. Hancock in London in 1842–1843. Various chemicals other than elemental sulfur are effective, for example, sulfur monochloride, selenium, and *p*-quinone dioxime.

Duerdon, F., *Thomas Hancock—An appreciation. Plas. Rubber Internat.*, 1986, **11**(3), 22.

W

Wacker (1) A general process for oxidizing aliphatic hydrocarbons to aldehydes or ketones by the use of oxygen, catalyzed by an aqueous solution of mixed palladium and copper chlorides. Ethylene is thus oxidized to acetaldehyde. If the reaction is conducted in acetic acid, the product is vinyl acetate. The process can be operated with the catalyst in solution, or with the catalyst deposited on a support such as activated carbon. There has been a considerable amount of fundamental research on the reaction mechanism, which is believed to proceed by alternate oxidation and reduction of the palladium:

$$CH_2{=}CH_2 + PdCl_2 + H_2O = CH_3CHO + Pd + 2HCl$$

$$Pd + 2HCl + \tfrac{1}{2}O_2 = PdCl_2 + H_2O$$

The naming of this process has been confused because of various corporate relationships. The basic invention was created in 1957 at the Consortium für Elektrochemische Industrie, Munich, a wholly owned subsidiary of Wacker-Chemie. It has therefore been called both the Wacker process and the Consortium process. But for many years, Wacker-Chemie has had a close relationship with Farbwerke Hoechst and the latter company has participated in some of the development and licensing activities, so two other names have come to be used: Wacker-Hoechst and Hoechst-Wacker. The five inventors (J. Schmidt, W. Hafner, J. Sedlmeier, R. Jira, and R. Rüttinger) received the Dechema prize in 1962 for this invention.

The acetaldehyde process was first operated commercially in 1960. In 1997, this process was used in making 85 percent of the world's production of acetaldehyde. Although Wacker-Chemie still makes vinyl acetate, it no longer uses the Wacker process to do so.

German Patents 1,049,845; 1,061,767.
Smidt, J., Hafner, W., Jira, R., Sedlmeier, J., Sieber, R., Rüttinger, R., and Kojer, H., *Angew. Chem.*, 1959, **71**(5), 176.
Chem. Eng. News., 1961, **39**(16), 52.
Jira, R., in *Ethylene and Its Industrial Derivatives*, Miller, S. A., Ed., Ernest Benn, London, 1969, 639.
Lowry, R. P. and Aquilo, A., *Hydrocarbon Process.*, 1974, **53**(11), 105.
Weissermel, K. and Arpe, H.-J., *Industrial Organic Chemistry*, 3rd ed., VCH Publishers, Weinheim, Germany, 1997, 165.

Wacker (2)   A process for making sodium salicylate by reacting sodium phenate with carbon dioxide.

Lindsey, A. S. and Jeskey, H., *Chem. Rev.*, 1957, **57**, 583.

Wacker-Hoechst   *See* Wacker (1).

Waelz   A process for extracting zinc and lead from lean ores, using a large rotary kiln. Developed by Metallurgische Gesellschaft and Fried. Krupp Grusenwerk at Magdeburg, Germany, in 1926. The process is used also for extracting zinc and lead from the dusts from electric arc furnaces. The pelletized dusts are mixed with 25 percent coke and 15 percent sand and heated in a rotary kiln to 1,200°C. Lead and zinc volatilize and are collected as dusts which, after bricketting, can be treated by the *Imperial Smelting process.

Cocks, E. J. and Walters, B., *A History of the Zinc Smelting Industry in Britain*, George G. Harrap, London, 1968, 61,150.
Morgan, S. W. K., *Zinc and Its Alloys and Compounds*, Ellis Horwood, Chichester, England, 1985, 141.

Wah Chang   *See* Benilite.

Walker   A process for partially oxidizing natural gas or LPG, forming a mixture of methanol, formaldehyde, and acetaldehyde. Air is the oxidant and aluminum phosphate the catalyst. Invented by J. C. Walker in the 1920s and operated by the Cities Service Corporation, OK, in the 1950s.

U.S. Patent 2,186,688.
Walker, J. C. and Malakoff, H. L., *Oil & Gas J.*, 1946, **45**(33), 59.
Meyer, R. E., *Oil & Gas J.*, 1955, **54**(7), 82.

Walterization   *See* metal surface treatment.

Walthal   An obsolete process for obtaining alumina from clay. The clay was roasted, extracted with sulfuric acid, and the aluminum sulfate dried and calcined.

Walther   Also called Walther Ammonia. A *flue-gas desulfurization process in which the gas is scrubbed with aqueous ammonia. Two scrubbing stages are used, operating at different pH values. The by-product is suitable for use as a fertilizer. Developed and licensed by Walther & Company. Two plants were operating in Germany in 1987, one of which was experiencing serious operational problems.

Merrick, D. and Vernon, J., *Chem. Ind. (London)*, 1989, 3, 55.

Warner   A novel process for extracting zinc from sulfide ores. Two linked furnaces are used. In the first, the ore is reacted with metallic copper:

$$ZnS + 2Cu = Zn + Cu_2S$$

The resulting zinc vapor is condensed to liquid and run off. The copper sulfide is oxidized to copper metal in the second furnace:

$$CuS + O_2 = Cu + SO_2$$

The molten copper is circulated back to the first furnace. The sulfur dioxide is converted to sulfuric acid for sale. Overall, the process uses much less energy than other processes. Developed in the 1980s by N. A. Warner at the University of Birmingham, UK, but not piloted as of 1992.

British Patent 2,048,309.
Gray, P. M. J., *Min. Mag.,* 1992, Jan, 14.

Washoe   A process for extracting silver from sulfide ores. The ore is heated with aqueous sodium chloride in an iron pot. The chloride dissolves the silver and the iron reduces it; addition of mercury gives silver amalgam:

$$2AgCl + Fe = 2Ag + FeCl_2$$

This is a variation of the *Patio and *Cazo processes. Invented around 1860 at the Comstock mines, Nevada, and named after the district where it was developed. Mark Twain described the operations in his autobiographical novel *Roughing It* (Vol. 1, Chap. 36).

Dennis, W. H., *A Hundred Years of Metallurgy,* Gerald Duckworth, London, 1963, 285.

water gas   A generic name applied to two processes and their products. The original process, dating to the end of the 18th century, makes a fuel gas by passing steam over a carbonaceous fuel which has been heated by partial combustion. The product is a mixture of carbon monoxide, carbon dioxide, and hydrogen. It is also known as "blue gas" because it burns with a blue flame. Enrichment of blue gas by adding hydrocarbons was invented by T. S. C. Lowe and first commercialized at Phoenixville, PA in 1874. The product, known as "carburetted water gas," was used mainly as an additive to coal gas; in 1931, 13 percent of the town gas distributed in Great Britain was made by this process. *See also* Blaugas.

Morgan, J. J., in *Chemistry of Coal Utilization,* Vol. 2, Lowry, H. H., Ed., John Wiley & Sons, New York, 1945, Chap. 37.
Peebles, M. W. H., *The Evolution of the Gas Industry,* Macmillan Press, London, 1980, 14.
Parker, A., *J. Soc. Chem. Ind.,* 1927, **46**, 72.

Water gas shift   *See* reforming.

W-D   *See* Woodall-Duckham.

WD-IGI   [Woodall-Duckham Il Gas Internazionale]   A two-stage, nonslagging coal gasification process. In the first stage, tar and volatile matter is removed; in the second, steam and air (or oxygen) gasify the coke, producing a mixture of carbon monoxide, hydrogen, and nitrogen (if air is used). The process is based on a design by Il Gas Integrale, Milan, Italy, developed in the 1950s. In 1984, over 100 plants had been installed in Europe, South Africa, and Australia.

Jones, D. M., in *Handbook of Synfuels Technology,* Meyers, R. A., Ed., McGraw-Hill, New York, 1984, 3-169.

Weber   *See* carbonization.

Weissenstein   An electrolytic process for making hydrogen peroxide by the electrolysis of sulfuric acid. Peroxodisulfuric acid, $H_2S_2O_8$, is formed first and this is then hydrolyzed via peroxomonosulfuric acid:

$$2H_2SO_4 = H_2S_2O_8 + H_2$$

$$H_2S_2O_8 + H_2O = H_2SO_5 + H_2SO_4$$

$$H_2SO_5 + H_2O = H_2O_2 + H_2SO_4$$

First operated in 1908 at the Österreichische Chemische Werke, Weissenstein, Austria and then by Degussa, Germany. This, and the other electrolytic processes, was made obsolete by the invention of the *AO proces.

Schumb, W. C., Satterfield, C. N., and Wentworth, R. L., *Hydrogen Peroxide,* Reinhold Publishing, New York, 1955, 132.

Weizmann A process for producing acetone and *n*-butanol by the fermentation of carbohydrates by bacteria isolated from soil or cereals. Later work has shown that effective bacteria are *Clostridium acetobutylicum* and *Bacillus granulobacter pectinorum*. Used in Britain in World War I for the manufacture of acetone, needed for the production of cordite. Subsequently operated by Commercial Solvents Corporation in Terre Haute, IN, and in two plants in Canada. Later abandoned in favor of synthetic processes. Invented by C. Weizmann in the University of Manchester in 1915, based on earlier work at the Pasteur Institute by A. Fernbach and E. H. Strange (hence the alternative name: Fernbach–Strange–Weizmann). The money that Weizmann obtained from royalties on this process was used in founding the State of Israel, of which he was the first president.

British Patents 21,073 (1912) (Fernbach and Strange); 4,845 (1915) (Weizmann).
U.S. Patent 1,315,585.
Goodman, P., Ed., *Chaim Weizmann: A Tribute on his Seventieth Birthday,* Victor Gollancz, London, 1945.
New Ency. Brit., 1988, **12,** 565.
Benfey, T., *Chem. Ind. (London),* 1992, (21), 827.

Weldon An early process for making chlorine by oxidizing hydrochloric acid (from the *Leblanc process) with manganese dioxide. The mixture was heated with steam in stone tanks. Manganese was recovered from the liquor by precipitation with calcium hydroxide and subsequent oxidation by air:

$$4HCl + MnO_2 = MnCl_2 + Cl_2 + 2H_2O$$

$$MnCl_2 + Ca(OH)_2 = Mn(OH)_2 + CaCl_2$$

$$2Mn(OH)_2 + O_2 = 2MnO_2 + 2H_2O$$

The process was complicated by the formation of calcium manganite, $CaMn_2O_6$, known as Weldon mud. Invented by W. Weldon in 1866 and developed at St. Helens from 1868 to 1870. Operated in competition with the *Deacon process until both were overtaken by the electrolytic process for making chlorine from brine. Weldon mud has been used as a catalyst for oxidizing the hydrogen sulfide in coal gas to elemental sulfur.

British Patents 1,948 (1866); 133 (1867).
Hardie, D. W. F., *A History of the Chemical Industry in Widnes,* Imperial Chemical Industries, Widnes, England, 1950, 66.

Welland A process for making nitroguanidine, an explosive. Cyanamide dimer is reacted with ammonium nitrate to form guanidine nitrate, which forms nitroquanidine when dehydrated by heating with 96 percent sulfuric acid. *See also* Marquerol and Loriette.

Smith, G. B. L., Sabetta, V. J., and Steinbach, O. F., Jr., *Ind. Eng. Chem.,* 1931, **23,** 1124.

Wellman A coal gasification process, widely used since its introduction in the 1950s. Air and steam, at atmospheric pressure, are passed through a fixed bed of coal supported on a rotating ash bed. *See also* Riley-Morgan.

Hebden, D. and Stroud, H. J. F., in *Chemistry of Coal Utilization,* 2nd Suppl. Vol., Elliott, M. A., Ed., John Wiley & Sons, New York, 1981, 1616.

Wellman-Galusha   A coal gasification process using a fixed bed; the dry ash is removed through a revolving grate.

van der Hoeven, B. J. C., in *Chemistry of Coal Utilization,* Vol. 2, Lowry, H. H., Ed., John Wiley & Sons, New York, 1945, 1659.

Wellman-Lord   The most widely used regenerable *flue-gas desulfurization process. The sulfur dioxide is absorbed in sodium sulfite solution in a wet spray scrubber, forming sodium bisulfite:

$$Na_2SO_3 + H_2O + SO_2 = 2NaHSO_3$$

The solution is regenerated by heat to provide a sulfur-rich gas which can be used to make elemental sulfur, sulfuric acid, or sulfur dioxide. A small amount of sodium sulfate is produced, which must be crystallized out and disposed of. Initially the process used the potassium salts. Developed in the late 1960s. Licensed by Davy McKee and used in 22 plants in the United States, Japan, Germany, and Austria.

British Patent 1,557,295.
Hydrocarbon Process., 1975, **54**(4), 111.
Kohl, A. L. and Riesenfeld, F. C., *Gas Purification,* 4th ed., Gulf Publishing, Houston, TX, 1985, 351.
Ford, P. G., in *The Problem of Acid Emissions,* Institution of Chemical Engineers, Rugby, England, 1988, 151.

Welsh   The Welsh process is the general name given to the complex copper smelting operations carried out in Swansea, South Wales, from around 1800 until the introduction of larger smelters at the end of the 19th century. The heart of the Welsh process was a reverboratory furnace in which all the operations of roasting, fusing, and refining were conducted. It was superseded by the development of much larger furnaces, initially in the United States, and by the use of Bessemer-type converters for the final stages.

Dennis, W. H., *A Hundred Years of Metallurgy,* Gerald Duckworth, London, 1963, 129.

Wendell Dunn   A family of chlorine beneficiation processes based on selective chlorination of ores in a fluidized bed. Developed by W. E. Dunn of Chlorine Technology in Australia in the 1970s, primarily for beneficiating ilmenite. The first such commercial ilmenite beneficiation plant, completed in 1991, was that of Bene-Chlor Chemicals Private, Madras.

U.S. Patents 3,699,206; 4,349,516.
German Patents 2,103,478; 2,220,870; 2,221,006.

Western Gas   One of the processes for making water-gas, in which heavy oil was introduced. This enabled the cost to be optimized, depending on the relative prices of coal and oil. In 1937, 45 installations used the process in the United States. *See also* water gas, Willien-Stein.

Hartzel, F. W. and Lueders, C. J., *Proc. Am. Gas Assoc.,* 1932, **14**, 882 (*Chem. Abstr.,* **27**, 3319).

Westinghouse   A proposed thermochemical process for decomposing water to oxygen and hydrogen by electrolysis, coupled with the high-temperature decomposition of sulfuric acid:

$$2H_2O + SO_2 \rightarrow H_2 + H_2SO_4$$

$$H_2SO_4 \xrightarrow{870°C} H_2O + SO_2 + \frac{1}{2}O_2$$

Demonstrated only on the laboratory scale. It was developed in the 1970s as a potentially economic method of obtaining hydrogen fuel from a high-temperature source.

Williams, L. O., *Hydrogen Power: An Introduction to Hydrogen Energy and Its Applications,* Pergamon Press, Oxford, 1980, 85.

Westvaco (1) A variation of the *Claus process for removing hydrogen sulfide from gas streams, in which the sulfur dioxide is catalytically oxidized to sulfur trioxide over activated carbon at 75 to 150°C. The adsorbed sulfur trioxide is hydrated to sulfuric acid and then converted back to sulfur dioxide by reaction with the hydrogen sulfide at a higher temperature.

Ball, F. G., Brown, G. N., Davis, J. E., Repik, A. J., and Torrence, S. L., *Hydrocarbon Process.,* 1972, **51**(10), 125.
Oil & Gas J., 1978, **76**(37), 88.
Sulphur, 1974, (111), 51.

Westvaco (2) A process proposed for making chlorine by electrolyzing aqueous copper chloride. Invented in 1928 by F. S. Low at Chlorine Products, New York. Piloted by Westvaco in the 1940s, but not commercialized.

U.S. Patent 1,746,542.
Roberts, C. P., *Chem. Eng. Prog.,* 1950, **46**(9), 456.
Berkey, F. M., in *Chlorine, Its Manufacture, Properties and Uses,* Sconce, J. S., Ed., Reinhold Publishing, New York, 1962, 220.

Wetherill *See* American.

wetox [**wet ox**idation] A generic name for processes for oxidizing organic wastes, based on the use of hydrogen peroxide and a catalyst. Examples are *WINWOX, *WOX.

Wet Process A process for making phosphoric acid by treating phosphate rock with an acid. The acid is usually sulfuric acid, but hydrochloric and nitric acids are used commercially in special circumstances — in Israel where by-product hydrochloric acid is available, and in Norway and Switzerland where nitric acid is made by cheap hydroelectric power. The basic process with sulfuric acid is:

$$3H_2SO_4 + Ca_3(PO_4)_2 = 3CaSO_4 + 2H_3PO_4$$

The calcium sulfate by-product separates as either the dihydrate or the hemihydrate, depending on the conditions. The process originates from the work of J. B. Lawes in 1842 who patented a method of making a fertilizer by treating bones with sulfuric acid. Many variations are practiced today. *See also* Dorr and Haifa.

Childs, A. F., in *The Modern Inorganic Chemicals Industry,* Thompson, R., Ed., The Chemical Society, London, 1977, 386.
Becker, P., *Phosphates and Phosphoric Acid,* 2nd. ed., Marcel Dekker, New York, 1989.
McCoubry, J. C., in *Industrial Inorganic Chemicals: Production and Use,* Thompson, R., Ed., Royal Society of Chemistry, Cambridge, 1995, 379.

Wiberg-Soderfors A direct reduction process for extracting iron. *See* DR.

Wiewiorowski A process proposed for removing hydrogen sulfide from industrial gases by reacting it with sulfur dioxide in molten sulfur in the presence of an amine catalyst. Invented by T. K. Wieriorowski at the Freeport Sulfur Company, but not known to have been commercialized.

U.S. Patent 3,447,903.
Chem. Eng. News, 1970, **48**(18), 68.
Kohl, A. L. and Riesenfeld, F. C., *Gas Purification,* 4th ed., Gulf Publishing, Houston, TX, 1985, 545.

Wilbuschewitsch *See* Normann.

Willhoft A proposed process for making aluminum chloride from the solid waste from paper mills. The waste is mainly a mixture of clay with cellulose. It is dried and calcined in an inert atmosphere, giving a mixture of clay and carbon which chlorinates readily. Conceived by E. M. A. Willhoft and briefly examined by the Research Association for the Paper and Board, Printing and Packaging Industries (PIRA) in England in 1977, but not piloted.

British Patent 1,472,683.

Willien-Stein   A method for increasing the hydrogen content of water-gas by introducing gas-oil at one stage in the process. Invented by L. J. Willien and L. Stein in 1929. Piloted at three locations in the United States in the 1930s, but apparently not adopted on a large scale. *See also* water gas, Western Gas.

Canadian Patent 305,227.
Morgan, J. J., in *Chemistry of Coal Utilization,* Vol. 2, Lowry, H. H., Ed., John Wiley & Sons, New York, 1945, 1741.

Willson   A process for making calcium carbide by heating calcium oxide with tar or carbon in an electric furnace:

$$CaO + 3C = CaC_2 + CO$$

Invented by T. L. Willson in 1892 and first practised commercially at Niagara Falls in 1896.

Chem. Eng. (N.Y.), 1950, **57**(6), 129.

Wilman   A metallurgical process for removing manganese from steel scrap. Developed at the Electricity Council Research Centre, Capenhurst, United Kingdom, and first commercialized in 1988.

Winkler   Also called Fritz Winkler. A process for gasifying coal, using oxygen (or air) and steam in a fluidized bed at atmospheric pressure. Introduced by F. Winkler of IG Farbenindustrie, Germany, in 1922. It was developed in the 1920s and used mainly in Germany; the first plant was built at Leuna in 1926; by 1979, 36 units had been built. The largest was 33 m high, 6 m in diameter. The Flesch-Winkler process is a modification which permits the use of relatively unreactive coals which produce ash having a low melting-point. *See also* HTW.

German Patent 437,970.
Dainton, A. D., in *Coal and Modern Coal Processing,* Pitt, G. J. and Millward, G. R., Eds., Academic Press, London, 1979, 138.
Cornils, B., in *Chemicals from Coal: New Processes,* Payne, K. R., Ed., John Wiley & Sons, Chichester, England, 1987, 13.

Winkler-Koch   A early mixed-phase petroleum cracking process.

WINWOX   [Winfrith wet oxidation]   A process for oxidizing hazardous organic wastes by wet oxidation with hydrogen peroxide and a catalyst containing a transition-metal such as iron or copper. Developed in 1987 by the Winfrith Technology Centre of the UK Atomic Energy Authority, originally for destroying ion-exchange resins containing radioactive isotopes, but now proposed for hazardous organic wastes generally. A pilot plant was built in 1989.

Eur. Chem. News, 1989, **53**(1389), 24.
Wilks, J. P. and Holt, N. S., *Waste Manag.,* 1990, **10**, 197.

WIP   [Waste Immobilization Plant]   A process for immobilizing nuclear waste by incorporation in a borosilicate glass for long-term disposal. Developed in the 1970s in India for use at the waste immobilization plant at Tarapur.

Lutze, W., in *Radioactive Waste Forms for the Future,* Lutze, W. and Ewing, R. C., Eds., North-Holland, Amsterdam, 1988, 11.

Wisconsin A thermal process for fixing atmospheric nitrogen. Air is heated to over 2,000°C by contact with a bed of magnesia pebbles, and then cooled rapidly by contact with a bed of cold pebbles. The resulting air, containing 1 to 2 percent of nitric oxide, is passed through beds of silica gel to dry it; to permit the nitric oxide to be oxidized to dinitrogen tetroxide; and to concentrate the dinitrogen tetroxide before desorbing it and dissolving it in water. Developed by F. Daniels at the University of Wisconsin during World War II. Piloted in 1953 but subsequently abandoned.

Gilbert, N. and Daniels, F., *Ind. Eng. Chem.,* 1948, **40,** 1719.

Ermenc, E. D., *Chem. Eng. Prog.,* 1956, **42**(4), 149.

Chem. Eng. Prog., 1956, **42**(11), 488.

Wisner *See* carbonization.

Witten A process for making dimethyl terephthalate by the concurrent oxidation and esterification of *p*-xylene. Similar to the *Imhausen and *Katzschmann processes. Developed by Chemische Werke Witten and subsequently operated by Dynamit Nobel in Germany. World production capacity for this process in 1993 was 3 million tonnes.

Katzschmann, E., *Chem. Ing. Tech.,* 1966, **38,** 1.

Landau, R. and Saffer, A., *Chem. Eng. Prog.,* 1968, **64**(10), 20.

Weissermel, K. and Arpe, H.-J., *Industrial Organic Chemistry,* 3rd ed., VCH Publishers, Weinheim, Germany, 1997, 393.

WLP [Wasserstoff-Lichtbogen-Pyrolyse, German, meaning hydrogen arc pyrolysis] A process for converting gasoline into a mixture of acetylene and ethylene by injecting a jet of it into a hydrogen plasma. Piloted by Knapsack-Griesheim in Germany in the 1960s.

Miller, S. A., *Acetylene: Its Properties, Manufacture and Uses,* Vol. 1, Ernest Benn, London, 1965, 407.

Wohlwill An electrolytic process for refining gold. The crude gold, which may be made by fusing the anode slimes from the *Balbach process, is used as the anode, the cathode is of pure gold, and the electrolyte is a solution of gold chloride in hydrochloric acid. Gold deposits on the cathode. Silver deposits as a sediment of silver chloride. The process is relatively slow, so the interest lost on the inventory of metal in process is significant. Developed by E. Wohlwill at the Norddeutsche Affinerie in Hamburg in 1874, it became the principle method of gold refining in the world. It was largely superseded by the *Miller chlorine process at the end of the 19th century.

Dennis, W. H., *A Hundred Years of Metallurgy,* Gerald Duckworth, London, 1963, 281.

Yannopoulos, J. C., *The Extractive Metallurgy of Gold,* Van Nostrand Reinhold, New York, 1991, 243.

Woodall-Duckham Also called the Babcock W-D process. A process for recovering hydrochloric acid and metal oxides from spent metal chloride solutions, such as those obtained from metal pickling and ilmenite beneficiation. The liquor is first concentrated by evaporation, and then atomized in a heated spray-tower. Water evaporates from the droplets in the upper part of the tower, and chlorides are converted to oxides in the hotter, lower part. Developed by Woodall-Duckham in the 1960s; by 1992, over 150 installations were in use worldwide. Now offered by Babcock Woodall-Duckham, United Kingdom.

Woolwich [Named after the British Government laboratory at Woolwich Arsenal, where it was invented] A process for making the explosive RDX by nitrating hexamethylene tetramine.

WORCRA [Worner Conzinc Rio-Tinto of Australia] A family of continuous smelting and refining processes developed by Conzinc Riotinto of Australia in the 1960s. Invented by

H. K. Worner. The copper smelting process was piloted in Port Kembla, New South Wales, in 1968, but later abandoned.

Worner, H. K., in *Advances in Extractive Metallurgy,* Institution of Mining & Metallurgy, London, 1968, 245.
Worner, H. K., *Eng. Min. J.,* 1971, **172**(8), 64.
Worner, H. K., Reynolds, J. O., Andrews, B. S., and Collier, A. W. G., in *Advances in Extractive Metallurgy and Refining,* Institution of Mining & Metallurgy, London, 1972, 18.

Workman   *See* Dual-Spectrum.

WOX   [Wet oxidation]   A process for destroying organic materials by catalyzed oxidation with hydrogen peroxide. Developed by ASEA Atom, Sweden. *See also* WINWOX.

WSA   [Wet gas sulphuric acid]   A process for recovering sulfur from flue-gases and other gaseous effluents in the form of concentrated sulfuric acid. It can be used in conjunction with the *SCR process if oxides of nitrogen are present too. The sulfur dioxide is catalytically oxidized to sulfur trioxide, and any ammonia, carbon monoxide, and carbonaceous combustibles are also oxidized. The sulfur trioxide is then hydrolyzed to sulfuric acid under conditions which produce commercial quality 95 percent acid. Developed by Haldor Topsoe; 15 units were commissioned between 1980 and 1995. *See also* SNOX.

WSA-2   A variation of the WSA process, developed in 1989 but abandoned in 1994 in favor of the original WSA process.

WSA-SNOX   A combined flue-gas treatment process which converts the sulfur dioxide to sulfuric acid and the nitrogen oxides to nitrogen. Developed by Snamprogetti and Haldor Topsoe, based on the *WSA process. A large demonstration unit was under construction in 1989.

Wulff   A two-stage process for making acetylene by the pyrolysis of saturated aliphatic hydrocarbons. The feed gas is first pyrolyzed at approximately 1,300°C and then passed into a refractory brick reactor at below 400°C. Developed by R. G. Wulff in California in 1927. Operated in the United States, Brazil, and Europe until the end of the 1960s. *See also* Ruhr Chemie.

U.S. Patents 880,308; 917, 627; 1,843,965.
Bixler, G. H. and Coberly, C. W., *Ind. Eng. Chem.,* 1953, **45**, 2596.
Bogart, M. J. P. and Long, R. H., *Chem. Eng. Prog.,* 1962, **58**(7), 90.
Miller, S. A., *Acetylene: Its Properties, Manufacture and Uses,* Vol. 1, Ernest Benn, London, 1965, 384.
Tedeschi, R. J., *Acetylene-based Chemicals from Coal and Other Natural Resources,* Marcel Dekker, New York, 1982, 25.
Weissermel, K. and Arpe, H.-J., *Industrial Organic Chemistry,* 3rd ed., VCH Publishers, Weinheim, Germany, 1997, 96.

Wünsche   An electrolytic process for liberating bromine from a bromide solution. It uses carbon electrodes and a porous clay separator. Developed in Germany in 1902. *See also* Kossuth.

German Patent 140,274.
Yaron, F., in *Bromine and Its Compounds,* Jolles, Z. E., Ed., Ernest Benn, London, 1966, 16.

WWT   *See* Chevron WWT.

Wyandotte   A process for making a mixture of ethylene and propylene glycols, for use as antifreeze, from propane. The propane is cracked to a mixture of ethylene and propylene, which are not separated but converted to the corresponding glycols by *chlorohydrination. Developed by the Wyandotte Chemicals Corporation.

X

XIS [Xylene isomerization]   A process for isomerizing *p*-xylene to the equilibrium mixture of C_8 aromatic hydrocarbons. Developed by Maruzen Oil in the United States.

Weissermel, K. and Arpe, H.-J., *Industrial Organic Chemistry,* 3rd ed., VCH Publishers, Weinheim, Germany, 1997, 332.

Xylenes-plus   A catalytic process for isomerizing toluene to a mixture of benzene and xylenes. A silica/alumina catalyst is used in a moving bed. It is unlike the related *Tatoray process, in that no hydrogen is required. Developed by Sinclair Research in 1964 and then licensed by Atlantic Richfield.

U.S. Patents 3,116,340; 3,350,469; 3,437,709.
Verdol, J. A., *Oil & Gas J.,* 1969, **67**(23), 63.

Xylofining [Xylol refining]   A process for isomerizing a petrochemical feedstock containing ethylbenzene and xylenes. The xylenes are mostly converted to the equilibrium mixture of xylenes; the ethylbenzene is dealkylated to benzene and ethylene. This is a catalytic, vapor-phase process, operated at approximately 360°C. The catalyst (Encilite-1) is a ZSM-5–type zeolite in which some of the aluminum has been replaced by iron. The catalyst was developed in India in 1981, jointly by the National Chemical Laboratory and Associated Cement Companies. The process was piloted by Indian Petrochemicals Corporation in 1985 and commercialized by that company at Baroda in 1991.

Indian Patent 155,892.

Z

Zadra   A process for extracting gold from its ores. After *cyanidation and adsorption on activated carbon, the gold is re-extracted into a hot alkaline cyanide solution and stripped from it by electrolysis using a steel wool cathode.

Yannopoulos, J. C., *The Extractive Metallurgy of Gold,* Van Nostrand Reinhold, New York, 1991, 201.

Zenith   A process for refining vegetable oils by passing droplets of them down a column of dilute aqueous sodium hydroxide.

Braae, B., *J. Am. Oil Chem. Soc.,* 1976, **53**, 353.

Z-forming   A process for making aromatic hydrocarbons from aliphatic hydrocarbons. Developed jointly by Chiyoda and Mitsubishi Oil and operated in a demonstration plant in Kawasaki until it was closed in 1992.

Eur. Chem. News, CHEMSCOPE, 1994, **61**, 7.

Zeoforming   A process for converting light paraffinic feedstocks to high-octane gasoline components. The catalyst is zeolite ZSM-5. Developed in the CIS, engineered by KTI, and first installed by Lurgi in Gorlice, Poland in 1997.

Ziegler (1)   A process for polymerizing ethylene under moderate temperatures and pressures, catalyzed by a mixture of titantanium tetrachloride and a trialkyl aluminum such as tri-ethyl aluminum. Invented in 1953 by K. Ziegler at the Max Planck Institut für Kohlenforschung, Mülheim/Ruhr, Germany. Operated worldwide on a very large scale. *See also* Ziegler-Natta.

Belgian Patent 533,362.
Ziegler, K., Holzkamp, E., Breil, H., and Martin, H., *Angew. Chem.*, 1955, **67**, 426.
Natta, G., *Angew. Chem.*, 1956, **68**, 393.
Ziegler, K., *Angew. Chem.*, 1959, **71**, 623.
Ziegler, K., *Angew. Chem.*, 1960, **72**, 829.
Raff, R. A. V., in *Ethylene and Its Industrial Derivatives,* Miller, S. A., Ed., Ernest Benn, London, 1969, 335.

Ziegler (2)   A process proposed for making tetraethyl lead by electrolyzing the molten complex of ethyl potassium with triethyl aluminum, $KAl(C_2H_5)_4$, using a lead electrode. Invented in 1963 by K. Ziegler and H. Lehmkühl but not commercialized.

U.S. Patent 3,372,097.

Ziegler-Natta   Also called Z-N. A general name for the family of olefin polymerization processes invented by K. Ziegler and G. Natta in the 1950s. Ziegler and Natta were jointly awarded the Nobel Prize for Chemistry in 1963 for their discoveries. *See* Natta, Ziegler (1).

Boor, J., Jr., *Ziegler-Natta Catalysts and Polymerizations,* Academic Press, New York, 1979.
James, L. K., Ed., *Nobel Laureates in Chemistry 1901–1992,* American Chemical Society and Chemical Heritage Foundation, Washington, D.C., 1993, 442, 449.
Fink, G., Mülhaupt, R., and Brintzinger, H. H., Eds., *Ziegler Catalysts: Recent Scientific Innovations and Technical Improvements,* Springer Verlag, Berlin, 1995.

Zimmermann   *See* Zimpro.

Zimpro   [**Zim**mermann **pro**cess]   Also called the Zimmermann process, and wet-air oxidation. A thermal process for oxidizing organic wastes in aqueous solution, and for conditioning sewage sludge. Raw sewage sludge is pressurized with air and heated with steam to 150 to 250°C in a pressure vessel; the product is sterile and easy to filter. Invented by J. F. Zimmermann in the United States in 1954, first operated in Chicago in 1957, and now offered by U.S. Filter/Zimpro. As of 1991, more than 200 units had been installed worldwide. *See also* SCWO.

Teletzke, G. M., *Chem. Eng. Prog.*, 1964, **60**(1), 33.
Pradt, L. A., *Chem. Eng. Prog.*, 1972, **68**(12), 72.
Metcalf and Eddy, Inc., *Wastewater Engineering: Treatment, Disposal, Re-use,* 2nd. ed., McGraw-Hill, New York, 1979, 636.
Hydrocarbon Process., 1996, **75**(8), 109.

Zincex   [**Zinc ex**traction]   A process for extracting zinc from pyrite cinder leachate, using organic solvents. The chloride leachate is first extracted with a secondary amine, and then with di(2-ethylhexyl)phosphoric acid to remove iron. Developed by Tecnicas Reunidas, first commercialized in 1976, and now used in Spain and Portugal.

Nogueira, E. D., Regife, J. M., and Arocha, A. M., *Eng. Min. J.,* 1979, **180**(10), 92.
Nogueira, E. D., Regife, J. M., and Blythe, P. M., *Chem. Ind. (London),* 1980, (2), 63.

Zinclor   A development of the *Zincex process which uses di-pentylpentylphosphonate (DPPP) as the extractant. Developed by Tecnicas Reunidas.

Cox, M., in *Developments in Solvent Extraction,* Alegret, S., Ed., Ellis Horwood, Chichester, England, 1988, 181.

Zincote *See* metal surface treatment.

Zirpro A process for flame-proofing textiles by treating them with aqueous solutions of zirconium complexes. Wool is treated with aqueous potassium hexafluorozirconate and citric acid. Developed by the International Wool Secretariat, Yorkshire, now based in Melbourne, Australia.

Benisek, L., *J. Textile Inst.,* 1974, **65,** 102.
Ingham, P. E. and Benisek, L., *J. Textile Inst.,* 1977, **68,** 176.

Z-N *See* Ziegler-Natta.

Z-Sorb A process for removing hydrogen sulfide and other sulfur compounds from gas streams by absorption in a proprietary granular absorbent containing zinc oxide. The process can be operated at temperatures between 315 and 555°C.

Chem. Eng. (N.Y.), 1998, **105**(2), 25.

BIBLIOGRAPHY

Gowan, J. E. and Wheeler, T. S., *Name Index of Organic Reactions,* Longmans, London, 1960.

Krauch, H. and Kunz, W., *Organic Name Reactions,* John Wiley & Sons, New York, 1964.

Encyclopedia of Chemical Processing and Design, McKetta, J. J. and Cunningham, W. A., Eds., Marcel Dekker, New York, from 1976.

Rompps Chemie-Lexicon, 8th ed., Neumuller, O.-A., Frank'sche Verlagshandlung, Stuttgart, 1979.

Tottle, C. R., *An Encyclopedia of Metallurgy and Materials,* The Metals Society and MacDonald & Evans, London, 1984.

Ullmann's Encyclopedia of Industrial Chemistry, 5th ed., VCH Publishers, Weinheim, Germany, 1985–1996.

Mundy, B. P. and Ellerd, M. G., *Name Reactions and Reagents in Organic Synthesis,* John Wiley & Sons, New York, 1988.

Kirk-Othmer's Encyclopedia of Chemical Technology, 4th ed., John Wiley & Sons, New York, 1991–1996.

Hassner, A. and Stumer, C., *Organic Syntheses Based on Name Reactions and Unnamed Reactions,* Elsevier Science, Oxford, 1994.

East, M. B. and Ager, D. J., *Desk Reference for Organic Chemists,* Krieger Publishing, Malabar, FL, 1995, Chap. 4.

BIBLIOGRAPHY

APPENDIX: KEY TO PRODUCTS

acetaldehyde Grünstein, Hoechst-Wacker, Wacker (1), Walker.

acetic acid Cativa, DF, Llangwell, Monsanto (3), Rhône-Poulenc/Melle Bezons, Shawinigan, Suida.

acetic anhydride Hoechst-Shawinigan.

acetone Cumene, Cumox, Hock, Roka, Weizmann.

acetylene Avco, CCOP, Hoechst HTP, Hoechst-WLP, Ruhrchemie, Sachsse, SBA, Schoch, WLP, Wulff.

acrylic acid Reppe.

acrylic esters Reppe.

acrylonitrile ammoxidation, Andrussov, Knapsack, Kurtz, OSW, PETROX, SNAM (2), SOHIO.

adiponitrile CANDID, EHD, Hydrocyanation, Monsanto, UCB-MCI.

aggregate Neutralysis.

alcohols, aliphatic Alfol, Bashkirov, Epal, Octamix.

aldehydes Consortium, OXO, RCH/RP.

alkanes *See* hydrocarbons, aliphatic.

alkenes *See* olefins.

alkyl anilines Thoma.

Alnico Magnicol.

alumina AEROSIL, Alumet, Anortal, Bayer, Blanc, Deville-Pechiney, Grzymek, Hall (4).

aluminium alloys Cowles.

aluminium chloride Toth.

aluminium chlorohydrate ACH (2).

aluminium extraction Aloton, Bretsznajder, Büchner, Calsinter, Hichlor, H-Plus, Kalunite, Nuvalon, Pechiney H⁺, Pedersen, Peniakoff, Reynolds Metal, Séailles-Dyckerhoff, Toth, VAW Lurgi, VAW-Sulfite, Walthal.

aluminium metal AIAG Neuhausen, ALCOA, ALUREC, Compagnie AFC, Deville (1), Grätzel, Hall-Héroult, Hoopes, Netto, Pechiney (1).

aluminium nitride Serpek, SHS.

aluminium oxide *See* alumina.

aluminium trichloride Hichlor, Toth, Willhoft.

amines, aromatic Béchamp.

amino acids Acylase.

o-**aminothiophenols** Herz.

ammonia   AMV, Braun, BYAS, Casale, Claude (1), Claude-Casale, Fauser, Frank-Caro, Haber, KAAP, LCA, LEAD, Mond Gas, Mont Cenis, NEC, PARC, Purifier, Serpak.

ammonium alum   Aloton.

ammonium chloride   Engechlor.

ammonium nitrate   NSM, Stengel, Uhde-Hibernia, Victor.

ammonium phosphates   Cros, Gardinier, Minifos, Norsk-Hydro, Swift.

ammonium sulfate   Cominco, Erdölchemie, GEESI, Merseburg, OSAG.

t-**amyl methyl ether**   NExTAME.

antimony extraction   BRGM.

argon   HARP.

ascorbic acid   Bertrand.

Bakelite   Baekeland.

benzene   Aromax (1), Benzorbon, DETOL, Dynaphen, HDA, Houdry-Litol, Hydeal, hydrodealkylation, Hytoray, Kellogg Hydrotreating, Lignol, Litol, MHC, MHD, MSTDP, MTDP, Newton Chambers, PX-Plus, Pyrotol, Tatoray, T2BX, TDP, THD, THDA, Xylenes-plus.

benzole   BASF/Scholven, Ibuk, Thiofex.

benzonitrile   ammoxidation.

beryllium extraction   Copaux, Copaux-Kawecki, Fuse-Quench, Kjellren-Sawyer, Perosa, Schwenzfeier-Pomelée.

beryllium metal   Kjellgren.

bisphenol-A   CT-BISA.

brass   cementation (2).

bromine   Dow bromine, Kossuth, Kubierschky, Wünsche.

BUNA   Buna.

butadiene   Aldol, CAA, Catadiene, DIFEX, GPB, KLP, Lebedev, Ostromislenski, O-X-D, Oxo-D, Phillips (3), Reppe.

butane   Krupp-Koppers (2).

1,4-butanediol   Geminox, Linde/Yukong.

butanetriol trinitrate   Biazzi.

n-**butanol**   Weizmann.

t-**butanol**   Oxirane.

1-butene   Alphabutol, Idemitsu, IFP-SABIC, Isopol, SHP, Sorbutene.

2-butene   Arco, BUTACRACKING, ISOMPLUS, Isopol, Trolefine.

butenes   Krupp-Koppers (2), MTO.

butyraldehyde   RCH/RP.

cadmium recovery Tetronics.

calcium ammonium nitrate CAN.

calcium carbide Willson.

calcium cyanamide Frank-Caro, Polzeniusz-Krauss.

calcium hypochlorite Mathieson (2), Pennsalt/Pennwalt, Perchloron, PPG, Thann.

calcium nitrate Cerny.

caprolactam HPO, HSO, PNC.

carbides, cemented Coldstream, ROC.

carbon black Ayers, Acetylene Black, Channel Black, Enasco, Furnace Black, Jones, Lampblack, Thermal Black, Thermatomic.

carbon dioxide Backus, Reich (1).

carbon disulfide Folkins.

carbon monoxide Calcor, COPISA, COPSA, COSORB.

carboxylic acids Amoco, Armour, Bernardini, SOT.

cellulose acetate Acetate.

cellulose fibre Alceru, Bemberg, Carbacell, Chardonnet, Cross-Bevan-Beadle, Cuprammonium, Newcell, Viscose, Sini.

cement Grzymek.

ceramics, non-oxide SHS.

ceramics, oxide Pechini, Sol-Gel, Stöber.

charcoal, activated Urbain.

chlorine Airco, Castner-Kellner, Chlor-Alkali, Deacon, De Nora, Diaphragm cell, Downs, Glanor, Griesheim (1), Grosvenor-Miller, Hargreaves-Bird, Hasenclever, Hoechst-Uhde (1), Kel Chlor, LeSeur, Membrane cell, Salt, Schroeder, Shell Deacon, Weldon, Weldon-Pechiney, Westvaco (2).

chlorine dioxide Erco, Fröhler, Holst, Kesting, Munich, Mathieson (1), Persson, R-2, SVP.

chlorobenzene Auger.

chlorosulfonic acid Schweizerhall.

chromium metal Thermite, van Arkel and de Boer.

coal tar Ab der-Halden, Heinrich Koppers.

cobalt extraction Caron.

cobalt metal Thermite.

copper extraction AMAR, Anatread, Arbiter, CLEAR, Cuprex, Cymet (1), Cymet (2), Electroslurry, Falconbridge, Henderson, LM, Longmaid-Henderson, Mannhés, Mansfield, MECER, Noranda, Orford, Outokumpu, RLE, Sherritt-Cominco, Sirosmelt, Treadwell, Welsh, Worcra.

copper metal   AMAR, Actimag.

cotton   mercerization, Zirpro.

cresols   Cresex.

cumene   CD-Cumene, Mobil/Badger Cumene, Q-Max.

cyclohexane   HB Unibon, Hydrar.

cyclohexane oxime   Nixan.

cyclohexanol   Halcon.

cyclohexanone   Cyclopol.

cyclohexanone oxime   ammoximation, Nixan.

cyclopropane   Hass (2).

p-**cymene**   Cymex.

cymenes   Attisholz.

detergents   Chemithon, Detal, Detergent Aklylate.

deuterium   G-S.

1,2-dichloroethane   CER, HTC, LTC, oxychlorination.

1,2-dichloropropane   Hass-McBee.

1,4-dicyanobutane   hydrocyanation.

dimethyl ether   TIGAS.

dimethyl terephthalate   Katzschmann, Hercules (2), Imhausen, Witten.

di-isopropyl ether   Oxypro (1).

di-isopropyl naphthalene   Kureha.

dinitrogen pentoxide   Dipen.

EDTA   Geigy.

emeralds   Espig.

EPDM rubber   Lovacat.

ethanol   Biostil, Emert, ENSOL, Gulf, Keyes, Van Ruymbeke (2).

ethyl acetate   Tischenko.

ethylbenzene   Albene, Alkar, Ebex, Ebmax, Ethyl benzene, Mobil/Badger.

ethylene   ARS, Benson, CCOP, Dianor, Ethoxene, Hoechst HTP, SBA-Kellogg, Triolefine, WLP.

ethylene chlorohydrin   chlorohydrination.

ethylene glycol   GO, Wyandotte.

ethylene oxide   Lefort, Scientific Design.

ethylene propylene copolymer   Catalloy, Flexomer.

fatty acids   *See* carboxylic acids.

formaldehyde   Adkins-Peterson, Formox, Gutehoffnungshütte, Hibernia, Walker.

fructose   Sarex.

furfural   Attisholz.

gallium extraction   Beja, De la Breteque.

glass   CRISFER.

glucose   Sarex, Scholler.

glue   Scheidemandel.

glycerol   Garrigue, Neuberg, Shell Glycerol, Twitchell, Van Ruymbeke.

gold extraction   APOL, Arseno, AuPLUS, Betts, Calmet, Cashman, CGA, CIL, CIP (1), Cyanide, K-Process, GOLDOX, Magchar, Merrill-Crowe, Moebius, PAL, Plattner, Rose (2), Salsigne, Zadra.

gold metal   Miller, parting, Thum, Wohlwill.

graphite   Acheson (2), Castner (1).

grout   Joosten, Siroc.

gypsum   Donau Chemie, Cerphos, Guillini, Knauf.

hexachloro cyclopentadiene   Lidov.

hexane   Hexall.

hexenes   Dimersol G.

hydrazine   Bayer Ketazine, Hoffman, Ketazine, Raschig (1).

hydrocarbons, aliphatic, C_4   BUTENEX.

hydrocarbons, aliphatic, $C_5 - C_8$ Isomate.

hydrocarbons, aliphatic, iso   Hysomer, Penex.

hydrocarbons, aliphatic, linear   ENSORB, IsoSiv, Molex, MS2, N-ISELF, Nurex, Parex (2), TSF.

hydrocarbons, alkyl aromatic   Detal, Detergent Alkylate.

hydrocarbons, aromatic   Alpha, Aroforming, Carom, Cyclar, hydroforming, Hall, Koch, MTA, Platforming, Rittman, Z-Forming.

hydrochloric acid   Aman, Hargreaves, Hargreaves-Robinson, Mannheim (1), Woodall-Duckham.

hydrogen   Hoechst-Uhde (1), Hydrogen Polybed PSA, Hypro, HYSEC, HyTex, Lane, LO-FIN, MRH, Proximol, Rincker-Wolter, RKN, SMART H_2, Sumitomo-BF, Westinghouse.

hydrogen cyanide   ammoxidation, Andrussov, BMA, Degussa, Fluohmic.

hydrogen peroxide   AO, Barium, Huron-Dow, Krutzsch, Loewenstein-Riedel, Pietzsch and Adolph, Weissenstein.

hydroquinone   Hock.

α-hydroxy isobutyric acid   Escambia (1), Lonza (1).

hydroxylamine   Olin Raschig, Raschig (3).

ilmenite beneficiate   Becher, Benilite, ERMS, Musro/Murso, SREP, TSR (2), Wah Chang.

iron and steel   Bloomery, Brassert, Catalan, Corex, DR, Finmet, Hall (1), Hoerde, Larkin, NSC, Pedersen, Puddling; *see also* steelmaking.

iron oxide pigment   Laux, Penniman-Zoph.

isobutane   Butamer, Butomerate, I-Forming.

isobutene   Adib, Arco, BLISS, BUTACRACKING, C_4 Butesom, Cold Acid, Isofin, Isomplus, OlefinSiv, Olex, SKIP.

isopentenes C_5 Pentesom.

isoprene   IDAS.

isopropyl alcohol   Ellis.

di-isopropyl ether   Oxypro.

Lanxide   Lanxide.

L-DOPA   Monsanto (2).

lead carbonate, basic   Bischof, Carter, Dutch, Thénard, Thompson-Stewart.

lead extraction   Boliden (2), Carinthian, Flintshire, Imperial Smelting, Isasmelt, Kaldo, Kivcet, QSL, Waelz.

lead metal   Betterton (1), Betterton (2), Betterton-Kroll, Betts, Brittania, Davey, Harris, liquation, Parkes, Pattinson, Tetronics.

lead monoxide   Barton.

lead tetraethyl   Ziegler (2).

leather   Liritan, TAL.

lignosulfonates   Howard, Magnefite.

l-lysine   Toray (1).

magnesium extraction   Dow Seawater, MAGRAM, Pattinson (2).

magnesium metal   Elektron, Gardner, Hansgirg, Magnetherm, Norsk-Hydro, Pidgeon, Radenthein.

maleic anhydride   ALMA, Petro-Tex.

malononitrile   Lonza (2).

manganese extraction   Dean, MHO.

manganese metal   Chemetals, Pidgeon, Thermite.

mercury fulminate   Chandelon.

methacrylates   Escambia.

methacrylic acid   Lonza.

methane   ANTHANE/ANODEK, Binax, Biogas, HCM, Hydrane, Kryosol, Laran.

methanol   Attisholz, DMO, ICI Low Pressure Methanol, LCM, LPMEOH, MAS, Pier-Mittasch, POX, Remet, Walker.

methyl ethyl ketone   Hoechst-Wacker.

methyl methacrylate   ACH (1), MIGAS.

methyl *t*-butyl ether   catalytic distillation, CDETHEROL, Ethermax, Etherol, Isotex, NExEthers.

mica   Bardet, Samica.

moulds, foundry   Isocure, Pep Set, SAPIC.

mustard gas   Levinstein.

naphthalene   Brodie, Sulzer-MWB, UNIDAK, Unionfining.

naphthaquinone   Kawasaki Kasei.

nickel catalyst   Raney, SURECAT.

nickel extraction   Caron, Falconbridge, Hybinette, INCO, Mond Nickel, Nicaro, Orford, Outokumpu, Sherritt-Gordon.

niobium carbide   Menstruum.

niobium metal   Balke, Kroll (1).

nitrate esters   Biazzi, Gyttorpp.

nitric acid   Arc, Birkeland-Eyde, CNA, CONIA, DSN, DWN, HOKO, Hycon (2), Ostwald, SABAR, Wisconsin.

nitrogen   Bergbau-Forschung, KURASEP, Linde, MOLPSA-Nitrogen, NitroGEN, Serpek.

nitroglycerine   Biazzi, Nitro Nobel, Schmidt.

nitroguanidine   Marqueyrol and Loriette, Welland.

nitro-compounds, organic   Bofors, Hass (1), Hercules.

nuclear waste   AVM, ESTER, FINGAL, FIPS, HARVEST, PAMELA, PHOTHO, PIVER, Synroc, VERA, WINWOX, WIP.

nylon   EFFOL.

octenes   Dimersol X, Octol.

oils, lubricating   Bensmann, Chlorex (1), Mohawk.

oils, vegetable   Behr, Zenith.

olefins   Alfene, CATOFIN, Catpoly, DCC, Define, Linear-1, MTO, Oleflex, Olex, Pacol, Polynaphta Essence, PYROCAT, SHOP, SOR, Superflex, TRC, Vniios.

oleum   Squire and Messel.

organofluorine compounds   Halex, La-Mar, Simons.

oxalic acid   Rhône-Poulenc.

oxygen   Bosch, Brin, DWO, Linde, Mallet, Moltox, OxyGEN, Oxy-Rich, OXYWELL.

ozone   MEMBREL, POZONE.

palmitic acid   Emersol.

paper and pulp   Acedox, Ahlstage, Alcell, ASAM, BFR, Burkheiser, Celdecor, DegOX, ECF, Green Liquor, GreenOx, Kraft, Kramfors, Lignox, Lyocell, Magnefite, Macrox, Middox, Milox, MOXY, NSSC, Organocell, Organosolv, Oxypro (2), Prenox, Sapoxal, SAPPI-Air Liquide-Kamyr, SCA-Billerand, Sirola, Solvay (2), Solv-X, SRP, Stora, Sulfate, Sulfite, TCF.

paraffin wax   Krupp-Kohlechemie, Sharple.

pentanes   Pentafining.

perchloric acid   Pernert.

phenol   Cumene, Cumox, Dennis-Bull, Dow-Phenol, Dynaphen, Guyot, Halcon, HDA, Hercules-BP, Hock, Lignol, Phenolsolvan, Phenoraffin, Raschig (2), Tyrer.

phenols   Noguchi.

phosphate fertilizer   Davidson, Den, LETS, Oberphos, Odda.

phosphoric acid   Adex, Central-Prayon, CFB, Dorr, Dorr-Oliver, Gulf-Swenson, Haifa, HDH, Jacobs-Dorr, KPA, Nordac, Ozark-Mahoning, Phorex, Prayon, Progil, SIAPE, Wet.

phosphorus   Furnace.

phthalic acid   Kawasaki Kasei.

phthalic anhydride   Gibbs, Heyden-Wacker, LAR, Sapper, von Heyden.

platinum metals   Deville and Debray, Footing, Leidie, Matthey, Moebius, parting, Slow-Cooled Matte, Sobolevsky, Solvex.

plutonium   Butex, Purex, Recuplex, Redox, Tramex, Truex.

polybutene   Mobil-Witco-Shell.

polyethylene   ATOL, Borstar, Catalloy, CP, CX, Exxpol, High Productivity, Innovene, Naphtachimie, Phillips (1), Sclair, Sclairtech, Selexsorb, Spherilene, Standard Oil, UNIPOL, Ziegler (1).

polyisobutene   Cosden.

polymer gasoline   Selectopol.

polypropylene   Addipol, GPB, Hypol, LIPP-SHAC, Natta, MPC, Novolen, Sherpol, Spheripol, Rexene, UNIPOL.

polystyrene   NORSOLOR, SDS, Styro-Plus.

polyvinyl chloride   MSP3.

poly-*p*-xylene   Paralene.

potassium carbonate   Engel, Engel-Precht, Precht.

potassium cyanide   Beilby, Erlenmeyer, Rodgers.

potassium hydroxide Griesheim (1).

potassium metal Griesheim (2).

potassium permanganate Carus.

potassium sulfate Alumet.

propylene MTO, META-4, OCT.

propylene glycol Wyandotte.

propylene oxide Cetus, Daicel, Escambia (2), Montoro, Oxirane, Propylox, SMPO, Pekilo, Provesteen, Pruteen, Scholler-Tornesch, SMPO.

pulp *See* paper and pulp.

quartz Heraeus.

rayon *See* cellulose fibre.

RDX Bachmann, KA, Woolwich.

resorcinol Hock.

rubber, foam Dunlop, Talalay.

rubber, natural Kaysam, Peachy, Vulcanization.

rubber, synthetic Aldol, Alfin, Buna, GRS.

rutile, synthetic *See* ilmenite beneficiate.

silica AEROSIL, HAL, Soot, Stöber.

silicon carbide Acheson (1).

silicon metal Siemens.

silk, artificial *See* cellulose fibre.

silver extraction Boss, Brittania, Cazo, Clandot, Kiss, Patera, Patio, Russell, Washoe.

silver metal Balbach, Betts, cupellation, Davey, liquation, Luce-Rozan, Miller, Moebius, Parkes, parting, Pattinson (1), Rozan, Thum, Wohlwill.

soap Armour, Bradshaw, Clayton, Crosfield, De Laval Centripure, Kettle, Mazzoni, Mon Savon, saponification, Sharples, Twitchell.

sodium carbonate ammonia-soda, Fresnel, Leblanc, Schloesing-Rolland, Solvay (1).

sodium chloride Alberger, Grainer, Recrystallizer, Salex.

sodium cyanide Bucher, Castner (2), neutralization, Raschen, Röhm, Schlempe.

sodium dithionite Amalgam, Formate.

sodium hydroxide Castner-Kellner, causticization, Chlor-Alkali, De Nora, Diaphragm cell, Ferrite, Glanor, Hargreaves-Bird, Kiflu, Lime-soda, Löwig, Membrane cell, De Nora, Glanor.

sodium metal Castner (3), Castner (4), Deville (2), Downs, Tekkosha.

sodium nitrate Guggenheim, Salt, Shanks.

sodium perborate   Acid, Duplex.

sodium salicylate   Wacker (2).

sodium sulfate   Climax, Hargreaves-Robinson, Mannheim, saltcake.

sorbitol   Creighton, Dynatol.

stannic oxide   Tin Sol.

stearic acid   Emersol.

steel   *See* iron and steel.

styrene   Fina/Badger, Mark and Wulff, Montoro, Oxirane, SMART SM, Styro Plus.

styrene co-polymers   POSTech.

sucrose   Bergius (2), Bergius-Rheinau, Boivan-Loiseau, Deguide, Harloff, Madison, NRS, Quentin, Scheibler, Scholler-Tornesch, Steffen, Sucro-Blanc, Talafloc, Talodura.

sulfur   Boliden (1), Catasulf, Clinsulf, Claus, Hysulf, Mond, Resox, Selectox, SPOR, Sulpel, Trail.

sulfur extraction   Chemico, Frasch, Orkla.

sulfuric acid   Bayer-Bertrams, BOSAC, CAT-OX, Chamber, Contact, IPA, Knietsch, Mannheim (2), Müller-Kühne, NoTICE, Pauling-Plinke, OSW/Krupp, PERCOS, SAR, SARP, Schröder-Grillo, Satco, Schaffner, Tenteleff, Winkler.

Superphosphate   *See* phosphate fertilizer.

syngas   AGC-21, ATR (1), ATR (2), CAR, Electropox, Fauser-Montecatini, GasCat, Hycar (1), Hydrocol, Hytex, ICAR, ICI Low Pressure methanol, Koppers Hasche, KRES, MRG, Octamix, Onia-Gegi, POX, PRENFLO, reforming, SGP, SBA-HT, Selox, SGP, SMDS, THGP.

Synroc   Synroc.

tantalum   Kroll (1).

tantalum carbide   Menstruum.

tar acids   Phenolsolvan.

terephthalic acid   Amoco, Henkel (1), Henkel (2), Maruzen (1), Mid-Century, Raecke, Teijin, Toray (2).

thiophene   Socony Vacuum.

tin extraction   Ashcroft-Elmore, Berzelius, Caveat, Goldschmidt, liquation.

titanium carbide   SHS.

titanium diboride   SHS.

titanium dioxide   AEROSIL, Blumenfeld, Chloride, Fletcher, ICON, Monk-Irwin, Sulfate.

titanium extraction   *See* ilmenite beneficiate.

titanium metal   Hunter, Kroll (1).

titanium tetrachloride   Chloride, Wendell Dunn.

toluene   Aromax (1).

transuranic elements   Tramex, Truex; *see also* plutonium.

triethylene glycol nitrate   Biazzi.

trimethylol ethane trinitrate   Biazzi.

trinitro toluene   Meissner, Selleting.

tungsten extraction   Fan Steel, Shoppler.

tungsten metal   Cooledge, van Arkel and de Boer.

ultramarine   Guimet.

uranium dioxide   ADU, AUC, Comurhex, IDR (2), Sol-Gel.

uranium extraction   AMEX, Bufflex, Cogema, Dapex, DEPA-TOPO, Eluex, Purlex, Thorex, Tramex.

uranium metal   Ames (1).

uranium tetrafluoride   Excer.

urea   Bosch-Meiser, HR, IDR (1), Mitsui-Toatsu, Pechiney (2), SNAM (1), Urea 2000plus, UTI.

vanadium metal   McKechnie-Seybolt, van Arkel and de Boer.

vanillin   Howard, Riedel.

vinyl acetate   Bayer-Hoechst, Hoechst-Uhde (2), Wacker-Hoechst.

vinyl chloride   BPR, DOC, Transcat.

wax, paraffin   *See* paraffin wax.

white lead   *See* lead carbonate, basic.

wood   Bethell, Empty-Cell, Full-Cell, Heiskenkjold, Iotech, Rueping.

wood pulp   *See* paper and pulp.

wool   Chlorine/Hercosett, Hercosett, Kroy, Zirpro.

***m*-xylene**   JGCC, MGCC, MS Sorbex.

***p*-xylene**   Aromax (2), Chevron (2), Eluxyl, Krupp-Koppers (1), Maruzen (2), MHTI, MSTDP, MTPX, MVPI, Octafining, Parex (1), Phillips (2), PROABD MSC, PX-Plus.

xylenes, mixed   Isarom, Isolene II, Isomar, MLPI, MTDP, Octafining, STDP, Tatoray, T2BX, TDP, Transplus, XIS, Xylenes-plus, Xylofining.

xylitol   Creighton.

zinc extraction   Champion, English, Goethite, Haematite, Jarosite, Imperial Smelting, Lacell, liquation, New Jersey, QSL, St Joseph, Sherritt-Gordon, Tetronics, Waelz, Warner, Zincex, Zinclor.

zinc oxide   American, Belgian, Direct, French, Fricker, Wetherill.

zircon   HAL.

zirconium diboride   SHS.

zirconium metal   Kroll (1).